ADVANCES IN GENETICS

VOLUME 25

ADVANCES IN GENETICS

Edited by

E. W. CASPARI

Department of Biology
University of Rochester
Rochester, New York

JOHN G. SCANDALIOS

Department of Genetics
North Carolina State University
Raleigh, North Carolina

VOLUME 25

ACADEMIC PRESS, INC.
Harcourt Brace Jovanovich, Publishers
San Diego New York Berkeley Boston
London Sydney Tokyo Toronto

ACADEMIC PRESS, INC.
1250 Sixth Avenue
San Diego, California 92101

United Kingdom Edition published by
ACADEMIC PRESS INC. (LONDON) LTD.
24-28 Oval Road, London NW1 7DX

LIBRARY OF CONGRESS CATALOG CARD NUMBER: 47-30313

ISBN 0-12-017625-4 (alk. paper)

PRINTED IN THE UNITED STATES OF AMERICA
88 89 90 91 9 8 7 6 5 4 3 2 1

CONTENTS

Transferrin: Evolution and Genetic Regulation of Expression

Barbara H. Bowman, Funmei Yang, and Gwendolyn S. Adrian

The *Drosophila* Alcohol Dehydrogenase Gene–Enzyme System

Geoffrey K. Chambers

Genetics Research on Braconid Wasps

Daniel S. Grosch

The Variable Mitochondrial Genome of Ascomycetes: Organization, Mutational Alterations, and Expression

Klaus Wolf and Luigi Del Giudice

TRANSFERRIN: EVOLUTION AND GENETIC REGULATION OF EXPRESSION

Barbara H. Bowman, Funmei Yang, and Gwendolyn S. Adrian

Department of Cellular and Structural Biology,
The University of Texas Health Science Center at San Antonio,
San Antonio, Texas 78284

I. Introduction

Human plasma proteins serve as excellent models for (1) identifying evolutionary mechanisms important in the origin and diversification of gene families and for (2) analyzing the multiregulatory circuits responsible for gene expression. Characterization of the DNAs encoding transferrin, haptoglobin, and the group-specific component has illustrated examples of "molecular tinkering" (Jacob, 1982), where bits and pieces of several genes were melded together to produce plasma proteins with novel structures and diverse functions (Bowman and Yang, 1987). Transferrin (TF) is a member of a conserved family of genes that have remained linked on the same chromosome for hundreds of millions of years. It is a plasma protein of biological interest, not only because of its evolutionary history, but also because of its role as a growth factor required for the proliferation of normal and malignant cells.

Excellent reviews have been written about transferrin and the transferrin protein family (Putnam, 1984; Morgan, 1983; Aisen and Listowsky, 1980). Our laboratory's work, in identifying, characterizing, and chromosomally mapping the human *TF* gene, has recently been summarized in *The Plasma Proteins, Volume V* (Bowman and Yang, 1987). The evolution of transferrin, tissue sites of expression of the *TF* gene, and regulatory circuits that may influence its expression *in vivo* and *in vitro* will be reviewed in the following discussion.

II. Properties of Human Transferrin

Transferrin, a major plasma protein (Holmberg and Laurell, 1945; Schade and Caroline, 1946), migrates in alkaline electrophoresis with the β-globulins. Purified transferrin displays a red coloration because of its capacity to bind ferric iron. Transferrin's concentration in human plasma is approximately 2.5 mg/ml (Putnam, 1975). The glycoprotein has an approximate molecular weight of 76,500 (Putnam, 1984), 75,157 of which is contributed by a polypeptide chain of 679 amino acid residues (MacGillivray *et al.,* 1983; Yang *et al.,* 1984). The carbohydrate moiety of the transferrin molecule is composed of two identical polysaccharide chains which contribute to the protein's heterogeneity (Spik *et al.,* 1975). Transferrin is the product of an ancient intragenic duplication that led to homologous carboxyl and amino domains, each of which binds one ion of ferric iron. Transferrin carries iron from the intestine, reticuloendothelial system, and liver parenchymal cells to all proliferating cells in the body.

Transferrin carries iron into cells by receptor-mediated endocytosis. Iron is dissociated from transferrin in a nonlysosomal acidic compartment of the cell. Provision of intracellular iron for synthesis of ribonucleotide reductase, an enzyme that catalyzes the first step leading to DNA synthesis, is required for cell division. After dissociation of iron, transferrin and its receptor return, undegraded to the extracellular environment and the cell membrane, respectively.

III. Evolution of the Transferrin Gene

Evolutionary mechanisms contributing to the elongation and refinement of the plasma protein genes have included intragenic amplification, gene conversion, unequal crossing-over, reverse transcription, point mutation, and site replacement (Bowman and Yang, 1987). In the evolution of primitive to complex organisms, gene duplications contribute most heavily to the generation of new biochemical mechanisms. In addition to generating multiple genes with the same function, duplication of genetic material can also produce intragenic amplifications which increase the function of one protein product by increasing the number of its active sites.

A. Intragenic Duplication

The internal homology found in the amino and carboxyl halves of vertebrate transferrin support the notion that the protein evolved from a smaller ancestor. Transferrin gained an additional iron binding site following its intragenic duplication. The evolutionary ancestor of transferrin, 200 to 500 million years ago, was probably a polypeptide chain one-half the size of the contemporary vertebrate protein (Williams, 1982). The ancient intragenic duplication that produced the amplified transferrin gene occurred at a time when other plasma proteins are also thought to have expanded their lengths and increased their active sites by intragenic duplications and triplications (Doolittle, 1985).

The proposed scheme of evolution of the vertebrate transferrin gene is given in Fig. 1. The ancient *TF* duplication produced two homol-

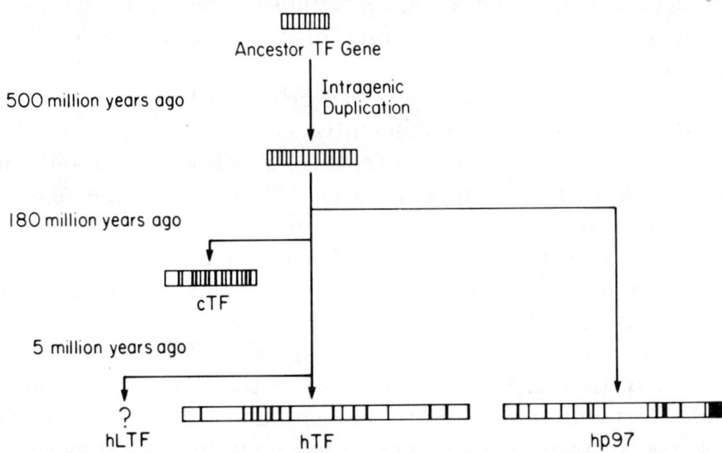

FIG. 1. A scheme for the evolution of the genes encoding the transferrin family. The relative intron–exon lengths in the genes of chicken transferrin (cTF), human p97 (hp97), and human transferrin (hTF) are based on data discussed in the text. Exons and introns of the ancestor genes are not drawn to scale. The genomic structure for human lactoferrin (hLTF) is currently not known.

ogous domains in the amino acid sequence of the protein (MacGillivray *et al.*, 1983; Yang *et al.*, 1984; Park *et al.*, 1985) and homologous exon–intron arrangements of the DNA (Plowman, 1986; Schaeffer *et al.*, 1985). Additional gene duplications of these DNA sequences occurred in vertebrates. Mammalian protein descendants include serum transferrin, the major plasma iron transport protein; p97, a melanoma antigen; and lactoferrin, a secreted protein present in milk and other body fluids. Amino acid and DNA sequences of human transferrin (MacGillivray *et al.*, 1983; Yang *et al.*, 1984; Park *et al.*, 1985), chicken transferrin (Williams *et al.*, 1982; Jeltsch and Chambon, 1982), p97 (Brown *et al.*, 1985; Plowman, 1986), and lactoferrin (Metz-Boutigue *et al.*, 1981; Pentecost and Teng, 1987) reflect the ancient intragenic duplication.

The determination of the nucleotide sequences encoding the amino and carboxyl domains of human transferrin indicated that during evolution selection acted more strongly on some regions of the *TF* gene exons than on others. When the nucleotide sequences of amino and carboxyl domains were compared, identity of 72, 64, and 62% was found in three regions, which therefore qualified as intragenic homology blocks (Sargent *et al.*, 1981; Yang *et al.*, 1984). Six pairs of tyrosine residues and three pairs of histidine residues conserved in the amino and carboxyl domains of both human and hen transferrin are encoded by nucleotides in the three homology blocks. These three regions appear, therefore, to be conserved in evolution because of the presence of codons for tyrosine and histidine residues predicted to be functional sites important in iron binding (MacGillivray *et al.*, 1983; Williams *et al.*, 1982).

During the evolution of the transferrin family, elongation of the genes occurred, not only by the intragenic duplication but also by increases of intron size that occurred in the period after divergence of the birds and mammals (approximately 150 to 180 million years ago). The coding regions of the transferrin gene in birds and serum transferrin in mammals are approximately the same length, 2.3 kb; however, the mammalian transferrin gene is 33.5 kb long while the chicken transferrin is only 10.5 kb. The difference is due to the significant expansion of several introns in the mammalian *TF* gene (Schaeffer *et al.*, 1985). Introns 2, 9, 13, 14, and 16 in the human *TF* gene have expanded to a length 6- to 16-fold that found in the chicken transferrin gene (Figs. 1 and 2). Sequence determination also suggests that unequal intragenic crossover occurred between exon 9 of one nonamplified gene and exon 2 of the other to produce the amplified ancestor gene (Schaeffer *et al.*, 1985). This is illustrated in Fig. 2. Insertion of an

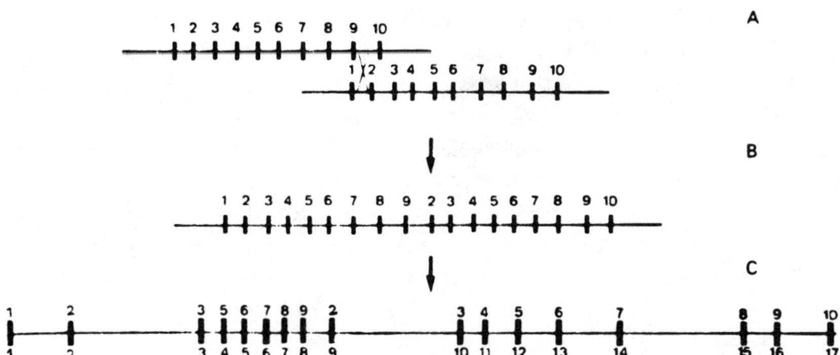

FIG. 2. A scheme for the origin of vertebrate *TF* gene by intragenic crossing-over. In A and B the distance between exons is arbitrary. In C, the different sizes of the exons are not drawn to the scale, but the lengths of the introns are proportional to the actual intron size. The upper numbers in C correspond to the numbers in A and B and the lower ones correspond to the exons of the present-day gene. (Modified from Schaeffer *et al.,* 1985.)

additional exon (exon 11) occurred in human and chicken transferrin genes but not in the gene for human p97 antigen (Plowman, 1986).

B. THE PRESENCE AND EXPRESSION OF THE *TF* GENE IN A PROCHORDATE

One of the most puzzling questions arising from the studies of vertebrate plasma proteins is why so many of them appear to be represented in the most primitive fish but have not been identified among the invertebrates or prochordates (Doolittle, 1984). Although an estimation can be made about common ancestry of human transferrin and chicken transferrin, and a rough approximation of the relative age of transferrin is possible, timing of the duplicative events leading to the elongation of primitive transferrin has not been precisely identified. However, the sensitivity and specificity of cDNA probes for specific plasma proteins are providing important information about the primal events leading to the contemporary plasma proteins.

The prochordate *Pyura,* a sea squirt having a notochord in its larval development, appears to possess in its genome unamplified versions of several of the plasma proteins that later duplicated and triplicated during evolution (Martin *et al.,* 1984; Yang *et al.,* 1985; Lum, 1987). In *Pyura stolonifera,* there appears to be an iron-binding transferrin-like protein, one-half the molecular weight of vertebrate transferrin (Martin *et al.,* 1984). We, in collaboration with Drs. Martin, Hubers, and Finch of the University of Washington in Seattle, have studied a related sea squirt, *Pyura haustor*. Hybridization analysis of *Pyura* DNA was carried out with radiolabeled human TF cDNA on Southern filters. Genomic DNA (Fig. 3) from *Pyura* was fragmented with 11 specific restriction endonucleases (Yang *et al.*, 1985). The filter, containing an impression of the *Pyura* genomic fragments that had been separated by electrophoresis, was exposed to a radiolabeled human TF cDNA probe. The bands shown in Fig. 3 are DNA fragments of *Pyura* that hybridized with human TF cDNA. At first, hybridization of

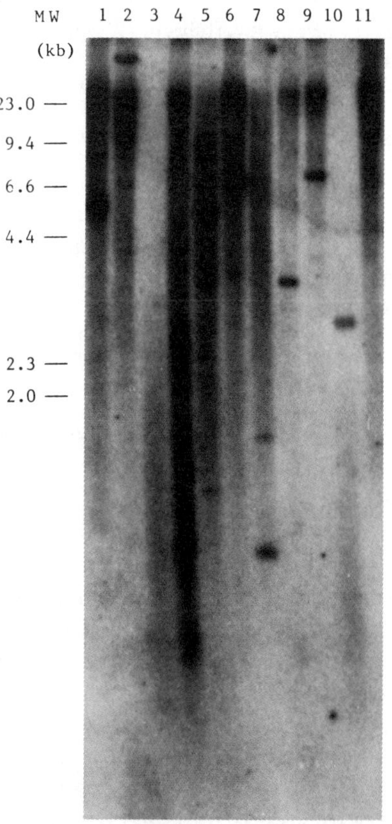

FIG. 3. Autoradiogram of *Pyura* DNA analysis by Southern hybridization with human TF cDNA. Each lane contains 5 μg of *Pyura* DNA digested with a restriction endonuclease. Restriction endonucleases used are as follows: lane 1, *Bam*HI; lane 2, *Bgl*II; lane 3, *Hpa*II; lane 4, *Dde*I; lane 5, *Eco*RI; lane 6, *Kpn*I; lane 7, *Pst*I; lane 8, *Pvu*II; lane 9, *Sal*I; lane 10, *Taq*I; lane 11, *Xho*I. Extensive homology of *Pyura* and human *TF* sequences is illustrated. (From Yang *et al.*, 1985.)

human TF cDNA with *Pyura* DNA was carried out at low temperature (30°C) to encourage cross-hybridization. Surprisingly, most of the hybridization signals could also be detected even when hybridization was performed at highly stringent conditions; this indicated that the *TF* gene sequence had been significantly conserved throughout the evolutionary period separating prochordates and humans, about 400–500 million years (Fig. 1). Judging from the sizes of hybridized fragments obtained from 11 restriction digests (Fig. 3), the *Pyura TF* sequence appears to be significantly smaller than the human counterpart which has been demonstrated to be over 30 kbp (Schaeffer *et al.*, 1985).

To determine whether the amino or carboxyl domain of the human *TF* gene was more homologous to *Pyura* DNA, a series of hybridizations were carried out using radiolabeled probes from the coding regions of the amino- and carboxyl-terminal domains after dissecting human TF cDNA into several fragments (Yang *et al.*, 1985). Hybridization of the cDNA fragments to *Pyura* genomic DNA demonstrated strong homology throughout the entire human cDNA sequence.

One point of interest was whether the regulatory signals that specify the tissue-specific expression of the *TF* gene were also conserved during evolution. Vertebrate transferrin is expressed mainly in liver although extrahepatic sites of expression exist and will be discussed below. To study the tissue-specific expression of the *Pyura TF* gene, *in situ* hybridization was performed (Lum, 1987) with radiolabeled human cDNA and *Pyura* tissue. Hybridization, which indicated sites of TF mRNA transcripts, was observed in unfertilized eggs and the primitive digestive glands of *Pyura*. The digestive gland cells constitute the vertebrate liver anlagen (Fig. 4). However, no hybridization was detected in other tissues such as muscle or stomach. Therefore, the highly conserved human *TF* homolog in *Pyura* appears to be expressed in a tissue-specific fashion similar to the human *TF* gene (Yang *et al.*, 1985).

C. Evolutionary Relationships of Transferrin and Its Receptor

Interesting relationships in the amino acid and nucleotide sequences of some ligands and the complement of their corresponding receptor mRNA have been indicated (Bost *et al.*, 1985). Short regions have been identified in the nucleotide and amino acid sequences of epidermal growth factor, interleukin-2, and transferrin that match short regions

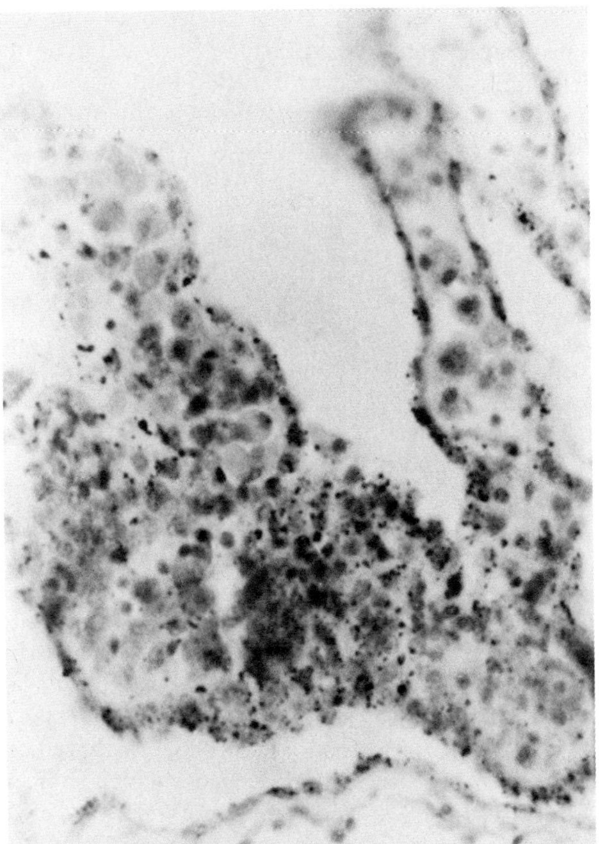

FIG. 4. *In situ* hybridization of primitive digestive gland of *Pyura*. Hybridization was carried out with human TF cDNA to localize TF mRNA transcripts. (From Lum, 1987.)

in their respective receptor complements (antisense strand) and their deduced amino acid sequences. In each case the region of homology of the receptor was in sequences external to the cytoplasmic membrane that might qualify for the ligand binding site. In Fig. 5, two of the seven examples reported for transferrin and its receptor are shown. Although the homology observed in transferrin and its receptor is faint, and any evolutionary basis unknown, the observation remains intriguing.

TF						TF RECEPTOR						TF RECEPTOR COMPLEMENT					
130					135	219					214	219					214
Ile	Pro	Ile	Gly	Leu	Leu	Tyr	Gly	Gly	Pro	Asn	Glu	Ile	Pro	Pro	Gly	Leu	Leu
AUC	CCC	AUA	GGC	UUA	CUU	TAT	GGT	GGG	CCT	AAT	GAG	AUA	CCA	CCC	GGA	UUA	CUC
281					286	375					370	375					370
Glu	Phe	Gln	Leu	Phe	Ser	Leu	Lys	Val	Asn	Lys	Ser	Glu	Phe	His	Leu	Phe	Ser
GAA	UUC	CAA	CUA	UUC	AGC	CTC	AAG	GTG	AAT	AAG	AGC	GAG	UUC	CAC	UUA	UUC	UCG

FIG. 5. Regions of homology between TF and its receptor complement. Two 18-nucleotide regions of human TF cDNA have homology with 18-nucleotide regions in the complement of human TF receptor cDNA. Nucleotide homologies for these regions are 61 and 67% for the upper and lower nucleotide sequences, respectively. Amino acid homologies are both 83%. The numbers above the amino acid abbreviations indicate amino acid positions. (From Bost *et al.*, 1985.)

D. Chromosomal Location of the *TF* Gene Family

Chromosomal mapping studies have been predictive of evolutionary relationships and have provided valuable information about families of related genes, some of which have existed together on the same chromosome for millions of years and other which have been separated over time by chromosomal translocations and inversions. Comparative studies of gene linkage have established that many linkage groups are conserved in widely divergent species. Plasma protein genes have been mapped on 18 of the 22 human autosomes, as well as on the X chromosome (Bowman *et al.*, 1987). The genes for transferrin and its receptor (TFR) are linked on chromosome 3. The *TF* gene was mapped to human chromosome 3q by hybridization analysis of radiolabeled cDNA to DNA of mouse–human hybrid cells including human parental lines having specific rearrangements of chromosome 3 (Yang *et al.*, 1984). Hybridization of radiolabeled TF cDNA with metaphase spreads of human chromosomes revealed the *TF* gene was located in the region of 3q21-25 (Yang *et al.*, 1984). Genes coding for the transferrin receptor, lactoferrin, and the melanoma antigen (p97) are also located on the long arm of chromosome 3. The transferrin receptor gene had previously been mapped at 3q22-qter and 3q26.2, respectively (Miller *et al.*, 1983; Rabin *et al.*, 1985) while *p97* had been located on chromosome 3 by somatic cell hybrid analysis (Plowman *et al.*, 1983). Comparison of *in situ* hybridization patterns of chromosomes from a normal and a malignant cell line containing a rearranged chromosome 3 identified the chromosomal locations of these related proteins to be *TF* on 3q21, *TFR* on 3q26, and *p97* within

3q28-29 (LeBeau *et al.,* 1985). Utilization of the cDNA encoding mouse lactoferrin (LTF) on DNA of somatic cell hybrids with human and rodent parental lines established its chromosomal location to be on human chromosome 3q (Teng *et al.,* 1987); *in situ* hybridization of the same cDNA on human chromosomal spreads established the exact *LTF* gene location to be 3q21-23 (McCombs *et al.,* 1987). Recent gene assignments to chromosome 3 are illustrated in Fig. 6.

Information available points to the presence of a single copy of the *TF* gene in the haploid chick (Lee *et al.,* 1978) and human (Yang, unpublished) genomes. However, a human transferrin pseudogene was recently described by Schaeffer *et al.* (1987). The pseudogene is likely to be a nonprocessed gene because differences in its reading frame produce several stop codons and modify intron–exon boundaries. When gaps were introduced in the pseudogene's sequence to maximize homology there was 65% identity with the amino acid sequence and 72% identity with the nucleotide sequence of exons 7, 8, 9, 10, and 12 of the human *TF* gene. The pseudogene, like the *TF* gene, was mapped to human chromosome 3.

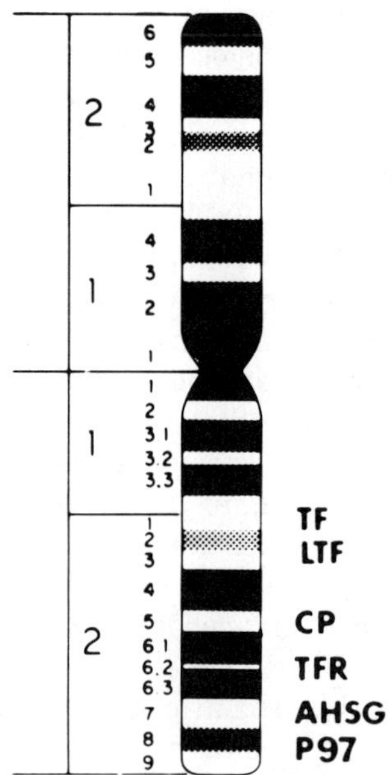

FIG. 6. Regional assignment of transferrin-related genes on human chromosome 3. TF, Transferrin; LTF, lactoferrin; P97, oncofetal protein p97; TFR, transferrin receptor. Two other linked metal-binding proteins, α_2-HS-glycoprotein (AHSG) and ceruloplasmin (CP), are encoded by genes on chromosome 3.

The *TF* gene is linked to genes encoding two other proteins that bind heavy metals, ceruloplasmin (CP) that binds copper and α_2-HS-glycoprotein (AHSG) that binds calcium and barium. Neither protein is homologous to transferrin. Linkage of *TF* and *CP* was found earlier by family studies; the number of recombinations between the *TF* and *CP* loci predicted that the two genes were separated by an approximate distance of 10 centimorgans (Weitkamp, 1983). Ceruloplasmin is thought to serve as an oxidant that converts ferrous to ferric iron (Frieden, 1986). Only the ferric form of iron can specifically be bound by transferrin. Human cDNA encoding CP has recently been cloned; utilization of the probe has mapped the *CP* gene to 3q25 (Yang *et al.*, 1986). Furthermore, the gene encoding α_2-HS-glycoprotein has been mapped to human chromosome 3q by cDNA hybridization with DNA panels of somatic cell hybrids (Lee *et al.*, 1987) and to 3q27-29 by *in situ* hybridization (Magnuson *et al.*, 1987). The locations of these plasma protein genes are shown in Fig. 6.

In mice the transferrin gene is linked to genes encoding lactoferrin (Teng *et al.*, 1987), aminoacylase, and β-galactosidase on chromosome 9, a homolog of human chromosome 3p (Naylor *et al.*, 1982). A rearrangement of chromosome 3 involving a pericentric inversion separated the transferrin gene during primate evolution from other markers of 3p (de Grouchy *et al.*, 1978; Yunis and Prakash, 1982). Humans, gorillas, and chimpanzees have the rearranged chromosome 3, while orangutans do not.

E. Homology of Promoter Regions of Transferrin-Related Genes

The availability of a 3.6-kb region of the 5′ region of the human *TF* gene enabled Adrian *et al.* (1986) to define the transcription start site, the translation start codon, and a 5′ region of the *TF* gene that directs tissue-specific expression in transfected cells (Adrian *et al.*, 1987). The transcription start site is located 50 bp 5′ to the translation start codon as determined by S_1 nuclease and primer extension (Adrian *et al.*, 1987). Striking homology was demonstrated when the human *TF* gene promoter was aligned with the chicken transferrin promoter (Cochet *et al.*, 1979). Figure 7 demonstrates the conservation of sequence and the similarity of positions of initiation site, TATA box, and two reverse CAAT boxes within the 5′ region of both genes.

Examination of the promoter regions of the human *TF* and *TFR* genes indicates some similarities in orientation. Because of the coordinated expression of *TF* and *TFR* in activated T lymphocytes, both

```
            -132                                                    -96
HUMAN TF:   TGGGACGAGTAA-GGAAGGGGGG----TTGGGAGAGGGGCGATTGGGCAA
            ****  ***    **********    *******      * **** * **
CHICK TF:   TGGGCTGAGCCGGGGAAGGGGGGCAACTTGGGAG-----CTATTGAG-AA

                                                -58
HUMAN TF:   CCC-GGCTGCACAAACACGGG-AGGTCAAAGATTGC----GCCCAGCCC
             *  **  * *******  *  *****     *****    **  *  *
CHICK TF:   ACAAGGAAGGACAAACAGCGTTAGGTC----ATTGCTTCTGCAAACACA

                              -27                            +1
HUMAN TF:   GCCCAGGCCGGGAATGGAATAAA-GGG-ACGC--GGGGCGCCGGACGTG
            ***  *** *    *   ***** *** * *   * *** *** *
CHICK TF:   GCCAGGGCTGCTCCTC-TATAAAAGGGGAAGAAAGAGGCTCCGCAGCCA
                                                             +1
```

FIG. 7. Conservation of sequence of chicken and human *TF* promoters. The human *TF* promoter (Adrian *et al.*, 1986) is compared with the chicken *TF* promoter (Cochet *et al.*, 1979). When arranged as above, there is a 64% nucleotide identity between these two promoters. Conserved sequences include two CAAT boxes (on the complement strand) and TATAA boxes which are boxed. Asterisks denote identical bases in the two promoters.

genes may be expected to share sequences that are common recognition sites for premitotic signals. The signals, in turn, may be responsible for expression of the genes in proliferating cells (Lum *et al.*, 1986). For example, Miskimins *et al.* (1986) have demonstrated the presence of GGGGC nucleotide sequences in a group of genes, including the *TFR* gene, that are expressed at elevated levels in proliferating cells. The human *TF* and *TFR* gene promoter regions are compared in Fig. 8. Similarities include GC-rich regions, Sp1 recognition sites (Dynan and Tjian, 1983) 5′ to the TATA sequence, and GGGGC repeats. Future studies will be required to determine whether the *TF* and *TFR* genes contain DNA sequences that respond to the same premitotic signals responsible for cell proliferation.

IV. Expression of *TF* Gene

A. Tissue-Specific Expression

Like most plasma proteins, transferrin is synthesized primarily in the liver; however, important extrahepatic sites of transferrin expression also have been discovered recently. Synthesis of transferrin in nonhepatic tissue probably contributes to the homeostasis of cells that are separated by blood barriers from plasma transferrin and to requirements of specific tissues deprived of transferrin because of

```
TF promoter:

                  Sp1                                              +1
ATTGCGCCCAGCCCGCCCAGGCCGGGAATGGAATAAAGGGACGCGGGGCGCCGGACGTGCA
             ******                                 +++++

TFR promoter:
                       Sp1
GCCCCCCTGGGGGCCGGGGCGGGGCCAGGCTATAAACCGCCGGTTAGGGGCCGCCATCCCCTCAG
                  ******                                 +++++           +1
```

Percentage GC in above Sequences

	5' to TATA	**3' to TATA**
TF:	**74**	**87**
TFR:	**93**	**75**

FIG. 8. A comparison of promoter regions of genes encoding TF and the TF receptor. Portions of the human TF promoter (Adrian *et al.*, 1986) and human TF receptor (TFR) promoter (Miskimins *et al.*, 1986) are compared. Both regions are GC rich, contain TATA boxes, Sp1 recognition sites 5′ to TATA, and one GGGGC pentanucleotide 3′ to TATA. In addition, the two gene promoters contain similar heptanucleotides, GGCCGGG and GGCCAGG, between the Sp1 sites and TATA.

physiological restrictions following inflammation or iron overload. Cellular signals appear to elicit expression of the *TF* gene in specific tissues during times in which there is a requirement for its function as a growth factor.

Work by McKnight and co-workers (1983) suggests that transcription of the *TF* gene is controlled by organ-specific and gene-specific factors. The TF mRNA content of various tissue of normal rats indicated at least five sites of transcription: in the liver (6500 molecules/cell), testis (114/cell), brain (83/cell), spleen (11/cell), and kidney (5/cell) (Idzerda *et al.*, 1986).

1. *Hepatic Sites*

The liver is the site of synthesis of essentially all of plasma transferrin, although minute contributions to this pool are undoubtedly made by other cells that have been shown to synthesize transferrin (Morgan, 1983; Lum *et al.*, 1986). Studies in humans have demonstrated that the rate of transferrin synthesis by liver was 10.5% of the plasma pool per day (Wochner *et al.*, 1968; Kernoff and Baker, 1980).

This value corresponds closely to determinations of the catabolic rate of plasma transferrin, 6–14% of the plasma pool per day (reviewed by Morgan, 1983). Following liver transplantations, the plasma transferrin type of the donor is reflected in the recipient's plasma (Alper *et al.*, 1980).

2. *Extrahepatic Sites*

a. Central Nervous System. The *TF* gene is expressed in the central nervous system at specific locations. Transferrin mRNA has been detected in choroid plexus (Aldred *et al.*, 1987) and in oligodendrocytes, a type of glial cell, in adult rat brain (Bloch *et al.*, 1985). Dickson *et al.* (1985) reported the presence of TF mRNA in extracts from spinal cord and spinal cord ependyma, but not in extracts from pia mater. TF mRNA was also observed in thalamus and medulla. Demonstration of synthesis of transferrin in the choroid plexus (Dickson *et al.*, 1985) implied that the transferrin concentration of the cerebrospinal fluid, formerly believed to originate from plasma, arises, in part if not entirely, from synthesis in this tissue.

Transferrin mRNA level in the rat brain is very low before birth, gradually increases during postnatal development, and reaches a maximum in the 60-day-old rat adult (Levin *et al.*, 1984).

b. Testis Sertoli Cells. Thorbecke *et al.* (1973) have reported that transferrin is synthesized by both testis and ovary. Skinner and Griswold (1980) demonstrated synthesis and secretion of transferrin by Sertoli cells *in vitro*. Spermatids, spermatozoa, and dividing spermatocytes obtain their nutrients from Sertoli cells which are responsible for secretion of most of the fluid components of the seminiferous tubules; transferrin and the androgen-binding protein represent a significant amount of the proteins secreted by the Sertoli cells and appear to be under independent hormonal control (Perez-Infante *et al.*, 1986). In cultured Sertoli cells retinoids, insulin, epidermal growth factor, and testosterone significantly increased transferrin secretion. Androgen-binding protein secretion was stimulated by follicle-stimulating hormone, testosterone, and insulin. Hormonal modulation of transferrin expression by follicle-stimulating hormone, insulin, retinol, and testosterone was shown to be on a transcriptional level (Huggenvik *et al.*, 1987).

Levels of transferrin synthesis and transcription in Sertoli cells *in vivo* appear to be significantly lower when compared to Sertoli cells grown *in vitro* (Lee *et al.*, 1986). This may reflect the correlation of levels of *TF* transcription and ambient concentrations of hormones and vitamins *in vivo*. *In situ* hybridization of human testicular tissue with radiolabeled TF cDNA probes demonstrated TF mRNA transcripts in some testicular cells *in vivo* although the identification of the cell types containing TF mRNA transcripts was not possible in this study (Lum, 1986). Since transferrin is probably required for proliferation of germ cells in the testis, a local supply is beneficial to the host. The tight junctions between Sertoli cells constitute a blood–testis barrier (Fawcett, 1975) that may prevent adequate transferrin from being supplied by the circulation.

c. Mammary Gland. Levels of transferrin in mammary glands of virgin mice are significantly lower than in pregnant or lactating animals. The rate of transferrin synthesis in lactating mammary glands of rat and rabbit has been reported to be 1.5- to 3-fold higher than that of liver (Morgan, 1968; Jordan and Morgan, 1969, 1970). Synthesis of transferrin by mammary epithelium of mice accounts for its presence as a major protein in milk. The *TF* gene expression appears to be regulated independently from that of other major milk proteins (Lee *et al.*, 1987).

d. Lymphocytes and Thymus Gland. Lum *et al.* (1986) demonstrated the transcription and synthesis of transferrin by human helper–inducer T lymphocytes at a period after lymphocyte activation and before proliferation. Soltys and Brody (1970) reported that human peripheral mononuclear cells synthesized transferrin and that transferrin was localized in the T cell compartment. The *TF* gene appears to be expressed in 60–75% of peripheral T lymphocytes of the $T4^+$ subset (Lum *et al.*, 1986). Furthermore, the transferrin protein was identified by immunoprecipitation of radiolabeled proteins from supernatants and cell extracts of phytohemagglutinin-activated human peripheral mononuclear cells. Studies of *in situ* hybridization of radiolabeled cDNA TF probes with lymphocytes revealed the presence of TF mRNA transcripts only in the activated helper–inducer ($T4^+$) subset, and not in B lymphocytes, the $T8^+$ subset, or resting $T4^+$ lymphocytes (Lum *et al.*, 1986). As discussed below, transferrin may participate in an autocrine scenario in lymphocytes that serves as a typical solution to the requirements of proliferating cells in regions spatially isolated from the circulation or during periods in which transferrin-bound iron is being sequestered to protect against bacterial or malignant cellular invasion.

Lum (1987) observed TF mRNA transcripts in cells of the fetal thymus gland and in some, but not all, malignant cell lines arising from the helper–inducer T lymphocyte subset. TF mRNA transcripts were identified in HPB-ALL and Molt-4 cell lines by *in situ* hybridization (Lum, 1987).

e. Lung, Heart, Spleen, Kidney, Muscle, and Intestine. Aldred *et al.* (1987) analyzed tissue-specific expression of transferrin using TF cDNA as a probe. These studies revealed TF mRNA at very low levels in rat placenta, spleen, kidney, muscle, and heart (Fig. 9). No transferrin mRNA was detected in total RNA of yolk sac, small intestine, and adrenal glands. The same workers, however, reported the presence of TF mRNA in yolk sac and placenta after analyzing polyadenylated RNA (Schreiber, 1987).

Levin *et al.* (1984) measured transferrin expression by RNA blot hybridization in adult rat liver, total brain, muscle, kidney, and spleen. The transferrin gene was found to be transcribed in muscle before birth and in brain after birth. Gene expression in muscle is maximal before birth during day 19 to day 20 of gestation, then

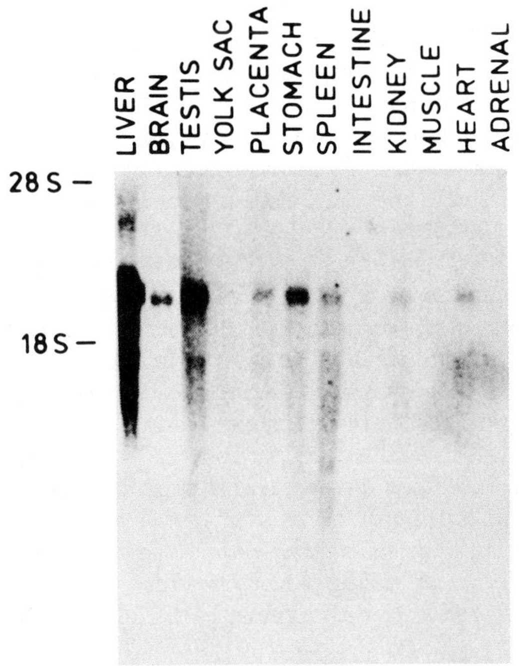

FIG. 9. Tissue distribution of transferrin mRNA by Northern blot analysis. (From Aldred *et al.*, 1987.)

decreases markedly during postnatal development. Maximum expression in the nonhepatic tissues reaches approximately one-tenth that of adult liver.

Transferrin mRNA was observed in intestine of newborns but was not detectable in adult intestine (Levin *et al.,* 1984). The role of transferrin in intestinal iron absorption remains a controversial topic (reviewed by Osterloh *et al.,* 1987). Mucosal transferrin has been described in the intestine, but whether it is derived from the plasma pool or is distinct remains to be settled. *In situ* hybridization with TF cDNA probes would be the ideal analytical method to apply to the solution of this problem.

3. Transferrin in Transgenic Mice

The transgenic mouse model presents an opportunity to analyze *in vivo* the expression of a foreign gene within a developing mammalian system (Mintz and Illmensee, 1975; Brinster and Palmiter, 1982). Following the physical insertion of a gene into fertilized mouse eggs, the foreign DNA is frequently incorporated into one or more of the mouse chromosomes. This usually enables the foreign gene to be transmitted to the progeny of transgenic mice in a Mendelian manner. Experimental results obtained from using the transgenic model attest to the value of the technique and demonstrate that the expression of a foreign gene can be studied in every cell type during every stage of mammalian development—from embryonic, fetal, and mature stages into the aging process.

Experiments by McKnight *et al.* (1983) demonstrated that the *TF* gene is directed by a robust promoter that leads to transferrin expression in transgenic mice in a tissue-specific manner. The chicken *TF* gene microinjected into the male pronucleus of fertilized mouse eggs was carried in the genome of, and expressed in livers of, six of seven transgenic mice studied and demonstrated a concentration 4- to 10-fold greater than the concentration of transferrin in other tissues. Five of the mice secreted chicken transferrin into their serum. Some offspring of transgenic mice expressed the chicken *TF* gene. The study demonstrated great promise of the transgenic mouse model for studying *in vivo* the hormonal and environmental factors regulating gene expression.

Modulation of expression of a 17-kb DNA fragment of the chicken *TF* gene inserted into the genome of transgenic mice was carried out by introducing estrogen as a single injection of 100 μg of estradiol benzoate or by estrogen administration for 10 days (Hammer *et al.,* 1986). After 10 days of estrogen administration, there was a 2-fold increase in transferrin mRNA and transcription. A single injection of estradiol led to a 4-fold increase in TF mRNA synthesis after 4 hours. Tissue specificity and gene modulation have been shown to result from the high affinity binding of specific nuclear trans-acting proteins with conserved cis-acting DNA elements frequently clustered in the promoter region of each gene (Dynan and Tjian, 1985). Maintenance of tissue specificity and estrogen responsiveness of the chicken transferrin gene in transgenic mice indicated that the cis-acting elements were highly conserved in species that diverged 150 to 180 million years ago (Hammer *et al.,* 1986).

A line of transgenic mice containing 11 copies of the complete 17-kb chicken transferrin gene on the X chromosome was used to study the process of inactivation of the X chromosome. The results found by Goldman *et al.* (1987) demonstrated that expression of the chicken transferrin failed to be suppressed even when the gene was located on the X chromosome that was inactivated. The results suggested that the inactivation process cannot spread over 187 kb of DNA in the absence of specific signal sequences, possibly CpG-rich islands (Cullen *et al.,* 1986), thought to be required for X inactivation.

B. Expression of Transferrin during Proliferation, Development, and Aging

The regulation of transferrin expression appears to be under temporal control during development as summarized above. For example, transferrin is synthesized in rat muscle, kidney, and intestine before birth and diminishes dramatically after birth, while transferrin synthesis in brain appears to be negligible before birth and increases significantly during development. The coordinated signals responsible for temporal control are not well understood but can be predicted to involve hormonal, autocrine, and mitotic signals that act *in trans* upon regions of DNA in the *TF* gene to control gene expression positively and negatively.

1. *Synthesis of Transferrin during Ontogeny*

Transferrin, which is probably the immediate source of iron needed for hemoglobin production in the developing red cells in blood islands of the visceral yolk sac, can be detected in fetal serum and in amniotic fluid. In mouse, hemopoiesis starts at about day 8 of gestation (Bateman and Cole, 1971). Transferrin is synthesized by mouse embryos in an undersialated form as early as day 13 of gestation, being gradually replaced by the sialated adult type by day 17 (Gustine and Zimmerman, 1972, 1973). In rats transferrin is synthesized by the liver by day 15 of gestation (Yeoh and Morgan, 1974).

The visceral yolk sac of the human fetus synthesizes transferrin, α-fetoprotein, α_1-antitrypsin, prealbumin, and albumin; the organ starts to degenerate after the sixth week of gestation (Gitlin and Pericelli, 1970).

The expression of the *TF* gene during early mammalian development was analyzed by following the incorporation of radiolabeled amino acids into the transferrin protein in postimplantation mouse embryos. Transferrin synthesis could be detected in egg cylinders as early as on day 7 of gestation onward and in the visceral yolk sac at all stages (Adamson, 1982). The major sites of synthesis in mouse fetuses are the yolk sac and liver.

Work in mouse development by Meek and Adamson (1985) strongly supports transferrin's role in the stimulation of fetal growth and differentiation. A survey of murine fetal tissues identified transferrin in embryonal cells that have special demands for transferrin. Since there appears to be a barrier to the passage of transferrin from maternal circulation (Faulk and Galbraith, 1979), the transferrin required in development would predictably be produced as an early product of the murine embryo. Meek and Adamson (1985) found transferrin synthesis in murine fetal lung, spleen, spinal cord, and rib cage. Very low transferrin levels were found in brain, muscle, and pancreas. The authors concluded that these sources of newly synthesized transferrin are significant for the development or maturation of tissues and demonstrated that transferrin synthesis is switched off in most adult tissues when it is no longer needed. Their results suggest that during gestation transferrin levels are highest when the proliferative demands of the fetal tissue are very great.

2. *Growth Factor and Autocrine Function*

Transferrin has been shown to be a growth factor required for both normal and malignant proliferating cells (Barnes and Sato, 1980; Brock and Mainou-Fowler, 1983; Seligman, 1983). Recognition of the requirement for transferrin during cell growth and proliferation has accompanied the development of serum-free media. Although the need for proliferating cells to acquire iron is sufficient to account for the transferrin requirement, it has been suggested that transferrin may fulfill an additional role unrelated to its iron-donating properties (Brock and Mainou-Fowler, 1983; Seligman, 1983). Recent work by Brock *et al.* (1986), however, indicates that in lymphocyte cultures transferrin is a growth factor simply because it can provide iron needed for cell growth and division.

a. Transferrin and T Lymphocyte Proliferation. Utilization of *TF* cDNA with the *in situ* hybridization technique provided the opportunity for identifying transferrin transcription, synthesis, and secretion after activation of human lymphocytes (Lum *et al.,* 1986). Transcription of *TF* mRNA was noted only in the $T4^+$ subset, but not in the $T8^+$ subset or B cells, specifying *TF* expression in the helper–inducer T lymphocytes.

The expansion of clones of lymphocytes is an essential feature of the immune response. T lymphocyte proliferation is controlled by a series of premitotic signals generated by interaction of growth factors and their membrane receptors. When T cells are activated, the series of events known to lead to DNA synthesis and cell proliferation include the expression of the receptor for interleukin 2 (IL-2; T cell growth factor) (Robb *et al.,* 1981; Leonard *et al.,* 1982), IL-2 production (Ruscetti *et al.,* 1977; Gillis and Smith, 1977), and TF receptor production (Neckers and Cossman, 1983; Trowbridge and Omary, 1981). Our laboratory has shown that the *TF* gene is also expressed, in a specific stage during the premitotic events of T lymphocyte differentiation, by the $T4^+$ "inducer" subset of T lymphocytes (Lum *et al.,* 1986). Figure 10 demonstrates the appearance of TF mRNA transcripts after activation of T lymphocytes. We demonstrated that *TF*

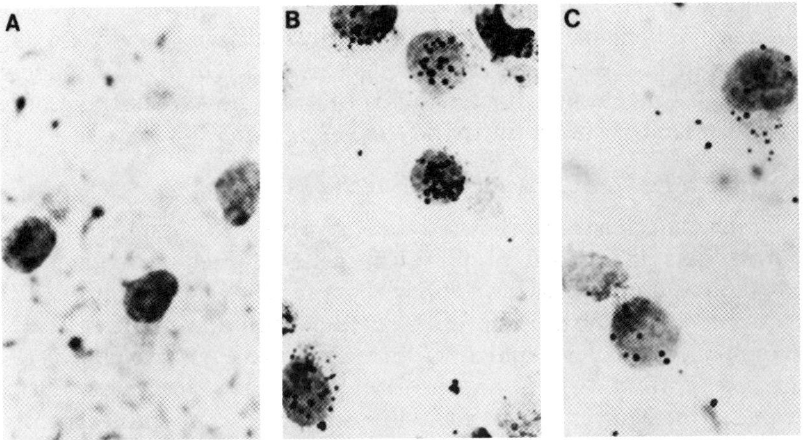

FIG. 10. TF mRNA transcription by a human $T4^+T8^-$ T cell clone. (A) Localization of TF mRNA in $T4^+T8^-$ T cell clones before activation, day 0; (B) day 1 after activation; and (C) day 2 after activation. (From Lum *et al.,* 1986.)

transcription is an intermediate event in the IL-2 autocrine cycle and occurs after IL-2 transcription and prior to expression of the IL-2 and the TF receptor. A proposed scheme of transferrin synthesis and cell proliferation is shown in Fig. 11. Transferrin synthesis by T lymphocytes appears to be part of a normal autocrine reaction in which growth factors and their respective receptors act as premitotic signals for DNA synthesis that is followed by cell proliferation. Expression of *TF* by helper–inducer T lymphocytes also predicted that human leukemia cells expressing the $T4^{+}$ marker might be frozen in a self-perpetuating automitogenic response and continually synthesize transferrin. Subsequent studies in our laboratory have revealed that the two human leukemia cell lines, HPB-ALL and Molt-4, indeed synthesize transferrin (Lum *et al.,* 1986; Lum, 1987).

The autocrine process in lymphocytes, where IL-2 and TF and their respective receptors serve as premitotic signals, is not clearly understood. Little is known about how proteins that serve as stimulatory signals react with each other. Neckers and Cossman (1983) demonstrated that an interaction of IL-2 with its receptor is required for the expression of TF receptors. Preliminary work by Lum (1987) indicates transferrin, synthesized by lymphocytes, must also interact with its receptor for an additional and successive stimulatory signal required for cellular proliferation.

The immune system requires the delivery of transferrin-bound iron for proper lymphocyte function (Brock, 1981; Mainou-Fowler and Brock, 1985). Therefore, during situations where iron is being withheld from the circulation, the capacity of a subset of human T lymphocytes to synthesize transferrin may provide a source of available iron to support localized proliferation of lymphocytes required in the host's immune response.

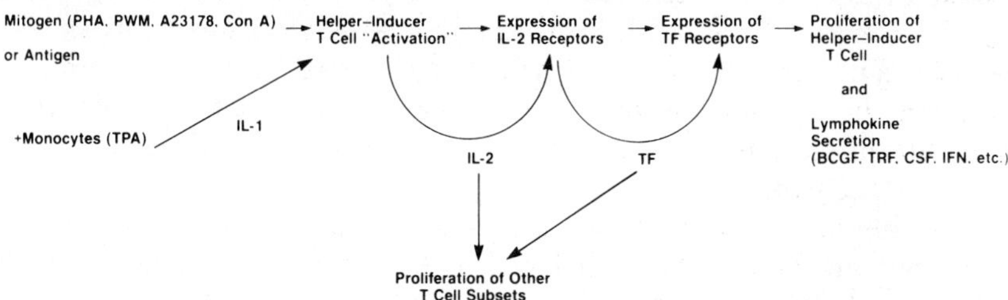

FIG. 11. Proposed role of transferrin (Tf) synthesis by helper–inducer T cells in T cell activation, proliferation, and IL-2 secretion. Additional abbreviations used in this figure: PWM, pokeweed mitogen; Con A, concanavalin A; TPA, 12-*O*-tetradecanoylphorbol 13-acetate; BCGF, B cell growth factor; TRF, T cell-replacing factor; CSF, colony-stimulating factor; IFN, interferon-γ. (From Lum *et al.,* 1986.)

Of particular interest has been the discovery of a truncated transferrin protein synthesized by T cells. In addition to the 79-kDa TF found in serum, a 62-kDa TF protein was also observed to be synthesized by normal T cells (Lum *et al.,* 1986). This 62-kDa TF protein constituted more than 60% of the total TF immunoreactivity detected in T cells. The molecular weight of the normal TF polypeptide backbone is predicted to be at least 75,000; therefore, the 62-kDa TF species made by lymphocytes does not appear to correspond to a nonglycosylated or partially glycosylated form of transferrin. It is possible that the truncated TF protein is the product of altered transcription of the *TF* gene. Since the 62-kDa TF derivative has never been identified in serum, its production by lymphocytes may be the product of a T cell-specific component(s). The nature and origin of the 62-kDa TF as well as its impact on the autocrine pathway of T lymphocyte proliferation await further study.

Transferrin receptor (TFR) expression has been correlated with sensitivity of target cells to natural killer (NK) T cells and thus suggested to be an NK recognition site (Vodinelich *et al.,* 1983; Lazarus and Baines, 1985). Shau *et al.* (1986), however, concluded that TF and TFR function not during the execution of NK cytotoxicity but rather during the induction of cytotoxic potential in the T cells.

b. Transferrin and Differentiation. Transferrin has been observed to be necessary for proliferation and differentiation of mouse fetal kidney (Ekblom *et al.,* 1983), muscle (Ii *et al.,* 1982), and lymphocytes (Brock, 1981; Lum *et al.,* 1986). Transferrin is also required for the proper development of teeth (Partanen *et al.,* 1984).

In hepatocytes, heme biosynthesis is controlled by 5-aminolevulinic synthase. However, Ponka and Schulman (1985) have presented evidence that heme synthesis in reticulocytes is not entirely regulated by 5-aminolevulinic acid synthase but also by some step in the acquisition of iron from transferrin. They suggest that this controlling step may be through the activity or processing of transferrin receptors, the rate of release of iron from transferrin, or the reduction and/or transport of iron from transferrin to ferrochelatase.

Actively proliferating cells express a high density of surface receptors for transferrin, and transferrin is known to be a required factor when most cells are cultured in serum-free media. Shannon *et al.* (1986) examined the importance of transferrin receptor expression during the growth and differentiation of erythroid bursts. They demonstrated that erythroid progenitors are extremely sensitive to inhibition by monoclonal antibodies directed against transferrin or its receptor. The erythroid precursors are much more sensitive than myeloid precursors. Their findings suggest that transferrin plays a major regulatory role in the differentiation of burst-forming units into erythroid bursts.

3. Aging

Aging is accompanied by alterations in gene expression in somatic cells (Phillips and Cristofalo, 1985). Meaningful results have emerged from investigations in which levels of major plasma and urinary proteins were followed throughout development and aging in experimental animals. Roy and Chatterjee (1983) reported changes in levels of 17 hepatic proteins during the aging of rats; 5 proteins increased and 12 decreased in levels as the animals aged. Some of the observed differences in gene expression occur in the transcription step and are attributed to decreased levels of mRNAs (Roy *et al.,* 1983a,b; Richardson *et al.,* 1985). Among the effects of the aging processes is a significant decrease in transcription of interleukin 2, the T lymphocyte growth factor, in aged mice, rats, and humans (Gillis *et al.,* 1981; Thoman and Weigle, 1981; Chang *et al.,* 1982; Cheung *et al.,* 1983; Rabinowich *et al.,* 1985). Decrease in IL-2 and observations of the decreased levels of IL-3 mRNA in mitogen-stimulated lymphocytes of aged mice (Wu *et al.,* 1986) have helped explain some of the events associated with the decline of the cellular immune response during aging.

The TF mRNA levels in the livers of different aged mice have been examined in our laboratory. Preliminary results suggested that there may be a decrease in the mRNA levels in livers of aging animals (Fig. 12). These results are presently being confirmed in transgenic mice that carry human *TF* gene constructs in their genomes. It will also be possible to study *TF* gene regulation in other tissues, such as brain and testis, during development and aging of transgenic mice. The brain regulation of *TF* gene expression will be of particular relevance to the varying levels of iron present in brain.

Iron is unevenly distributed in the rat brain (Hill and Switzer, 1984); it is localized in neurons in some brain areas but most accumulation is in oligodendrocytes where transferrin has been shown to be expressed (Levin *et al.,* 1984). Since iron in brain increases with age (Hallgren and Sourander, 1958), the role of transferrin in brain tissue must be of extreme importance. Transferrin receptors in the brain are not necessarily located in regions of high iron distribution. Hill *et al.* (1985) have suggested, since iron-accumulating areas of the brain are efferent to areas of high transferrin receptor density, that iron is delivered to

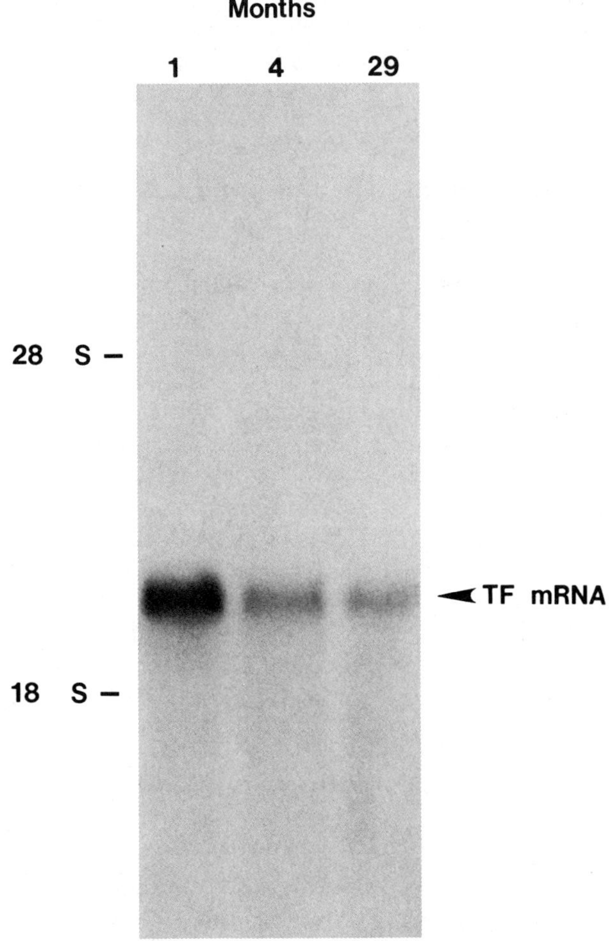

FIG. 12. Northern blot analysis of transferrin mRNA from livers of different aged mice. The ribosomal markers are shown on the left. Steady-state TF mRNA appears to be decreased in livers of aged mice (29 months).

these sites through neuronal transport. Expression of the *TF* gene by the choroid plexus and oligodendrocytes during aging has not yet been correlated with the reported increases of iron in brains of aging animals.

C. Multiregulatory Circuits

In humans, increased plasma transferrin levels are found in childhood, pregnancy, and iron deficiency anemia and after estrogen administration. In both chicken and rats, dietary iron deficiency increases transferrin synthesis in the liver (McKnight *et al.,* 1980b; Morton *et al.,* 1976). Estrogen and other steroid hormones, on the other hand, regulate *TF* gene expression in the oviduct and the liver of chicken (Lee *et al.,* 1978; McKnight *et al.,* 1980a) and in the testis and the liver of rats (Skinner and Griswold, 1982). The molecular mechanisms of gene regulation include trans-acting nuclear proteins, e.g., hormonal receptors, that bind with high affinity to cis-regulatory elements in the DNA sequence of each gene (Dynan and Tjian, 1985).

1. Iron Modulation

The induction of transferrin by iron deficiency appears to be organ and gene specific. In the chicken, where ovotransferrin (conalbumin) and serum transferrin are encoded by the same gene, an iron-deficient diet induces an increase of 2- to 4-fold in serum TF as well as in TF mRNA synthesis in the liver (McKnight *et al.,* 1980a). No detectable effect is observed in transferrin synthesis in the oviduct (McKnight *et al.,* 1980b). Rats raised on a low-iron diet showed a 2.4-fold increase in TF transcription activity in isolated liver nuclei, while the TF mRNA content of other TF-synthesizing tissues, including brain, testis, spleen, and kidney, remained unchanged (Idzerda *et al.,* 1986). Iron deficiency produced no changes in the synthesis of other proteins including albumin (McKnight *et al.,* 1980b; Idzerda *et al.,* 1986). When iron stores were rapidly replenished by the administration of iron-saturated ferritin, both the rate of TF synthesis and liver TF mRNA returned to control levels within 2–3 days (McKnight *et al.,* 1980b). In iron-loaded rats, a small decrease in liver TF mRNA was observed although the same rats did not show a change in total serum iron binding capacity (Idzerda *et al.,* 1986).

The molecular aspects of *TF* gene regulation by iron storage levels are unknown, although other examples of transcription regulation by metals have been defined. In mouse liver as well as in cultured liver cells, cadmium and zinc induce transcription of metallothionein genes (Mayo *et al.,* 1982; Yagle and Palmiter, 1985). On the molecular level, recent studies have shown that heavy metal induction of metallothionein gene expression involves multiple copies of "metal response DNA element" that reside within 200 bp upstream of the metallothionein transcription unit (Stuart *et al.,* 1984, 1985). The metal regulatory DNA element was shown to be functional, even when it was inserted into the promoter region of a gene that was normally not regulated by heavy metals (Searle *et al.,* 1985; Stuart *et al.,* 1984). It has not yet been demonstrated that isolated hepatocytes can respond to iron levels and regulate *TF* gene expression without the involvement of other tissues. Whether or not the effect of iron storage levels on liver TF synthesis is direct, it certainly involves a liver-specific component(s) (Idzerda *et al.*, 1986). Preliminary results obtained from transfecting rat hepatoma cells in our laboratory have indicated that gene expression directed by human *TF* gene promoter can be modulated by iron. This finding is consistent with the presence of metal regulatory elements in the promoter region of the *TF* gene (Adrian *et al.*, 1987). Table 1 specifies the locations of DNA sequences that match metal regulatory elements described by Stuart *et al.* (1985). The biological function of the DNA sequences in the *TF* gene awaits further testing. Yet, it is interesting to note that metallothionein synthesis is stimulated by cadmium and zinc, while *TF* expression is suppressed by increased iron storage levels. Modulation of *TF* gene expression by the interaction of cis-acting DNA elements with trans-acting protein factors responding to iron levels may be typical of the complicated regulatory processes responsible for positive and negative control of expression described by Zinn and Maniatis (1986).

2. *Hormonal Modulation*

Transferrin synthesis has been shown to be induced by estrogen, progesterone (Horne and Ferguson, 1972; Lee *et al.*, 1978; McKnight and Palmiter, 1979; McKnight *et al.*, 1980b), and glucocorticoid (Jeejeebhoy *et al.*, 1972a), in addition to thyroxine (Jeejeebhoy *et al.*, 1972b) and glucagon (Tavill *et al.*, 1972).

In chickens, both estrogen and progesterone induced a marked

TABLE 1
Sequences in Human TF Gene That Match Metal Regulatory Elements[a]

Source	Sequence	Location on gene (5′ end)
Metallothionein (Stuart *et al.*, 1985)	TGCRCYC	−51 to −147
Transferrin (Adrian *et al.*, 1986)	*TGCGCCC	−393
	TFCGCCC	−57
	TGCACTC	177
	TGCACTC	174
	TGCACCC	1546

[a] Heptanucleotides that may convey metal regulation in the human *TF* gene. Stuart *et al.* (1985) have identified a heptanucleotide consensus sequence that constitutes the core of a metal-dependent promoter element. R indicates purine and Y indicates pyrimidine in the consensus sequence. Five of these heptanucleotides have been identified in the human *TF* gene (Adrian *et al.*, 1986). Their location and sequence are shown above. The asterisk indicates a heptanucleotide that is on the complement strand. The other heptanucleotides are on the sense strand of the gene.

stimulation of ovotransferrin synthesis; the rate was elevated 6- to 10-fold after 4 days of hormone treatment (Lee *et al.*, 1978). Immediate induction of ovotransferrin mRNA synthesis by hormone was observed in isolated oviduct nuclei; a 2- to 3-fold stimulation of TF mRNA production was achieved within 30 minutes (McKnight and Palmiter, 1979). Estrogen also stimulated transferrin synthesis in the liver; however, the effect was less pronounced and there was a lag of at least 6 hours prior to the increase in TF mRNA. Progesterone, in contrast, had no effect on transferrin synthesis in liver. Although the estrogen-mediated stimulation of TF mRNA synthesis occurred slowly and could be an indirect response resulting from other changes, it appeared not to be a consequence of iron depletion produced by estrogen (McKnight *et al.*, 1980b). The estrogen-mediated induction of TF mRNA was not blocked when liver iron was maintained at high level. In addition, synergistic response of TF induction was observed in the combined treatment of iron deficiency and estrogen. This suggested that iron deficiency and estrogen interact with the *TF* gene in liver through separate mechanisms (McKnight *et al.*, 1980b). Tuil *et al.* (1985) showed no modulation of *TF* gene transcription or mRNA content by estrogen or iron content in the rat liver. These findings were in contrast to those of McKnight *et al.* (1980a) discussed above and may reflect a difference in the responsiveness of the chicken, rat, and mouse *TF* genes to estrogen. Hammer *et al.* (1986) have also examined the estrogen regulation of the complete chicken *TF* gene in transgenic mice, as summarized above.

The *TF* gene may be regulated by cyclic AMP. In studies of TF mRNA content and gene transcription rate in the liver of rats, Tuil *et al.* (1985) reported that glucagon and dibutyryl cyclic AMP reduced TF mRNA levels and transcription. Regulation of this gene by cAMP may also be supported by the identification of a potential cAMP regulatory region in the 5′ flanking region of the *TF* gene (Lucero *et al.*, 1986).

3. *Acute-Phase Reaction*

The acute-phase reaction in mammals follows injury or infection. It is characterized by a complex series of events that include a dramatic change in the pattern of protein synthesis rates in the liver. The rate of synthesis of certain proteins, the acute-phase reactants, increases rapidly while synthesis rates for some other proteins, the negative acute-phase reactants, decrease. Transferrin is not considered an acute-phase reactant protein (Putnam, 1984). However, Schreiber and Howlett (1983) reported an increase in serum transferrin levels in rats 3 days after a subcutaneous injection of turpentine. Since the increase of transferrin expression did not occur immediately after the injury, it was suspected of being a secondary response to some stimulatory factor such as iron reduction.

Sequestering of iron by transferrin during the acute-phase reaction in humans serves as a host defense mechanism against microbial pathogens and neoplasia (Letendre, 1985; Weinberg, 1984). Microbial growth is enhanced by adding exogenous iron or iron-saturated transferrin (Hunter *et al.*, 1984).

4. Extinction

Intertypic somatic cell hybrids have been a valuable tool in the investigation of the trans-acting control mechanisms which underlie tissue-specific gene expression. Hybrids formed by fusing a highly differentiated cell with a different cell type characteristically do not express the tissue-specific traits of either parental cell, a phenomenon called extinction (Davidson *et al.*, 1966). One of the best characterized examples of extinction in intertypic hybrids results when the highly differentiated rat hepatoma cell line (H4IIEC3) is fused to a variety of different cell types (Schneider and Weiss, 1971; Bertolotti and Weiss, 1972). The resultant intertypic hybrids are extinguished for the expression of liver-specific traits. Reexpression of tissue-specific traits may be observed in hybrids which have segregated chromosomes from the nonhepatic parental cell (Weiss and Chaplain, 1971; Bertolotti and Weiss, 1974), and, in fact, previously silent genes of the nonhepatic parental cell may become activated in particular segregant clones (Peterson and Weiss, 1972; Malawista and Weiss, 1974).

The genetic basis for extinction was determined initially using rat hepatoma × mouse fibroblast whole-cell hybrids and defined hepatoma cell clones containing single, specific mouse or human fibroblast chromosomes. Specifically, the extinction of tyrosine aminotransferase activity was correlated with the presence of a fibroblast-derived locus on murine chromosome 11 or its human homolog, chromosome 17 (Killary and Fournier, 1984). This locus, designated *Tse-1* for *tissue-specific extinguisher-1,* is active in fibroblasts (Peterson *et al.*, 1985) and acts across species barriers to negatively regulate the expression of a hepatic trait *in trans*.

Extinction of serum protein gene expression also has been observed in hepatoma × fibroblast hybrids (Szpirer and Szpirer, 1976; Fougere and Weiss, 1978; Cassio *et al.*, 1986). The presence of a particular mouse fibroblast chromosome in the hepatoma background resulted in extinction of serum albumin gene expression (Petit *et al.*, 1986; Lem *et al.*, unpublished results). Secretion of transferrin from intertypic hybrids was not completely extinguished; however, transferrin secretion was greatly reduced (Szpirer and Szpirer, 1976; Cassio *et al.*, 1986) and found only in a fraction of cells in the hybrid population (Cassio *et al.*, 1986).

The steady-state levels of TF mRNA have been examined in a karyotypically complete hepatoma × fibroblast hybrid. In this hybrid TF mRNA was extinguished (Chin and Fournier, 1987). Extinction of *TF* expression was presumably due to the presence of a chromosomal locus active in fibroblasts, since transferrin was reexpressed in a hybrid with a reduced complement of fibroblast chromosomes. In another study, 11 hepatoma × fibroblast hybrids were examined by Northern analysis for expression of TF mRNA. Five clones were extinguished for TF, four clones showed barely detectable levels of TF mRNA, and two clones clearly expressed TF mRNA. Detailed karyotypic analyses performed on two extinguished clones indicated that they were karyotypically complete, i.e., they retained at least one copy of each murine chromosome of the diploid complement at high frequency in the hepatoma background. The expressing clones were also virtually karyotypically complete with loss in each hybrid of a single, identical mouse fibroblast chromosome (Killary *et al.*, 1987). Transdominant negative regulation of transferrin gene expression, then, could be mediated by a locus or loci on a single fibroblast chromosome.

These results provide evidence that negative trans-acting factors appear to belong to the multiregulatory circuits modulating transferrin expression.

V. Conclusions

1. The *TF* gene is a member of a primitive family of genes which has remained chromosomally linked and structurally homologous. The ancestor gene doubled in length by intragenic amplification 300 to 500 million years ago. Descendants of this gene arose by subsequent gene duplications. Each gene in the transferrin family encodes conserved sequences of the proteins that probably contribute to the iron-binding functions and contains conserved chromosomal DNA in the promoter regions that account for tissue-specific expression.

2. In addition to synthesis in liver, tissue-specific expression of the transferrin gene has been described in normal and malignant cells of many types of tissue. The *TF* gene appears to be active in autocrine systems where a cell is stimulated by a factor which it synthesizes and to which it contains a receptor. In embryonic and fetal stages

transferrin appears to be synthesized by tissues requiring its stimulation in cell proliferation. Tissues in regions, or during periods, where a circulatory supply of transferrin is restricted, often produce their own transferrin.

3. Analysis of *TF* gene expression in every possible cell type throughout development into the aging process offers a promising model for learning more about gene modulation.

Acknowledgments

The research of the authors is supported in part by NIH Grants AG06872, AG06650, and GM33298; American Cancer Society Grant CD-303B; the American Heart Association (Texas Affiliate 86G-371); and the Meadows Foundation. We appreciate helpful discussions with Dr. Ann Killary and the expert manuscript preparation by Betty Russell and Judith Pride. We appreciate information from Drs. Evelyne Schaeffer, Mario Zakin, Tim Rose, Joe Brown, and Gregory Plowman about their unpublished results.

References

Adamson, E. D. (1982). The location and synthesis of transferrin in mouse embryos and teratocarcinoma cells. *Dev. Biol.* **91,** 227–234.

Adrian, G. S., Korinek, B. W., Bowman, B. H., and Yang, F. (1986). The human transferrin gene: 5′ region contains conserved sequences which match the control elements regulated by heavy metals, glucocorticoids and acute phase reaction. *Gene* **49,** 167–175.

Adrian, G. S., Yang, F., and Bowman, B. H. (1987). The 5′ flanking region of the human transferrin gene directs tissue-specific expression in human hepatoma cells. *Am. J. Hum. Genet.* **41,** A204 (Abstr. 605).

Aisen, P., and Listowsky, I. (1980). Iron transport and storage proteins. *Annu. Rev. Biochem.* **49,** 357–393.

Aldred, A. R., Dickson, P. W., Marley, P. D., and Schreiber, G. (1987). Distribution of transferrin synthesis in brain and other tissues in the rat. *J. Biol. Chem.* **262,** 5293–5297.

Alper, C. A., Raum, D., Awdeh, Z. L., Petersen, B. H., Taylor, P. D., and Starzl, T. E. (1980). Studies of hepatic synthesis *in vivo* of plasma proteins, including orosomucoid, transferrin, alpha 1-antitrypsin, C8, and factor B. *Clin. Immunol. Immunopathol.* **16,** 84–89.

Barnes, D., and Sato, G. (1980). Methods for growth of cultured cells in serum-free medium. *Anal. Biochem.* **102,** 255–270.

Bateman, A. E., and Cole, R. J. (1971). Stimulation of haem synthesis by erythropoietin in mouse yolk-sac-stage embryonic cells. *J. Embryol. Exp. Morphol.* **26,** 475–480.

Bertolotti, R., and Weiss, M. C. (1972). Expression of differentiated functions in hepatoma cell hybrids. II. Aldolase. *J. Cell. Physiol.* **79,** 211–224.

Bertolotti, R., and Weiss, M. C. (1974). Expression of differentiated functions in hepatoma cell hybrids. V. Reexpression of aldolase B *in vitro* and *in vivo*. *Differentiation* **2,** 5–17.

Bloch, B., Popovici, T., Levin, M. J., Tuil, D., and Kahn, A. (1985). Transferrin gene expression visualized in oligodendrocytes of the rat brain by using *in situ* hybridization and immunohistochemistry. *Proc. Natl. Acad. Sci. U.S.A.* **82,** 6706–6710.

Bost, K. L., Smith, E. M., and Blalock, J. E. (1985). Regions of complementarity between the messenger RNAs for epidermal growth factor, transferrin, interleukin-2 and their respective receptors. *Biochem. Biophys. Res. Commun.* **128,** 1373–1380.

Bowman, B. H., and Yang, F. (1987). DNA sequencing and chromosomal locations of human plasma protein gene. *In* "The Plasma Proteins: Structure, Function and Genetic Control" (F. W. Putnam, ed.), Vol. 5, 2nd Ed., pp. 1–49. Academic Press, Orlando, Florida.

Bowman, B. H., Adrian, G. S., and Yang, F. (1987). Expression of genes encoding the vitamin D binding protein and transferrin. *In* "Protides of the Biological Fluids" (H. Peetrs, ed.), Vol. 35, pp. 3–12. Pergamon, Oxford.

Brinster, R. L., and Palmiter, R. D. (1982). Induction of foreign genes in animals. *Trends Biochem. Sci. (Pers. Ed.)* **7,** 438–440.

Brock, J. H. (1981). The effect of iron and transferrin on the response of serum-free cultures of mouse lymphocytes to concanavalin A and lipopolysaccharide. *Immunology* **43,** 387–392.

Brock, J. H., and Mainou-Fowler, T. (1983). The role of iron and transferrin in lymphocyte transformation. *Immunol. Today* **4,** 347–351.

Brock, J. H., Mainou-Fowler, T., and Webster, L. M. (1986). Evidence that transferrin may function exclusively as an iron donor in promoting lymphocyte proliferation. *Immunology* **57,** 105–110.

Brown, J. P., Rose, T. M., and Plowman, G. D. (1985). Human melanoma antigen p97, a membrane-associated transferrin homologue. *In* "Proteins of Iron Storage and Transport" (G. J. Montreuil, R. R. Crichton, and J. Mazurier, eds.), pp. 39–46. Elsevier, Amsterdam.

Cassio, D., Rogier, E., Feldman, G., and Weiss, M. C. (1986). Plasma-protein production by rat hepatoma cells in culture, their variants and revertants. *Differentiation* **30,** 220–228.

Chang, M. P., Makinodan, T., Peterson, W. J., and Strehler, B. L. (1982). Role of T cells and adherent cells in age-related decline in murine interleukin 2 production. *J. Immunol.* **129,** 2426–2430.

Cheung, H. T., Twu, J. S., and Richardson, A. (1983). Mechanism of the age-related decline in lymphocyte proliferation: Role of IL-2 production and protein synthesis. *Exp. Gerontol.* **18,** 451–460.

Chin, A. C., and Fournier, R. E. K. (1987). A genetic analysis of extinction: Trans-regulation of 16 liver-specific genes in hepatoma–fibroblast hybrid cells. *Proc. Natl. Acad. Sci. U.S.A.* **84,** 1614–1618.

Cochet, M., Gannon, F., Hen, R., Maroteaux, L., Perrin, F., and Chambon, P. (1979). Organisation and sequence studies of the 17-piece chicken conalbumin gene. *Nature (London)* **282,** 567–574.

Cullen, C. R., Hubberman, P., Kaslow, D. C., and Migeon, B. R. (1986). Comparison of factor IX methylation on human active and inactive X chromosomes: Implications for X inactivation and transcription of tissue-specific genes. *EMBO J.* **5,** 2223–2229.

Davidson, R. L., Ephrussi, B., and Yamamoto, K. (1966). Regulation of pigment synthesis in mammalian cells, as studied by somatic hybridization. *Proc. Natl. Acad. Sci. U.S.A.* **56,** 1437–1440.

de Grouchy, J., Turleau, C., and Finaz, C. (1978). Chromosomal phylogeny of the primates. *Annu. Rev. Genet.* **12,** 289–328.

Dickson, P. W., Aldred, A. R., Marley, P. D., Tu, G. F., Howlett, G. J., and Schreiber, G.

(1985). High prealbumin and transferrin mRNA levels in the choroid plexus of rat brain. *Biochem. Biophys. Res. Commun.* **127,** 890–895.

Doolittle, R. F. (1984). Evolution of the vertebrate plasma proteins. *In* "The Plasma Proteins: Structure, Function, and Genetic Control" (F. W. Putnam, ed.), 2nd Ed., Vol. 4, pp. 317–359. Academic Press, New York.

Doolittle, R. F. (1985). The genealogy of some recently evolved vertebrate proteins. *Trends Biochem. Sci. (Pers. Ed.)* **10,** 233–237.

Dynan, W. S., and Tjian, R. (1983). The promoter-specific transcription factor Sp1 binds to upstream sequences in the VS40 early promoter. *Cell* **35,** 79–87.

Dynan, W. S., and Tjian, R. (1985). Control of eukaryotic messenger RNA synthesis by sequence-specific DNA-binding protein. *Nature (London)* **316,** 774–778.

Ekblom, P., Thesleff, I., Saxen, L., Miettinen, A., and Timpl, R. (1983). Transferrin as a fetal growth factor: Acquisition of responsiveness related to embryonic induction. *Proc. Natl. Acad. Sci. U.S.A.* **80,** 2651–2655.

Faulk, W. P., and Galbraith, G. M. (1979). Trophoblast transferrin and transferrin receptors in the host–parasite relationship of human pregnancy. *Proc. R. Soc. London. Biol.* **204,** 83–97.

Fawcett, D. W. (1975). Ultrastructure and function of the Sertoli cell. *In* "Handbook of Physiology: Male Reproductive System" (D. W. Hamilton and R. D. Grup, eds.), Vol. 5, pp. 21–56. Amer. Physiological Society, Washington, D.C.

Fougere, C., and Weiss, M. C. (1978). Phenotypic exclusion in mouse melanoma–rat hepatoma hybrids: Pigment and albumin production are not reexpressed simultaneously. *Cell* **15,** 843–854.

Frieden, E. (1986). Perspectives on copper biochemistry. *Clin. Physiol. Biochem.* **4,** 11–19.

Gillis, S., and Smith, K. A. (1977). Long term culture of tumour-specific cytotoxic T cells. *Nature (London)* **268,** 154–156.

Gillis, S., Kozak, R., Durante, M., and Weksler, M. E. (1981). Immunological studies of aging. Decreased production of and response to T cell growth factor by lymphocytes from aged humans. *J. Clin. Invest.* **67,** 937–942.

Gitlin, D., and Perricelli, A. (1970). Synthesis of serum albumin, prealbumin, alpha-foetoprotein, alpha-1-antitrypsin and transferrin by the human yolk sac. *Nature (London)* **228,** 995–997.

Goldman, M. A., Stokes, K. R., Idzerda, R. L., McKnight, G. S., Hammer, R. E., Brinster, R. L., and Gartler, S. M. (1987). A chicken transferrin gene in transgenic mice escapes X-chromosome inactivation. *Science* **236,** 593–595.

Gustine, D. L., and Zimmerman, E. F. (1972). Amniotic fluid proteins: Evidence for the presence of fetal plasma glycoproteins in mouse amniotic fluid. *Am. J. Obstet. Gynecol.* **114,** 553–560.

Gustine, D. L., and Zimmerman, E. F. (1973). Developmental changes in microheterogeneity of foetal plasma glycoproteins of mice. *Biochem. J.* **132,** 541–551.

Hallgren, B., and Sourander, P. (1958). The effect of age on the non-haemin iron in the human brain. *J. Neurochem.* **3,** 41–51.

Hammer, R. E., Idzerda, R. L., Brinster, R. L., and McKnight, G. S. (1986). Estrogen regulation of the avian transferrin gene in transgenic mice. *Mol. Cell. Biol.* **6,** 1010–1014.

Hill, J. M., and Switzer, R. C., III (1984). The regional distribution and cellular localization of iron in the rat brain. *Neuroscience* **11,** 595–603.

Hill, J. M., Ruff, M. R., Weber, R. J., and Pert, C. B. (1985). Transferrin receptors in rat brain: Neuropeptide-like pattern and relationship to iron distribution. *Proc. Natl. Acad. Sci. U.S.A.* **82,** 4553–4557.

Holmberg, C. G., and Laurell, C.-B. (1945). Studies of the capacity of serum to bind iron. A contribution to our knowledge of the regulation mechanism of serum iron. *Acta Physiol. Scand.* **10,** 307–319.

Horne, C. H. W., and Ferguson, J. (1972). The effect of age, sex, pregnancy, oestrogen and progestrogen on rat serum proteins. *J. Endocrinol.* **54,** 47–53.

Huggenvik, J. I., Idzerda, R. L., Haywood, L., Lee, D. C., McKnight, G. S., and Griswold, M. D. (1987). Transferrin messenger ribonucleic acid: Molecular cloning and hormonal regulation in rat Sertoli cells. *Endocrinology* **120,** 332–340.

Hunter, R. L., Bennett, B., Towns, M., and Vogler, W. R. (1984). Transferrin in disease. II: Defects in the regulation of transferrin saturation with iron contribute to susceptibility to infection. *Am. J. Clin. Pathol.* **81,** 748–753.

Idzerda, R. L., Huebers, H., Finch, C. A., and McKnight, G. S. (1986). Rat transferrin gene expression: Tissue-specific regulation by iron deficiency. *Proc. Natl. Acad. Sci. U.S.A.* **83,** 3723–3727.

Ii, I., Kimura, I., and Ozawa, E. (1982). A myotrophic protein from chick embryo extract: Its purification, identity to transferrin, and indispensability for avian myogenesis. *Dev. Biol.* **94,** 366–377.

Jacob, F. (1982). Molecular tinkering in evolution. *In* "Evolution from Molecule to Men" (D. S. Bendall, ed.), pp. 131–144. Cambridge Univ. Press, London.

Jeejeebhoy, K. N., Bruce-Robertson, A., Ho, J., and Sodtke, U. (1972a). The effect of cortisol on the synthesis of rat plasma albumin, fibrinogen and transferrin. *Biochem. J.* **130,** 533–538.

Jeejeebhoy, K. N., Bruce-Robertson, A., Ho, J., and Sodtke, U. (1972b). The comparative effects of nutritional and hormonal factors on the synthesis of albumin, fibrinogen and transferrin. *CIBA Found. Symp.* **9,** 217–247.

Jeltsch, J.-M., and Chambon, P. (1982). The complete nucleotide sequence of the chicken ovotransferrin mRNA. *Eur. J. Biochem.* **122,** 291–295.

Jordan, S. M., and Morgan, E. H. (1969). Plasma protein synthesis by tissue slices from pregnant and lactating rats. *Biochim. Biophys. Acta* **174,** 373–379.

Jordan, S. M., and Morgan, E. H. (1970). Plasma protein metabolism during lactation in the rabbit. *Am. J. Physiol.* **219,** 1549–1554.

Kernoff, L. M., and Baker, G. (1980). Direct measurement of transferrin synthesis rates in man using the ^{14}C carbonate method. *Anal. Biochem.* **106,** 529–534.

Killary, A. M., and Fournier, R. E. K. (1984). A genetic analysis of extinction: Trans-dominant loci regulate expression of liver-specific traits in hepatoma hybrid cells. *Cell* **38,** 523–534.

Killary, A. M., Riehl, E., Yang, F., Adrian, G. S., and Bowman, B. H. (1987). Trans-dominant tissue specific regulation of transferrin gene expression in intertypic hybrid cells (in preparation).

Lazarus, A. H., and Baines, M. G. (1985). Studies on the mechanism of specificity of human natural killer cells for tumor cells: Correlation between target cell transferrin receptor expression and competitive activity. *Cell. Immunol.* **96,** 255–266.

LeBeau, M. M., Diaz, M. O., Yang, F., Plowman, G. D., Brown, J. P., and Rowley, J. D. (1985). Molecular analysis of transferrin (TF), transferrin receptor (TFR) and p97 antigen genes in the INV(3) and T(3;3) in acute nonlymphocytic leukemia (ANLL). *Am. J. Hum. Genet.* **37,** 4, A31.

Lee, D. C., McKnight, G. S., and Palmiter, R. D. (1978). The action of estrogen and progesterone on the expression of the transferrin gene. A comparison of the response in chick liver and oviduct. *J. Biol. Chem.* **253,** 3494–3503.

Lee, N. T., Chae, C.-B., and Kierszenbaum, A. L. (1986). Contrasting levels of transferrin gene activity in cultured rat Sertoli cells and intact seminiferous tubules. *Proc. Natl. Acad. Sci. U.S.A.* **83,** 8177–8181.

Lee, C.-C., Bowman, B. H., and Yang, F. (1987). Human α_2-HS-glycoprotein: The A and B chains, with a connecting sequence, are encoded by a single mRNA transcript. *Proc. Natl. Acad. Sci. U.S.A.* **84,** 4403–4407.

Leonard, W. J., Depper, J. M., Uchiyama, T., Smith, K. A., Waldmann, T. A., and Greene, W. C. (1982). A monoclonal antibody that appears to recognize the receptor for human T-cell growth factor; partial characterization of the receptor. *Nature (London)* **300,** 267–269.

Letendre, E. D. (1985). The importance of iron in the pathogenesis of infection and neoplasia. *Trends Biochem. Sci. (Pers. Ed.)* **9,** 166–168.

Levin, M. J., Tuil, D., Uzan, G., Dreyfus, J.-C., and Kahn, A. (1984). Expression of the transferrin gene during development of non-hepatic tissues: High level of transferrin mRNA in fetal muscle and adult brain. *Biochem. Biophys. Res. Commun.* **122,** 212–217.

Lucero, M. A., Schaeffer, E., Cohen, G. N., and Zakin, M. M. (1986). The 5′ region of the human transferrin gene: Structure and potential regulatory sites. *Nucleic Acids Res.* **14,** 8692.

Lum, J. B. (1986). Visualization of mRNA transcription of specific genes in human cells and tissues using *in situ* hybridization. *BioTechniques* **4,** 32–40.

Lum, J. B. (1987). Role of transferrin in cellular proliferation. Ph.D. thesis, University of Texas Graduate School of Biomedical Science, San Antonio.

Lum, J. B., Infante, A. J., Makker, D. M., Yang, F., and Bowman, B. H. (1986). Transferrin synthesis by inducer T lymphocytes. *J. Clin. Invest.* **77,** 841–849.

McCombs, J. L., Teng, C. T., Pentecost, B. T., Manguson, V., Moore, C. M., and McGill, J. R. (1987). Localization of human lactotransferrin gene by *in situ* hybridization (submitted).

MacGillivray, R. T., Mendez, E., Shewale, J. G., Sinha, S. K., Lineback-Zins, J., and Brew, K. (1983). The primary structure of human serum transferrin. The structures of seven cyanogen bromide fragments and the assembly of the complete structure. *J. Biol. Chem.* **258,** 3543–3553.

McKnight, G. S., and Palmiter, R. D. (1979). Transcriptional regulation of the ovalbumin and conalbumin genes by steroid hormones in chick oviduct. *J. Biol. Chem.* **254,** 9050–9058.

McKnight, G. S., Lee, D. C., Hemmaplardh, D., Finch, C. A., and Palmiter, R. D. (1980a). Transferrin gene expression. Effects of nutritional iron deficiency. *J. Biol. Chem.* **255,** 144–147.

McKnight, G. S., Lee, D. C., and Palmiter, R. D. (1980b). Transferrin gene expression. Regulation of mRNA transcription in chick liver by steroid hormones and iron deficiency. *J. Biol. Chem.* **255,** 148–153.

McKnight, G. S., Hammer, R. E., Kuenzel, E. A., and Brinster, R. L. (1983). Expression of the chicken transferrin gene in transgenic mice. *Cell* **34,** 335–341.

Magnuson, V. L., Lee, C. C., Yang, F., McCombs, J. L., Bowman, B. H., and McGill, J. R. (1988). Human α_2-HS-glycoprotein (ASHG) localized to 3q27-29 by *in situ* hybridization *Cytogenet. Cell Genet.* (in press).

Mainou-Fowler, T., and Brock, J. H. (1985). Effect of iron deficiency on the response of mouse lymphocytes to concanavalin A: The importance of transferrin-bound iron. *Immunology* **54,** 325–332.

Malawista, S. E., and Weiss, M. C. (1974). Expression of differentiated functions in hepatoma cell hybrids: High frequency of induction of mouse albumin production in rat hepatoma–mouse lymphoblast hybrids. *Proc. Natl. Acad. Sci. U.S.A.* **71,** 927–931.

Martin, A. W., Huebers, E., Huebers, H., Webb, J., and Finch, C. A. (1984). A mono-sited transferrin from a representative deuterostome: The ascidian *Pyura stolonifera* (subphylum Urochordata). *Blood* **64,** 1047–1052.

Mayo, K. E., Warren, R., and Palmiter, R. D. (1982). The mouse metallothionein-I gene is transcriptionally regulated by cadmium following transfection into human or mouse cells. *Cell* **29,** 99–108.

Meek, J., and Adamson, E. D. (1985). Transferrin in foetal and adult mouse tissues: Synthesis, storage and secretion. *J. Embryol. Exp. Morphol.* **86,** 205–218.

Metz-Boutigue, M. H., Mazurier, J., Jolles, J., Spik, G., Montreuil, J., and Jolles, P. (1981). The present state of the human lactotransferrin sequence. Study and alignment of the cyanogen bromide fragments and characterization of N- and C-terminal domains. *Biochim. Biophys. Acta* **670,** 243–254.

Miller, Y. E., Jones, C., Scoggin, C., Morse, H., and Seligman, P. (1983). Chromosome 3q (22-ter) encodes the human transferrin receptor. *Am. J. Hum. Genet.* **35,** 573–583.

Mintz, B., and Illmensee, K. (1975). Normal genetically mosaic mice produced from malignant teratocarcinoma cells. *Proc. Natl. Acad. Sci. U.S.A.* **72,** 3585–3589.

Miskimins, W. K., McClelland, A., Roberts, M. P., and Ruddle, F. H. (1986). Cell proliferation and expression of the transferrin receptor gene: Promoter sequence homologies and protein interactions. *J. Cell Biol.* **103,** 1781–1788.

Morgan, E. H. (1968). Plasma protein turnover and transmission to the milk in the rat. *Biochim. Biophys. Acta* **154,** 478–487.

Morgan, E. H. (1983). Synthesis and secretion of transferrin. *In* "Plasma Protein Secretion by the Liver" (H. Glaumann, T. Peters, Jr., and C. Redman, eds.), Vol. 4, pp. 96–101. Academic Press, New York.

Morton, A., Hamilton, S. M., Ramsden, D. B., and Tavill, A. S. (1976). Studies on regulatory factors in transferrin metabolism in man and the experimental rat. *In* "Plasma Protein Turnover" (R. Bianchi, G. Mariani, and A. S. McFarlane, eds.), pp. 165–177. Univ. Park Press, Baltimore, Maryland.

Naylor, S. L., Elliott, R. W., Brown, J. A., and Shows, T. B. (1982). Mapping of aminoacylase-1 and beta-galactosidase-A to homologous regions of human chromosome 3 and mouse chromosome 9 suggests location of additional genes. *Am. J. Hum. Genet.* **34,** 235–244.

Neckers, L. M., and Cossman, J. (1983). Transferrin receptor induction in mitogen-stimulated human T lymphocytes is required for DNA synthesis and cell division and is regulated by interleukin 2. *Proc. Natl. Acad. Sci. U.S.A.* **80,** 3494–3498.

Osterloh, K. R., Simpson, R. J., and Peters, T. J. (1987). The role of mucosal transferrin in intestinal iron absorption. *Br. J. Haematol.* **65,** 1–3.

Park, I., Schaeffer, E., Sidoli, A., Baralle, F. E., Gowen, G. N., and Zakin, M. M. (1985). Organization of the human transferrin gene: Direct evidence that it originated by gene duplication. *Proc. Natl. Acad. Sci. U.S.A.* **82,** 3149–3153.

Partanen, A. M., Thesleff, I., and Ekblom, P. (1984). Transferrin is required for early tooth morphogenesis. *Differentiation* **27,** 29–66.

Pentecost, B. T., and Teng, C. T. (1987). Lactoferrin is the major estrogen inducible protein of mouse uterine secretions. *J. Biol. Chem.* **262,** 10134–10139.

Perez-Infante, V., Bardin, C. W., Gunsalus, G. L., Musto, N. A., Rich, K. A., and Mather, J. P. (1986). Differential regulation of testicular transferrin and androgen-binding protein secretion in primary cultures of rat Sertoli cells. *Endocrinology* **118,** 383–392.

Peterson, J. A., and Weiss, M. C. (1972). Expression of differentiated functions in hepatoma cell hybrids: Induction of mouse albumin production in rat hepatoma–mouse fibroblast hybrids. *Proc. Natl. Acad. Sci. U.S.A.* **69,** 571–575.

Peterson, T. C., Killary, A. M., and Fournier, R. E. K. (1985). Chromosomal assignment and trans regulation of the tyrosine aminotransferase structural gene in hepatoma hybrid cells. *Mol. Cell. Biol.* **5,** 2491–2494.

Petit, C., Levilliers, J., Ott, M. O., and Weiss, M. C. (1986). Tissue-specific expression of the rat albumin gene: Genetic control of its extinction in microcell hybrids. *Proc. Natl. Acad. Sci. U.S.A.* **83,** 2561–2565.

Phillips, P. D., and Cristofalo, J. S. (1985). A review of recent cellular aging research: The regulation of cell proliferation. *Rev. Biol. Res. Aging* **2,** 339–357.

Plowman, G. D. (1986). Ph.D. thesis, University of Washington, Seattle, Washington.

Plowman, G. D., Brown, J. P., Enns, C. A., Schroder, J., Nikinmaa, B., Sussman, H. H., Hellstrom, K. E., and Hellstrom, I. (1983). Assignment of the gene for human melanoma-associated antigen p97 to chromosome 3. *Nature (London)* **303,** 70–72.

Ponka, P., and Schulman, H. M. (1985). Acquisition of iron from transferrin regulates reticulocyte heme synthesis. *J. Biol. Chem.* **260,** 14717–14721.

Putnam, F. W. (1975). Transferrin. *In* "The Plasma Proteins" (F. W. Putnam, eds.), Vol. 1, 2nd Ed., pp. 265–316. Academic Press, New York.

Putnam, F. W. (1984). Alpha, beta, gamma, omega—the structure of the plasma proteins. *In* "The Plasma Proteins, Structure, Function, and Genetic Control" (F. W. Putnam, ed.), Vol. IV, 2nd Ed., pp. 45–146. Academic Press, Orlando, Florida.

Rabin, M., McClelland, A., Kuhn, L., and Ruddle, F. H. (1985). Regional localization of the human transferrin receptor gene to 3q26.2-qter. *Am. J. Hum. Genet.* **37,** 1112–1116.

Rabinowich, H., Goses, Y., Reshef, T., and Klajman, A. (1985). Interleukin-2 production and activity in aged humans. *Mech. Aging Dev.* **32,** 213–226.

Richardson, A., Rutherford, M. S., Birchenall-Sparks, M. C., Roberts, M. S., Wu, T. W., and Cheung, H. T. (1985). Levels of specific messenger RNA species as a function of age. *In* "Molecular Biology of Aging: Gene Stability and Gene Expression" (R. S. Sohal, L. S. Birnbaum, and R. G. Cutler, eds.), pp. 229–241. Raven, New York.

Robb, R. J., Munck, A., and Smith, K. A. (1981). T cell growth factor receptors. Quantitation, specificity, and biological relevance. *J. Exp. Med.* **154,** 1455–1474.

Roy, A. K., and Chatterjee, B. (1983). Sexual dimorphism in the liver. *Annu. Rev. Physiol.* **45,** 37–50.

Roy, A. K., Chatterjee, B., Demyan, W. F., Milin, B. S., Motwani, N. M., Nath, T. S., and Schiop, M. J. (1983a). Hormone and age-dependent regulation of alpha 2u-globulin gene expression. *Recent. Prog. Horm. Res.* **39,** 425–461.

Roy, A. K., Nath, T. S., Motwani, N. M., and Chatterjee, B. (1983b). Age-dependent regulation of the polymorphic forms of alpha 2u-globulin. *J. Biol. Chem.* **25,** 10123–10127.

Ruscetti, F. W., Morgan, D. A., and Gallo, R. C. (1977). Functional and morphologic characterization of human T cells continuously grown *in vitro*. *J. Immunol.* **119,** 131–138.

Sargent, T. D., Yang, M., and Bonner, J. (1981). Nucleotide sequence of cloned rat serum albumin messenger RNA. *Proc. Natl. Acad. Sci. U.S.A.* **78,** 243–246.

Schade, A. L., and Caroline, L. (1946). Iron binding component in human blood plasma. *Science* **100,** 14–15.

Schaeffer, E., Park, I., Cohen, G. N., and Zakin, M. M. (1985). Organization of the human serum transferrin gene. *In* "Proteins of Iron Storage and Transport" (G. J. Montreuil, R. R. Crichon, and J. Mazurier, eds.), pp. 361–364. Elsevier, Amsterdam.

Schaeffer, E., Lucero, M. A., Jeltsch, J.-M., Py, M.-C., Levin, M. J., Chambon, P., Cohen, G. N., and Zakin, M. M. (1987). *Gene* (in press).

Schneider, J. A., and Weiss, M. C. (1971). Expression of differentiated functions in hepatoma cell hybrids. I. Tyrosine aminotransferase in hepatoma–fibroblast hybrids. *Proc. Natl. Acad. Sci. U.S.A.* **68,** 127–131.

Schreiber, G. (1987). Synthesis, processing, and secretion of plasma proteins by the liver and other organs and their regulation. *In* "The Plasma Proteins: Structure, Function and Genetic Control" (F. W. Putnam, ed.), 2nd Ed., Vol. V, pp. 293–363. Academic Press, Orlando, Florida.

Schreiber, G., and Howlett, G. (1983). Synthesis and secretion of acute-phase proteins. *In* "Plasma Protein Secretion by the Liver" (H. Glaumann, T. Peters, Jr., and C. Redman, eds.), pp. 423–449. Academic Press, London.

Searle, P. F., Stuart, G. W., and Palmiter, R. D. (1985). Building a metal-responsive promoter with synthetic regulatory elements. *Mol. Cell. Biol.* **5,** 1480–1489.

Seligman, P. A. (1983). Structure and function of the transferrin receptor. *Prog. Hematol.* **13,** 131–147.

Shannon, K. M., Larrick, J. W., Fulcher, S. A., Burck, K. B., Pacely, J., Davis, J. C., and Ring, D. B. (1986). Selective inhibition of the growth of human erythroid bursts by monoclonal antibodies against transferrin or the transferrin receptor. *Blood* **67,** 1631–1638.

Shau, H., Shen, D., and Golub, S. H. (1986). The role of transferrin in natural killer cell and IL-2-induced cytotoxic cell function. *Cell. Immunol.* **97,** 121–130.

Skinner, M. K., and Griswold, M. D. (1980). Sertoli cells synthesize and secrete transferrin-like protein. *J. Biol. Chem.* **255,** 9523–9525.

Skinner, M. K., and Griswold, M. D. (1982). Secretion of testicular transferrin by cultured Sertoli cells is regulated by hormones and retinoids. *Biol. Reprod.* **27,** 211–221.

Soltys, H. D., and Brody, J. I. (1970). Synthesis of transferrin by human peripheral blood lymphocytes. *J. Lab. Clin. Med.* **75,** 250–257.

Spik, G., Bayard, B., Fournet, B., Strecker, G., Bouquelet, S., and Montreuil, J. (1975). Studies on glycoconjugates. LXIV. Complete structure of two carbohydrate units of human serotransferrin. *FEBS Lett.* **50,** 296–299.

Stuart, G. W., Searle, P. F., Chen, H. Y., Brinster, R. L., and Palmiter, R. D. (1984). A 12-base-pair DNA motif that is repeated several times in metallothionein gene promoters confers metal regulation to a meterologous gene. *Proc. Natl. Acad. Sci. U.S.A.* **81,** 7318–7322.

Stuart, G. W., Searle, P. F., and Palmiter, R. D. (1985). Identification of multiple metal regulatory elements in mouse metallothionein-I promoter by assaying synthetic sequences. *Nature (London)* **317,** 828–831.

Szpirer, C., and Szpirer, J. (1976). Extinction, retention and induction of serum protein secretion in hepatoma–fibroblast hybrids. *Differentiation* **5,** 97–99.

Tavill, A. S., East, A. G., Black, E. G., Nadkarni, D., and Hoffenberg, R. (1972). Regulatory factors in the synthesis of plasma proteins by the isolated perfused rat liver. *CIBA Found. Symp.* **9,** 155–179.

Teng, C. T., Pentecost, B. T., Marshall A., Solomon, A., Bowman, B. H., Lalley, P. A., and Naylor, S. L. (1987). Assignment of the lactotransferrin gene to human chromosome 3 and to mouse chromosome 9. *Somatic Cell Mol. Genet.* **13,** 689–693.

Thoman, M. L., and Weigle, W. O. (1981). Lymphokines and aging: Interleukin-2 production and activity in aged animals. *J. Immunol.* **127,** 2102–2106.

Thorbecke, G. J., Liem, H. H., Knight, S., Cox, K., and Muller-Eberhard, U. (1973). Sites of formation of the serum proteins transferrin and hemopexin. *J. Clin. Invest.* **52,** 725–731.

Trowbridge, I. S., and Omary, M. B. (1981). Human cell surface glycoprotein related to cell proliferation is the receptor for transferrin. *Proc. Natl. Acad. Sci. U.S.A.* **78,** 3039–3043.

Tuil, D., Vaulont, S., Levin, M. J., Munnich, A., Moguilewsky, M., Bouton, M. M., Brissot, P., Dreyfus, J.-C., and Kahn, A. (1985). Transient transcriptional inhibition of the transferrin gene by cyclic AMP. *FEBS Lett.* **189,** 310–314.

Vodinelich, L., Sutherland, R., Schneider, C., Newman, R., and Greaves, M. (1983). Receptor for transferrin may be a "target" structure for natural killer cells. *Proc. Natl. Acad. Sci. U.S.A.* **80,** 835.

Weinberg, E. D. (1984). Iron withholding: A defense against infection and neoplasia. *Physiol. Rev.* **64,** 65–102.

Weiss, M. C., and Chaplain, M. (1971). Expression of differentiated functions in hepatoma cell hybrids: Reappearance of tyrosine aminotransferase inducibility after the loss of chromosomes. *Proc. Natl. Acad. Sci. U.S.A.* **68,** 3026–3030.

Weitkamp, L. R. (1983). Evidence for linkage between the loci for transferrin and ceruloplasmin in man. *Ann. Hum. Genet.* **47,** 293–297.

Williams, J. (1982). The evolution of transferrin. *Trends Biochem. Sci. (Pers. Ed.)* **7,** 394–397.

Williams, J., Elleman, T. C., Kingston, I. B., Wilkins, A. G., and Kuhn, K. A. (1982). The primary structure of hen ovotransferrin. *Eur. J. Biochem.* **122,** 297–303.

Wochner, R. D., Weissman, S. M., Waldmann, T. A., and Berlin, N. I. (1968). Direct measurement of the rates of synthesis of plasma proteins in control subjects and patients with gastrointestinal protein loss. *J. Clin. Invest.* **47,** 3–12.

Wu, W., Pahlavani, M. A., Cheung, H. T., and Richardson, A. (1986). The effect of aging on the expression of interleukin 2 messenger ribonucleic acid. *Cell. Immunol.* **100,** 224–231.

Yagle, M. K., and Palmiter, R. D. (1985). Coordinate regulation of mouse metallothionein I and II genes by heavy metals and glucocorticoids. *Mol. Cell. Biol.* **5,** 291–294.

Yang, F., Lum, J. B., McGill, J. R., Moore, C. M., Naylor, S. L., VanBragt, P. H., Baldwin, W. D., and Bowman, B. H. (1984). Human transferrin: cDNA characterization and chromosomal localization. *Proc. Natl. Acad. Sci. U.S.A.* **81,** 2752–2756.

Yang, F., Lum, J. B., and Bowman, B. H. (1985). Molecular genetics of human transferrin. *In* "Protides of the Biological Fluids" (H. Peeters, ed.), pp. 31–34. Pergamon, Oxford.

Yang, F., Naylor, S. L., Lum, J. B., Cutshaw, S., McCombs, J. L., Naberhaus, K. H., McGill, J. R., Adrian, G. S., Moore, C. M., Barnett, D. R., and Bowman, B. H. (1986). Characterization, mapping, and expression of the human ceruloplasmin gene. *Proc. Natl. Acad. Sci. U.S.A.* **83,** 3257–3261.

Yeoh, G. C., and Morgan, E. H. (1974). Albumin and transferrin synthesis during development in the rat. *Biochem. J.* **144,** 215–224.

Yunis, J. J., and Prakash, O. (1982). The origin of man: A chromosomal pictorial legacy. *Science* **215,** 1525–1530.

Zinn, K., and Maniatis, T. (1986). Detection of factors that interact with the human β-interferon regulatory region *in vivo* by DNase I footprinting. *Cell* **45,** 611–618.

THE *Drosophila* ALCOHOL DEHYDROGENASE GENE–ENZYME SYSTEM

Geoffrey K. Chambers

Department of Biochemistry, Victoria University of Wellington, Private Bag, Wellington, New Zealand

ADVANCES IN GENETICS, Vol. 25

I. Introduction

November 28, 1985 marked the twenty-first anniversary of the earliest report of genetic variation at the alcohol dehydrogenase (*Adh*) locus of *Drosophila melanogaster* by Johnson and Denniston (1964). It seems fitting that we should celebrate the coming of age of this gene–enzyme system with an appreciative and perhaps even nostalgic review. This short period has seen a simple unmapped biochemical polymorphism develop to become a major genetic system of interest and use to disciplines so apparently different as ecology and molecular biology. The diversity of interest in *Adh* is reflected in the list of topics at the head of this article. The intensity of interest in *Adh* may be illustrated by comparing the entries in the first and second editions of *Genetic Variations of Drosophila melanogaster* (Lindsley and Grell, 1968; Lindsley and Zimm, 1985, respectively).

The existing literature on *Adh* for any single field of genetics is enormous. It is, therefore, hardly possible to do adequate justice to work in all areas in a single essay. Instead, I have tried to concentrate my efforts in those areas which are developing rapidly or which have been relatively lightly covered by previous authors. For genetic topics which have been reviewed recently only brief updates have been attempted. Several recent papers have reviewed genetic research on *Adh* and readers are encouraged to consult them to obtain a greater appreciation of the significance of the *Adh* gene–enzyme system for biology in general. Of particular interest are the articles by Clegg and Epperson (1985), McDonald (1983), and Lewontin (1986) on modern population genetics, by Laurie-Ahlberg (1985) on quantitative aspects of *Drosophila* gene expression, by Parsons (1983) on ecological genetics, and by Zera *et al.* (1985) on biochemical aspects of enzyme polymorphisms. For detailed background information on *Adh* itself, the reader is referred to Ashburner's *Adh* catalog in Lindsley and Zimm (1985). Van Delden's article "Selection at an enzyme locus" is a lucid and comprehensive history of biochemical and population genetics work on *Adh* (Van Delden, 1982). Other important research papers which serve as reviews of more limited areas are referenced in the appropriate sections.

In preparing this essay I not only increased my personal collection of reprints on *Adh* but also gained some understanding of the historical development of *Adh* research. It is a privilege to acknowledge the contributions made by those pioneers who have done so much to popularize this eminently amenable system. I feel that particular debts of gratitude are owed for the efforts of F. M. Johnson, S. B.

Pipkin, and co-workers in the area of population genetics and to M. Ashburner, W. Sofer, and colleagues for their painstaking classical genetic analysis of the locus and its subsequent molecular characterization. This is not to say that we should neglect the many other geneticists and biochemists who have each made their own valuable contributions.

II. Classical Genetics of *Adh*

A clear picture of the basic genetic features of any gene–enzyme system is of vital importance if one hopes to gain a full understanding of its function. Considerable efforts have been directed toward discovering the precise position of the *Adh* locus on the recombinational and cytological maps for chromosome arm 2L of *D. melanogaster*. A special feature of the analysis of the *Adh* locus has been the imaginative use of chemical screening procedures to select for *Adh* null mutants and revertants. The mutant lines obtained have been of value for a description of the mutagenic process itself and also in demonstrating the *in vivo* role of the *Adh* gene product, i.e., the enzyme alcohol dehydrogenase (ADH, EC 1.1.1.1). This work, *in toto,* is an excellent example of the power of a concerted biochemical and genetic approach for solving biological problems.

A. Map Position of the *Adh* Locus

The genetic map position of *Adh* on the left arm of chromosome 2 was established at 2-50.1 between *elbow* (*el,* 2-50.0) and *reduced-scraggly* (rd^S, 2-50.2) by Grell *et al.* (1965), and its location quickly confirmed by Courtright *et al.* (1966). Together these studies represent one of the first uses of electrophoretic variants for genetic mapping in *Drosophila.* The original placement has remained unchallenged since then, although the genetic structure of this interval is now known in considerably finer detail. Ashburner and colleagues used deletion mapping and saturation mapping of lethal and visible markers to obtain the map shown in Fig. 1 (see Chia *et al.,* 1985a). The closest loci with visible phenotypes are *noc* and rd^S which are separated from *Adh* by two and by eight lethal complementation groups, respectively. Their studies further revealed *l(2)br7* to be *Suppressor of Hairless* [*Su (H)*] and that the unusual properties of *Scutoid* (*Sco*) could be explained as a complex rearrangement of the *Adh* region (Ashburner *et al.,* 1982a,b; O'Donnell *et al.,* 1977). This model explains how genetic

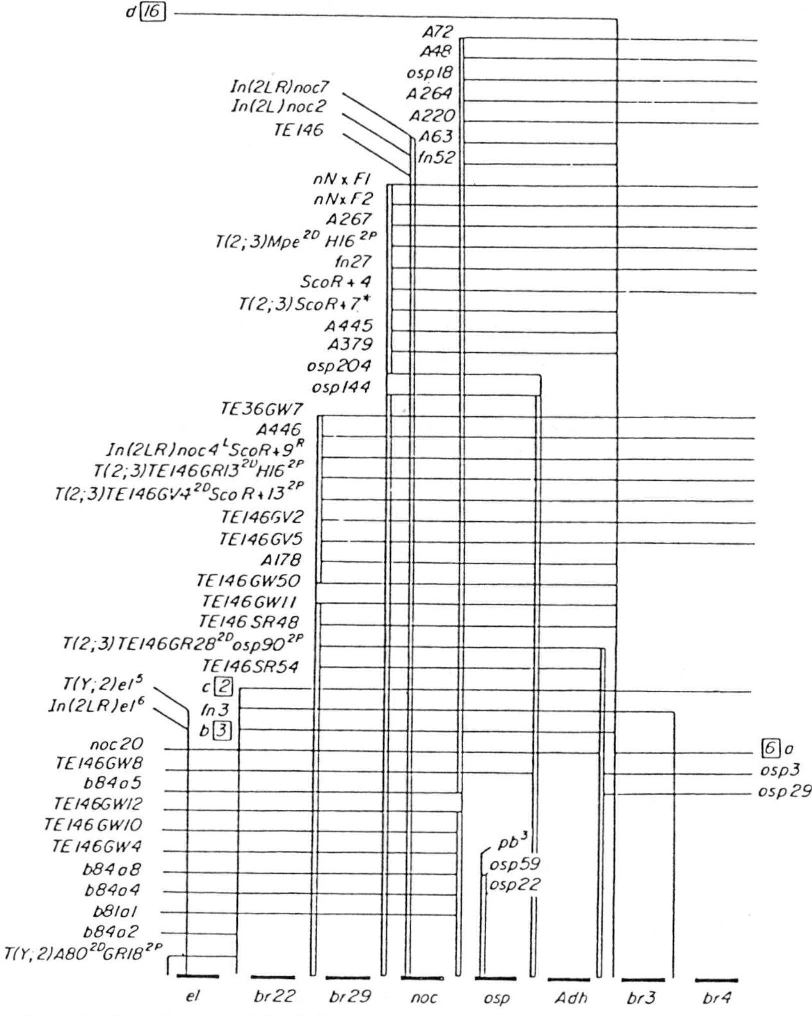

FIG. 1. Genetic map of the *Adh* region on the left arm of the second chromosome of *D. melanogaster*. The markers are *elbow* (*el*), *lethal(2)br22* (*br22*), *lethal(2)br29* (*br29*), *no ocelli* (*noc*), *outspread* (*osp*), *Alcohol dehydrogenase* (*Adh*), *lethal(2)br3* (*br3*), *lethal(2)br4* (*br4*). The marker *l(2)br29* is *not* a discrete locus but part of the *noc* complex (Chia *et al.*, 1985a; McGill, 1985). The genetic breakpoints of many different chromosomal aberrations are shown. The central seven positions (shown as double lines) are included in the chromosome "walk" of Chia *et al.* (1985a). Note that *osp* breakpoints fall to the left and right of *Adh* (see text). This is Fig. 1 of Chia *et al.* (1985a) and is reproduced by permission.

recombination between *Scutoid* and normal chromosomes can lead to duplication of the *Adh* locus (Maroni, 1980). The *Adh* region of the recombinational map was shown to lie within the polytene chromosome bands 35B1 to 35B3 (O'Donnell *et al.,* 1977; Woodruff and Ashburner, 1979a,b) by cytological analysis of deletion breakpoints. This location has since been confirmed directly by *in situ* hybridization using radioactive DNA probes for *Adh* (Benyajati *et al.,* 1980; Goldberg, 1980).

The region of the second chromosome 34F1 to 35F2 was found to correspond to approximately 2.7 map units and contains 29 lethal complementation groups and 9 visible loci (Asburner *et al.,* 1982b). It was noted that the 34-band region covered by deficiency Df(2L)64j (34D to 35C, 2.17 map units) contained approximately 34 such complementation groups (Woodruff and Ashburner, 1979a,b). The present estimate is 33 loci: 8 visibles, 21 lethals, and 4 steriles (M. Ashburner, personal communication). The physical and genetic structures of the region have since been described in very fine detail (Chia *et al.,* 1985a, and references therein). One of the more surprising findings of this work concerns the location of *Adh* with respect to *osp*. There seems to be no functional relationship between these two loci, and yet chromosomal breakpoints associated with *osp* phenotypes can occur both proximal (i.e., 3′) and distal (i.e., 5′) to *Adh*. These molecular mapping data imply that *Adh* is contained entirely within the *osp* region. A similar situation has been described at the *Gart* locus by Heinkoff *et al.* (1986). There are other possible but perhaps less likely interpretations of the genetic evidence as a whole and these are discussed in full by Chia *et al.* (1985a).

Genetic recombination around *Adh* seems to be quite normal despite the unusual cytological appearance of the region, which is due to intense ectopic pairing. Woodruff and Ashburner (1979b) found 0.07% recombination per polytene band (compare 0.057% for the *Drosophila* genome as a whole; Lefevre, 1976). The small size of the *Adh* gene has hindered conventional intragenic mapping of allelic substitutions (Ashburner *et al.,* 1979), but recombinational events within *Adh* itself have been reported (Maroni, 1978). Two novel *Adh* alleles, *A1* and *B7,* were recovered from crosses between two null mutants in this study.

B. Allelic Diversity at *Adh*

Many different *Adh* alleles have been described over the past two decades. They include electrophoretic mobility and heat stability variants from natural populations, and active and null alleles produced by mutagenesis and selection in the laboratory. A list of *Adh* alleles known to encode active enzyme forms is given in Table 1. The many Adh^{null} mutants are not listed here but can be found described in Ashburner (1985) and Kelley *et al.* (1985). However, a number of the better characterized null mutants which have played a crucial part in the development of the *Adh* system are mentioned elsewhere.

C. Chemical Selection for *Adh* Mutants

The screening procedure for *Adh* nulls invented by Sofer and colleagues (O'Donnell *et al.,* 1975; Sofer and Hatkoff, 1972) has advanced analysis of the genetics and biochemistry of *Adh* greatly. It has also made possible the study of mutagenesis per se in *Drosophila.* Their method is both simple and effective. Flies containing putative *Adh* null alleles (usually in heterozygous condition against an *Adh* deficiency) are exposed to an unsaturated secondary aliphatic alcohol such as 1-penten-3-ol (Sofer and Hatkoff, 1972) or 1-pentyn-3-ol (O'Donnell *et al.,* 1975). Such alcohols are poisonous for Adh^{+} individuals, probably because they are rapidly oxidized to highly toxic unsaturated ketone products. Thus, only those individuals which lack ADH activity entirely survive the screen. Conversely, rare Adh^{+} individuals (e.g., revertants) can be selected from an Adh^{-} population by exposure to low concentrations of ethanol (Grell *et al.,* 1968; Vigue and Sofer, 1976). This method is "leaky" and several repeats may be required to isolate the desired lines. It is presently applied to good advantage in selecting Adh^{+} progeny in P-element-mediated transformation experiments (Goldberg *et al.,* 1983; Posakony *et al.,* 1985).

III. Molecular Genetics of *Adh*

The *Adh* gene was one of the first *Drosophila* genes to be cloned and analyzed at the molecular level. As a consequence we have today an extremely detailed knowledge of *Adh* structure and expression for several *Drosophila* species. The picture which has emerged is one of a simple structural gene with a complex and fascinating mode of expression. Recent developments in molecular genetics, for instance P-factor-mediated transformation of *Drosophila* embryos (Rubin and Spradling, 1982; Spradling and Rubin, 1982), have made it possible to discover which DNA sequences in the vicinity of *Adh* exercise control over its expression. Thus, during the past 5 years *Adh* has become one

TABLE 1
Allelic Variants of the *Drosophila melanogaster Adh* Gene

Variant	Description of phenotype	Source	Amino acid substitution (relative to ADH-S)[a]
ADH-US	Ultra-slow mobility variant	Africa[b]	—
ADH-Ss	Heat-sensitive enzyme with ADH-S mobility	N. America[c]	—
ADH-S	Standard ADH-S	Worldwide[d]	Standard
ADH-F	Standard ADH-F	Worldwide[d]	Lys 192 → Thr
ADH-Fr	Heat-stable enzyme with ADH-F mobility	N. America[c]	—
ADH-71K		Laboratory stock[f]	Lys 192 → Thr[e] Pro 214 → Ser
ADH-FCh.D.		Australia[g]	Lys 192 → Thr[h] Pro 214 → Ser
ADH-Fs	Heat-sensitive enzyme with ADH-F mobility	N. America[c]	—
ADH-FR.City	Unstable enzyme	N. America[i]	—
ADH-F′	Stable enzyme with mobility slightly greater than ADH-F	Africa[b]	Ala 51 → Glu
ADH-UF	Ultra-fast mobility variant	Spain[j]	Asn 8 → Ala Ala 45 → Asp Lys 192 → Thr
ADH-D	Ultra-fast mobility mutant	Laboratory induced[k]	Lys 192 → Thr Gly 232 → Gly
ADH-A1	Recombinants	Laboratory cross[l]	—
ADH-B7			—

[a] After Ashburner (1985). References are also given in the text for amino acid replacement data.
[b] David *et al.* (1980).
[c] Sampsell (1977).
[d] Kreitman (1983).
[e] K. Th. Eisses (personal communication).
[f] Thörig *et al.* (1975).
[g] Gibson *et al.* (1980).
[h] Protein sequence data (Chambers *et al.*, 1981a) have been confirmed by DNA sequencing (C. Collet, personal communication).
[i] Fletcher (unpublished); see Chambers *et al.* (1984b).
[j] Malpica and Briscoe (unpublished); see Thatcher (1977).
[k] Grell *et al.* (1968).
[l] Maroni (1978).

of the most promising loci from which to learn how genes are regulated in eukaryotes. Techniques from molecular biology have now been adopted by many different fields of biological research and not surprisingly *Adh* figures prominently in the new molecular approach to the developmental and population biology of *Drosophila* species.

A. Cloning of the *Adh* Locus

The *D. melanogaster Adh* gene was cloned simultaneously and independently in two laboratories using different approaches. Goldberg (1980) obtained his *Adh* clone from a genomic DNA library (Maniatis *et al.*, 1978) by a double screening procedure. The two probes used were derived from poly(A)$^+$ RNA preparations extracted from a wild-type strain (Oregon R, control) and a strain *Df(2L)fn2*/*Df(2L)Sco*$^{R+4}$ which is deficient for the *Adh* locus but still viable. Benyajati *et al.* (1980) obtained their cDNA clone from *Adh* mRNA isolated by precipitation of polysomes with goat anti-ADH antibodies. The *Adh* gene structure was revealed by comparing the cDNA and genomic DNA sequences (Benyajati *et al.*, 1981).

B. The Structure of *Adh*

The *Adh* gene is a small genetic unit encoding a single short polypeptide. The coding region is split into three parts by two small (65 and 70 bp) intervening sequences. The *Adh* gene is unusual in that it is under the control of two separate promoters (Fig. 2). These were

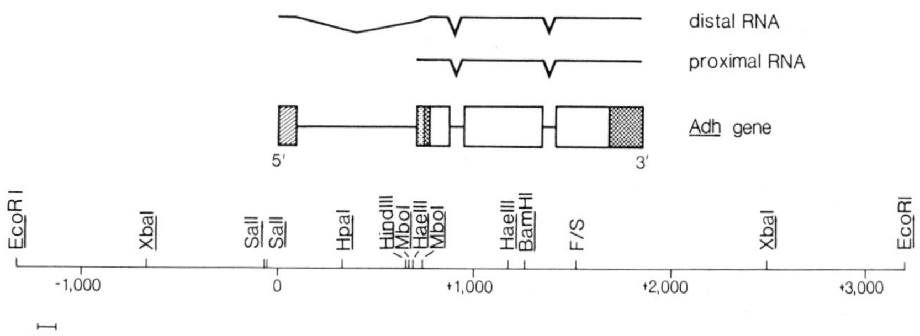

Fig. 2. Diagram of the molecular structure of the *Adh* locus showing the two major transcription products initiated at the proximal and distal promoters (above) and a restriction map of the region (below). In the illustration of the *Adh* gene itself, the boxes represent exons and the lines, intervening sequences. Untranslated exon segments are hatched. This figure was redrawn from Savakis *et al.* (1986) with permission.

revealed only after it was discovered that the major ADH mRNA species present in larvae and adult flies had different 5′ ends. The way in which these two promoters function in transcription was first proposed by Benyajati *et al.* (1983b) and quickly supported by results of exon mapping experiments (Henikoff, 1983) and extensive 5′ DNA sequence data (Kreitman, 1983).

In larvae, transcription begins from the proximal (to *Adh*) promoter 70 bp 5′ to the initiation codon, continues through the coding region, and ends with a short 3′ noncoding segment (121 bp from the TAA termination codon to the AATAAA polyadenylation signal). The mature mRNA is produced by normal RNA processing events including excision of the two introns, capping, and polyadenylation. In adults, *Adh* transcription begins upstream at the distal promoter and proceeds to the same 3′ terminus. Processing this long primary transcript to produce adult *Adh* mRNA thus requires the additional removal of a large (654 bp) intron which includes the larval promoter. The "adult" leader sequence is longer (123 vs 70 bp) than the larval leader. The two *Adh* mRNA species possess different 5′ ends but they do share a common 37-bp sequence immediately preceding the ATG translation start signal. There is considerable recent experimental evidence which helps to explain how the temporal and tissue-specific use of these two promoters is directed and this is described in detail in the next section.

C. DNA Sequences of *Adh* Variants

Several *Adh* variants have now been analyzed by DNA sequencing and Southern blotting. These are listed in full in Ashburner (1985), Kelley *et al.* (1985), and Martin *et al.* (1985). It is gratifying to note that much of the hard-won structural information concerning ADH variants, obtained initially by protein chemistry (e.g., Retzios and Thatcher, 1979; Schwartz and Jörnvall, 1976; Thatcher, 1980), has been substantiated by molecular biology. However, only from DNA sequence determinations (Benyajati *et al.*, 1980; Kreitman, 1983) do we know that the threonine 192 for lysine substitution, first discovered by Fletcher *et al.* (1978), is the *only* coding sequence difference between ADH-F and ADH-S. There are, of course, many noncoding differences between any two individual examples of Adh^F and Adh^S, including some intriguing insertions and deletions in the region between the two promoters (Kreitman, 1983). The proline for serine replacement at residue 214 has been confirmed as the only coding substitution in the heat-resistant $Adh^{FCh.D.}$ variant compared to *ADH-F* (C. Collett, personal communication). This is most interesting news in view of the remarkable resemblance in properties between this enzyme and ADH-S (see later).

Analysis of *Adh*^{null} mutants has revealed much about the effects of X rays and the chemical mutagens formaldehyde and ethyl methanesulfonate (EMS) on *D. melanogaster*. It has also provided some useful insights into the functional significance of various elements of the DNA sequence. Details of the latest experiments can be found in Benyajati *et al.* (1982, 1983a), Kelley *et al.* (1985), Kubli *et al.* (1982), and Martin *et al.* (1985) and a complete list of references is given in Ashburner (1985). Formaldehyde, a substance to which human populations are regularly exposed (see Benyajati *et al.*, 1983a), generates both small (6–11 bp) and large (17–34 bp) deletions at *Adh*, and also the removal of several genetic loci and polytene chromosome bands (mean value 21.09 bands; Ashburner *et al.*, 1982a; O'Donnell *et al.*, 1977). Deletions are sometimes accompanied by other genetic aberrations and when they fall inside intervening sequences they perturb mRNA processing by blocking intron excision (Benyajati *et al.*, 1982). The EMS-induced null mutant *Adh*^{nB} contains the first TGA termination codon reported in *Drosophila* (Martin *et al.*, 1985). It is suppressible *in vitro* with yeast opal suppressor tRNA (Kubli *et al.*, 1982). Kelley *et al.* (1985) examined the structures of 33 X-ray-induced *Adh* mutants obtained by Aaron (1979). Twenty-eight of these mutants were shown to be deletions caused by multiple DNA chain breaks, confirming the earlier classification of 23 of them made by genetic methods (Ashburner *et al.*, 1982a). The mutant *Adh*^{nLA248} obtained by Aaron (1979) deserves special mention because of its unusual structure. Chia *et al.* (1985b) present DNA sequence evidence and argue convincingly that it arose from two close, staggered single-stranded breaks in a haploid sperm. Replication following fertilization produced an *Adh* gene with an internal tandem duplication involving exons 2 and 3. This mutant produces a larger than normal *Adh* transcript from which the nonmutant introns have been removed in the usual fashion but which contains an extra 8-bp junction sequence and duplicated exon segments.

D. DNA Sequences of *Adh* Flanking Regions

The detailed genetic map of the *Adh* region obtained by Ashburner and colleagues (see Section II) has recently been extended to fine resolution by chromosome walking (Chia *et al.*, 1985a) and DNA sequencing (Kreitman and Aguadé, 1986a; Schaeffer and Aquadro,

1986). The chromosome 2 region in the vicinity of the *Adh* locus can now truly be said to be one of the best characterized parts of the entire *Drosophila* genome. A length of 165 kbp spans the polytene chromosome bands 35A4 to 35B1 and includes the morphological markers *osp* and *noc*. Both *osp* and *noc* are extremely large loci (minimum estimates are 52 and 50 kbp, respectively). As noted earlier, the *Adh* locus could be located inside the *osp* region. Kreitman and Aguadé (1986a) found an open reading frame (ORF) 4.2 kbp 5′ to *Adh* but could not demonstrate unambiguously that it produces transcripts. Could this perhaps represent the distal *osp* element? Schaeffer and Aquadro (1986) have sequenced 1.6 kbp of DNA 3′ to the *Adh* gene of *D. pseudoobscura*. Comparison of their sequence to the *D. melanogaster* sequence suggested to them that a new gene was lurking there incognito (see also Kreitman and Aguadé, 1986a). Their proposal, they claim, accounts for the high degree of sequence conservation 3′ to *Adh* observed by Kreitman (1983) in his survey of several Adh^F and Adh^S alleles. This putative coding region might then possibly be taken to represent a proximal section of *osp*. However, it is hard to reconcile Schaeffer and Aquadro's (1986) model for a 3′ gene which has both a TATA box and an ATG start codon, with the *osp* function, unless the *outspread* locus is transcriptionally complex. Also the *osp* breakpoints on the 3′ side of *Adh* fall beyond (i.e., 3′ to) this ORF. A further problem with the present speculations is that the two new genes 5′ and 3′ to *Adh* are separated by only 6 kbp. Alternatively, they may be like *Adh,* and reside within *osp* but be of unknown and dispensable function.

IV. Developmental Genetics of *Adh*

Developmental biology is concerned with describing and explaining how living organisms attain their characteristic morphology. The process of development is the result of complex interactions of genes with each other and between those genes and the environment. Geneticists who are interested in this process are, therefore, seeking to discover when and where particular genes are active, how genetic functions are turned on and off, and what interactions particular loci have with other genes or with hypothetical regulatory DNA elements. Given what has already been revealed of the *Drosophila* alcohol dehydrogenase gene–enzyme system in this article, one can readily see how well suited it is to such studies. However, it must be admitted that the *Adh* function in no direct way controls the morphology of the individual (see Aguadé *et al.*, 1981). Nor does it have any very well characterized interactions with other loci (Leibenguth, 1977; Libion-Mannaert *et al.*, 1976). In fact, under normal laboratory conditions flies can survive perfectly well without it. Despite these reservations, *Adh* has provided and will continue to provide fundamental insights into the ways in which genes in higher organisms are regulated.

A. The *Adh* Developmental Program in *D. melanogaster*

Expression of the *Adh* locus *in vivo* can be monitored by ADH activity measurements over the course of development. The pattern of expression shown by *D. melanogaster* is not typical of *Drosophila* species in general (Anderson and McDonald, 1983b; Batterham *et al.*, 1983b; Dickinson *et al.*, 1984; Maroni and Stamey, 1983; Ursprung *et al.*, 1970). In *D. melanogaster* ADH activity rises slowly during larval life, falls to a very low level during pupation, and rises to a new high early in adult life. The enzyme activity is high in larval and adult fat bodies and is also present in significant amounts in other tissues including Malpighian tubules and larval midgut (Korotchkin *et al.*, 1972; Ursprung *et al.*, 1970). Other *Drosophila* species differ in both the timing and tissue specificity of *Adh* expression and are described briefly later.

The developmental pattern of *Adh* transcription involves the use of both promoters at different times and in different tissues. The pattern of *Adh* mRNA production has recently been described in detail (Savakis and Ashburner, 1986; Savakis *et al.*, 1986). These workers recommend the use of the terms "proximal" and "distal" for the two *Adh* promoters and transcripts, rather than the slightly inaccurate descriptors "larval" and "adult." The proximal promoter is chiefly, but not exclusively, active in larvae and the distal promoter in adults. The level of proximal transcript mRNA rises slowly during larval development and dramatically falls to zero in mid-third instar (Fig. 3). Simultaneously there is a minor peak of transcription from the distal promoter (see also Benyajati *et al.*, 1983b). Transcription of *Adh* is apparently turned off altogether during pupation, and when *Adh* mRNA reappears at eclosion it is the longer message derived by distal transcription which predominates. In females, both proximal and distal transcripts are present in ovaries but they disappear quickly from newly laid (2–4 hours) eggs. Expression of paternal and maternal genes begins together some 4 or 5 hours later. After a brief and minor appearance of distal transcripts, the proximal transcript begins to predominate (see also Leibenguth *et al.*, 1979).

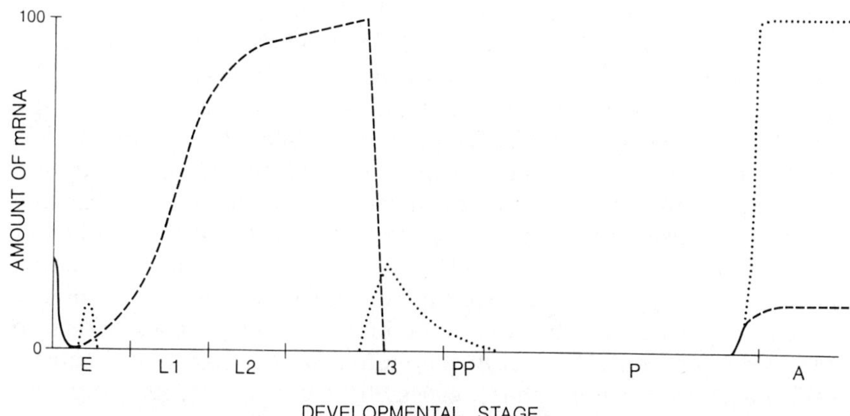

FIG. 3. Changes in amounts of *Adh* transcripts during development of *D. melanogaster*. Life stages are embryo (E); first, second, and third larval instars (L1–L3); prepupal (PP); pupal (P); and adult (A). The *y* axis is a subjective estimate of mRNA levels judged from Northern blots. (——) Maternal transcript, (- - -) proximal transcript, and (•••••) distal transcript. The diagram was redrawn from Savakis *et al.* (1986) with permission.

The cis-acting DNA sequences necessary for correct developmental programming of *Adh* expression are all contained within a short DNA region 5′ to the coding region. Goldberg *et al.* (1983) transformed the germ line of Adh^{null} flies with a P-element containing an 11.8-kbp *Sac*I fragment insert with a centrally located *Adh* gene. They recorded correct developmental expression of the genetically transplanted *Adh* locus, including promoter switching, regardless of the position in the genome into which the P-element had been inserted. A similar observation was made by Benyajati and Dray (1984), who reported correct transient transcription and translation of *Adh* introduced into nonexpressing Schneider line-2 *Drosophila* cells in tissue culture. This work has been taken to even finer levels of analysis in both *D. melanogaster* and *D. mulleri* (see Posakany *et al.*, 1985).

Other promising experiments (see Benyajati, 1984) are presently under way using *Adh* genes with modified 5′ upstream sequences. Their aim is to localize exactly those sequence segments essential to expression and to tissue specificity. Fusion of *Adh* with the promoter region of an entirely different gene (e.g., heat-shock protein 70, *HSP70*) produces a functional hybrid genetic unit. Such constructs confer altered patterns of *Adh* expression dictated by the foreign promoter on their recipients (Bonner *et al.*, 1984; Dudler and Travers, 1984; Klemenz *et al.*, 1985). The hybrid *Adh* genes used in these experiments were temperature shock inducible and expressed in almost all tissues with no apparent major ill effect on the organism. A particularly valuable *in vitro* transcription system derived from nuclei of cultured *Drosophila* Kc cells is giving some very interesting results too. Using this system Heberlein *et al.* (1985) demonstrated that sequences between −24 and −85 bp upstream from the distal start site were essential to mRNA production. They also fractionated a protein called Adf-1 from a nuclear extract and showed that it bound to DNA sequences partially overlapping this region. Results of similar deletion and footprinting analysis experiments gave less clear but equally fascinating results for the proximal promoter. Overall, it seems certain that work on control of *Adh* expression is entering a particularly exciting and rewarding phase, especially since regions flanking *Adh* and implicated in the control of expression have been shown to reside at nuclear scaffold binding regions (Gasser and Laemmli, 1986).

B. The Expression of *Adh* in Other *Drosophila* Species

In contrast to *D. melanogaster,* most *Drosophila* species have a rather low level of ADH activity as adults (Batterham *et al.,* 1983a,b; Dickinson *et al.,* 1984). There are a few exceptions such as *D. lebanonensis* (David *et al.,* 1979) and *D. hydei* (Imberski and Strömen, 1972). Studies on other members of the *melanogaster* subgroup are presently under way (see Savakis and Ashburner, 1986) and seem extremely promising in view of the report by Dickinson *et al.* (1984) that *Adh* genes are under strict cis-acting control in *D. melanogaster–D. simulans* hybrids.

A second species array which is currently attracting considerable attention is the *mulleri* subgroup. All members of this group examined to date possess a duplicated *Adh* locus (Batterham *et al.,* 1984; Mulley *et al.,* 1979; Oakeshott *et al.,* 1982a). The *Adh* genes have been cloned from several members of this group (see Savakis and Ashburner, 1986) and the *D. mulleri Adh* region has been fully sequenced (Fischer and Maniatis, 1985). The situation in this species is best summarized by saying that what *D. melanogaster* achieves with its proximal and distal promoters, *D. mulleri* accomplishes with its proximal and distal genes. This species has three copies of *Adh*: a pseudogene (transcribed but untranslated), *Adh-2* (adult specific), and *Adh-1* (larval specific) in order 5′ to 3′. The three genes are separated by 1.2 and 2 kbp of DNA, respectively. Each gene has its own proximal promoter but none of them has a distal promoter. It is remarkable that the *D. mulleri Adh* pseudogene is transcribed at low levels in adults, even though it lacks a recognizable 5′ TATA box. The 5′ regions of the *D. mulleri* genes show little sequence conservation with *D. melanogaster* 5′ *Adh* sequences. Still, they do possess cis-regulatory sequences which can be recognized by the *D. melanogaster* transcription machinery because both of the true *D. mulleri Adh* genes are expressed in a correct stage- and tissue-specific manner when introduced into the *D. melanogaster* genome on a P-element (Posakony *et al.,* 1985). Further analysis of this system using 5′ deletions has shown that most of the sequences required for expression of *Adh-1* are located within 350 bp immediately 5′ to the gene. However, although correct expression of *Adh-2* is also controlled by cis-acting sequences immediately 5′ to the gene, it further requires that extensive sequences upstream of the pseudogene also be intact! In their absence, *Adh-2* expression is very much (80–90%) reduced but correctly programmed. This report of apparent coordinate control in the *D. mulleri Adh* region is fascinating in the light of Schaeffer and Aquadro's (1986) suggestion concerning possible overlapping regulatory sequences in the vicinity of *Adh* in *D. pseudoobscura*. The correct functioning of a transplanted *Adh* gene from one species (*D. mulleri*) in another (*D. melanogaster*) shows that, different as DNA sequences may appear, important general developmental signals remain preserved within them. It is also evident that trans-acting regulatory factors (e.g., Heberlein *et al.,* 1985) must have conserved functional properties between *Drosophila* species.

The third group of species presently under intensive investigation is the Hawaiian *Drosophila* (Dickinson, 1980a–c, 1983). Among these species the developmental program for *Adh* seems to be evolutionary labile to a quite amazing degree but cis-regulation remains the rule (Dickinson, 1980a, 1983; Dickinson and Carson, 1979). The picture-winged group of Hawaiian *Drosophila* do have two *Adh* promoters. The *D. affinisdisjuncta Adh* gene, for which the full DNA sequence has been obtained, resembles that of *D. melanogaster* quite closely in general design (Brennan *et al.,* 1984b; Rowan and Dickinson, 1986a,b; Rowan *et al.,* 1986). It has been found that *D. melanogaster Adh* carried on P-elements could produce stable transformants in *D. hawaiiensis* (Brennan *et al.,* 1984a) and so it seems likely that analysis of *Adh* gene expression in these species may proceed quickly.

V. Quantitative Genetics of *Adh*

Knowledge is advancing rapidly concerning the genetic signals responsible for turning the *Adh* gene on and off and for determining which tissues shall have the potential to express alcohol dehydrogenase activity. However, much less is known about those genetic and environmental factors which determine the amount of ADH protein produced in organisms. The steady-state quantity of any enzyme in any tissue is decided by the point of balance between its rate of synthesis and its rate of degradation (see Anderson and McDonald, 1981a,b, 1983a, for discussion on *Adh* in this context). It is readily apparent that the ADH activity level in any line of *Drosophila* is a phenotypic character which can be influenced by a wide range of internal and external events. The dissection of the influences of genotype and environment on characters is properly the domain of that field of endeavor which we call quantitative genetics. And so it was that, following the earliest observations (Gibson, 1970, 1972) that $Adh^{F/F}$ homozygotes had higher ADH activity than $Adh^{S/S}$ flies, expression of the *Adh* locus began to be treated and analyzed as any other quantitative phenotypic character (Ward, 1974, 1975; Ward and Hebert, 1972). Subsequent studies have concentrated on either analysis of genotype alone (e.g., Maroni *et al.*, 1982) or on the influence of the environmental conditions: food source, temperature, etc. (e.g., Birley, 1984a). An alternative approach to unraveling the mysteries of quantitative gene expression is the selection experiment. Various selection regimes have been applied to the *D. melanogaster Adh* gene with rather diverse outcomes.

A. The Effect of Genotype on *Adh* Expression

For every experimental observation reported concerning *Adh* there almost always seems to be some piece of contrary evidence. However, there is one happy consensus which has stood the test of time: lines of *D. melanogaster* homozygous for Adh^F alleles have higher ADH enzyme activity than lines homozygous for Adh^S (see Birley and Marson, 1981; Day *et al.*, 1974; Lewis and Gibson, 1978; Maroni *et al.*, 1982; McDonald and Ayala, 1978, for examples and see Ashburner, 1985, for a full list of references). The same authors also agree that within lines homozygous for either allele, there is considerable ADH activity variation. Interest in this variability of expression has largely been as a possible example of regulatory (in its very broadest sense) gene variation and the significance of such variation in evolution (see Hedrick and McDonald, 1980). Variation in ADH activity has been subjected to detailed analysis by rigorous quantitative genetics methods and attempts have been made to localize major regulatory effects to individual chromosomes (Barnes and Birley, 1978; Maroni and Laurie-

Ahlberg, 1983; McDonald and Ayala, 1978). This whole area of research has been very well reviewed by Laurie-Ahlberg (1985) and little more should be added here.

The expression of the *Adh* locus is under complex polygenic control of loci on all of the *D. melanogaster* chromosomes. The first specific locus implicated in control of ADH activity levels to be mapped is *L* and lies approximately 0.01 centimorgan away from *Adh* toward rd^S (Thompson *et al.*, 1977). Detailed mapping experiments were carried out by Maroni and Laurie-Ahlberg (1983) on a selection of the Adh^F and Adh^S lines which they had assayed earlier (Maroni *et al.*, 1982). They discovered that cis-acting elements closely linked to Adh^S alleles were responsible for the observed activity variation. Their Adh^F alleles lacked this form of variation entirely and instead were controlled by trans-acting factors on remote regions of the second chromosome. These results are fully in accord with their later restriction mapping survey (Aquadro *et al.*, 1986) and the idea that Adh^S is the ancestral form.

There is hope that the molecular basis for the difference in expression between Adh^F and Adh^S alleles may soon be resolved. This difference was suggested as pretranslational in origin because Anderson and McDonald (1983a) showed that *Adh* mRNA levels reflect ADH CRM (cross-reacting material) levels. Rates of degradation for ADH-F and ADH-S do not seem to differ despite reported differences in their *in vitro* thermal stabilities. Two recent publications (Aquadro *et al.*, 1986; Birley, 1984b) have reported marked linkage disequilibrium between the allozyme substitution site and flanking restriction site markers. Analysis of alcohol dehydrogenase activity levels and DNA sequences among restriction site haplotypes within each allele class has produced strong evidence that characteristic 2-fold activity differences between certain ADH-F and ADH-S lines are associated with one (or more) of three single-base substitutions in the third exon, including the allozyme substitution site itself (Aquadro *et al.*, 1986). These findings have been confirmed (Laurie-Ahlberg and Stam, 1987) in a study using recombinant *Adh* constructs contained in P-elements to transform host flies lacking ADH activity.

B. The Effect of Environment on *Adh* Expression

All geneticists with an interest in phenotypic characters are well aware that the effects which they study can be strongly influenced by prevailing culture conditions. Any genotype has a norm of reaction and an environmentally induced variance of expression. With the *Adh* system, concerns of this nature are of great importance and certainly well founded. There is clear evidence that chemicals fed to *Drosophila* can markedly affect ADH activity levels and biological half-life (Anderson and McDonald, 1981a,b; Clarke *et al.,* 1979; Papel *et al.,* 1979; Schwartz and Sofer, 1976). The first report of this kind was a tantalizing and never-since-repeated experimental observation that *Drosophila* cells in tissue culture not expressing *Adh* under control conditions could be persuaded to do so by adding alcohol to the tissue culture medium (Fox *et al.,* 1967; Horikawa *et al.,* 1967). Later investigators (above) also concentrated particularly on feeding alcohols of various types. It is only relatively recently that the effects of other food components on *Adh* have been examined in the laboratory, although the chemical composition of natural *Drosophila* substrates has been under the scrutiny of field ecologists for some time; see, for example, the articles by Fogleman and by Starmer in Barker and Starmer (1982).

Contemporary laboratory studies, conducted under carefully controlled conditions, have shown that ADH activity levels can change dramatically when diet is altered (Birley, 1984a; Clarke and Whitehead, 1984; Geer *et al.,* 1985, 1986). Taken together, these studies clearly demonstrate how specific (e.g., ethanol and sucrose) and nonspecific (e.g., apples vs dates) food additives can increase alcohol dehydrogenase levels. Especially exciting is the finding of Clarke and Whitehead (1984) that increasing the amount of threonine available to *Drosophila* adults effected a repeatable and highly significant elevation of ADH activity. The authors are quick to point out that the enzyme itself consists of more than 10% threonine. As ADH is an extremely abundant intracellular protein it was suggested that ADH synthesis is in some way threonine limited. The explanation is ingenious but it cannot be extended to account for the overall greater rate of apparent synthesis of ADH-F (Anderson and McDonald, 1983a), since ADH-F contains one more threonine than ADH-S (28 and 27, respectively).

Genotype–environment interactions are extremely important to ecological and evolutionary studies. There is a regrettable shortage of such data for *Adh* apart from the few specific examples described here. Detailed quantitative genetic investigations of genotype–environment interactions for *Adh* are even scarcer and should perhaps be read in their original form (Birley, 1984a; Birley and Marson, 1981).

C. The Effect of Selection on *Adh* Expression

A very effective way of probing genetic variation in a quantitative character is to subject that character to an artificial selection regime intended to enhance its expression. This well-established technique of quantitative genetics has been applied to the *Adh* locus but with rather surprising infrequency. Three basic experimental designs have been tried: selection for changed ADH activity levels, selection for ethanol tolerance, and selection for ethanol utilization. The results obtained from all three types of experiment are filled with seeming contradictions. Fortunately interpretative help is at hand in the form of an excellent review (Van Delden, 1982).

Successful production of high ADH lines has been achieved after very few generations of selection (Ward, 1975; Ward and Hebert, 1972). This result is not surprising when one recalls how abundant *Adh* modifiers seem to be, and that McDonald and Ayala (1978) found certain high activity modifers to be dominant over low activity genes located elsewhere in the genome. Selection for increased ethanol tolerance has also been successful (Dorado and Barbancho, 1984; Gibson *et al.,* 1979; Sánchez-Cañete *et al.,* 1986; Van Herrewege and David, 1980) and occasionally spectacular (David and Bocquet, 1977). Some procedures caused increases in ADH activity (e.g., McDonald *et al.,* 1977) and some have failed to alter the starting activity level or *Adh* allele frequency significantly (e.g., Gibson *et al.,* 1979). These contrasting experiences serve to emphasize the distinction between alcohol tolerance and utilization (Van Herrewege and David, 1980).

A new form of tolerance-based selection experiment using ethanol vapor has just been developed. To challenge flies with ethanol fumes is not itself a new technique (Oakeshott *et al.,* 1980; Van Herrewege and David, 1984). The novel feature of the modern science of "inebriometry" is the development of an automated procedure used to select flies for resistance to ethanol anesthesia. Results show ethanol knock-down resistance to be selectable with high heritability (Cohan and Graf, 1985; Cohan and Hoffman, 1986; K. E. Weber personal communication). The effects of selection of *Adh* activity (Chambers and Weber, unpublished) or allele frequency (Cohan and Graf, 1985; K. E. Weber *et al.,* personal communication) are small in early generations. However, persistent selection over many generations leads to a marked elevation in frequency for *Adh*^{F} in some populations only (Cohan and Hoffman, 1986). This once again emphasizes the partial independence of "alcohol tolerance" and alcohol dehydrogenase.

However, resistance to anaesthesia is related to the ability to survive and reproduce on ethanol-enriched media (Oakeshott *et al.*, 1985). The above observations are in contrast with those of Oakeshott *et al.* (1980), whose single-generation knockdown test favored Adh^S. The real value of "inebriometry" may not lie in the part it currently plays in the *Adh* story, but that it offers an efficient and flexible way to do large-scale selection experiments.

The final category of experiments, selection for ethanol utilization, involves the establishment of long-term cage populations of *Drosophila* on alcohol-supplemented media and monitoring *Adh* allele frequencies, alcohol tolerance, etc., at intervals over many generations. During the 1970s these experiments were so popular that cage construction became something of a cottage industry for population geneticists. The products of their concerted labors are described later.

VI. The *Adh* Gene Product: *Drosophila* Alcohol Dehydrogenase

The *Adh* gene encodes a small polypeptide of some 255 amino acid residues. The translated product is processed intracellularly by an, as yet, uncharacterized mechanism which removes the N-terminal methionyl residue and leaves a new *N*-acetylserine terminus (Auffret *et al.*, 1978). The ultimate product of translation is a modified protein dimer (ADH) with alcohol dehydrogenase catalytic activity. A clear picture of the structural and functional properties of this enzyme and its metabolic role *in vivo* is essential for a full appreciation of the significance of *Adh* gene function for populations of *Drosophila*. Biochemical analysis of ADH catalysis is relatively undeveloped by comparison to the genetic analysis of *Adh* structure and expression. Nonetheless, the small amount of biochemical information available for ADH has provided plenty of food for thought.

A. Functional Properties of ADH

ADH is very abundant, forming up to 2% of the total protein in *D. melanogaster* adults. It is also quite stable over a broad range of temperature and pH. These two facts have made the task of obtaining the enzyme in a pure form fairly straightforward and several protocols have been published (see Chambers *et al.*, 1984a; Lee, 1982; Leigh-Brown and Lee, 1979; Moxon *et al.*, 1985; Thatcher, 1977, for recent examples). Purified ADH preparations consist of two or three isozymic forms (see Section VI,A,4 below) which complicate kinetic analysis. Differences in biochemical properties of ADH reported from time to time probably result from experiments conducted with enzyme preparations of differing isozymic composition. However, the value of analyzing ADH isozyme mixtures should not be underestimated because a similar multiplicity of enzyme forms occurs *in vivo* (Johnson and Denniston, 1964).

Most biochemical studies have concentrated on comparing the properties of the two most common allelic variants, ADH-F and ADH-S. This work is almost exclusively directed toward the development of an explanation of the worldwide *Adh* polymorphism in *D. melanogaster* at the molecular level. Recently there has been an awakening interest in the enzyme as a challenging biochemical topic in its own right (Hovik *et al.,* 1984; Winberg *et al.,* 1982a,b, 1983, 1985, 1986).

1. Substrate Specificity

Drosophila alcohol dehydrogenase is accorded the Enzyme Commission number EC 1.1.1.1. by all workers because of its ability to catalyze reaction (1) below.

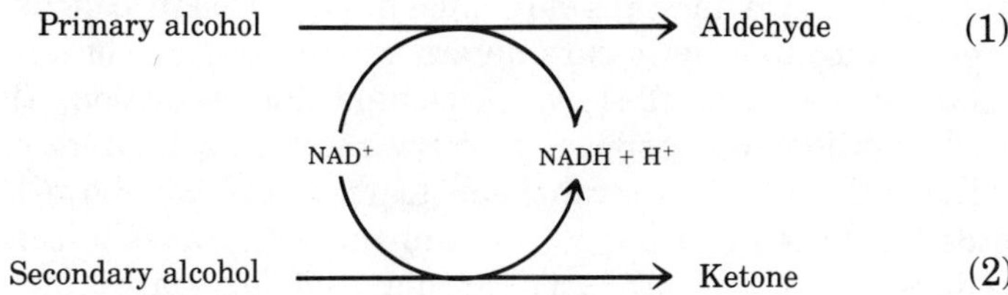

However, unlike the other members of the prestigious company of alcohol dehydrogenases which head the EC list, *Drosophila* ADH oxidizes secondary alcohols to ketones even more readily [reaction (2) above]. The degree to which secondary alcohols are favored substrates for ADH is extremely dependent on experimental conditions, especially pH (Chambers *et al.,* 1981b; Hovik *et al.,* 1984; Winberg *et al.,* 1982a). This property is common to the alcohol dehydrogenases from every *Drosophila* species so far examined (Atrian and Gonzàlez-Duarte, 1985a; Oakeshott *et al.,* 1982a).

To complicate matters further in *D. melanogaster* ADH-F and ADH-S differ in their substrate specificities, as indeed they differ in almost all of their other biochemical properties. The ADH-F allozyme has a greater preference for secondary alcohols than does ADH-S. This fact gains added significance when one selects an ADH assay system to monitor *Adh* expression. It is universally observed that $Adh^{F/F}$ lines have greater ADH enzyme activity than *Adh*$^{S/S}$ lines. This difference arises because (1) the *Adh*$^{F/F}$ lines produce more ADH protein and (2) the protein produced, ADH-F, has a higher specific activity than ADH-S. This phenomenon is clearly illustrated by the data of Lewis and Gibson (1978). Thus there is a real danger of overestimating *Adh* expression when using propan-2-ol based assays. An immunologically based assay system is therefore recommended for such purposes (Anderson and McDonald, 1981a,b, 1983a).

2. *Kinetic Constants and Reaction Mechanism*

Several experimental determinations of Michaelis constants (K_m) values for alcohols have been presented. The most extensive studies are those of Atrian and Gonzàlez-Duarte (1985a), Hovik *et al.* (1984), Juan and Gonzàlez-Duarte (1981), Winberg *et al.* (1982a,b), McDonald *et al.* (1980), and Vilageliu and Gonzàlez-Duarte (1980). There is general agreement that K_m values for short-chain secondary alcohols are around 1.0 m*M*, about 5-fold lower than those for the corresponding primary alcohols. The ADH-F allozyme has higher K_m values for all alcohols tested than ADH-S does, especially when high concentrations of cofactor are present. The reaction mechanism for the enzyme is not yet fully known and thus care must be exercised in attaching biological significance to kinetic parameters as one exchange of letters suggests (Hall and Koehn, 1981, vs McDonald and Anderson, 1981). Recent investigations have shown that the reaction catalyzed by *Drosophila* ADH with secondary alcohol substrates at pH 9.5 follows a compulsory ordered Theorell–Chance mechanism with rate-limiting release of reduced coenzyme product. Ternary complexes are kinetically insignificant when secondary alcohols serve as substrates. In contrast, it is the rate of interconversion of ternary complexes (hydride transfer) which is rate limiting in the case of primary alcohol substrates (Winberg *et al.* 1982a, 1985). These results provide an explanation for the substrate specificity of *Drosophila* ADH.

3. *Thermal Stability*

Population geneticists who are interested in *Adh* have made much of reports that ADH-S has greater thermal stability than ADH-F *in vitro* (see McKay, 1981; Johnson and Powell, 1974; and Sampsell and Sims, 1982). Their arguments offer a direct causal role for environmental temperature in maintaining this polymorphism. Attractive as such ideas may be, some nagging doubts remain. For instance, ADH thermostability *in vitro* is enhanced considerably when purified and depends on experimental conditions (Anderson *et al.*, 1980; Gibson *et al.*, 1980; Thatcher, 1977). However, when investigated with the sophisticated technique of thermal gradient electrophoresis (Thatcher and Hodson, 1981), ADH-S was found to possess greater conformational stability than ADH-F (Thatcher and Sheikh, 1981).

4. *Isozymic Forms*

Electrophoresis of extracts of homozygous flies of any *Adh* genotype produces two or three isozymic staining regions (ADH-5, ADH-3, and ADH-1, in order of anodal migration). The origin of these forms is well understood and arises from the binding of zero, one, or two molecules of an NAD^+–carbonyl adduct per ADH dimer (Schwartz *et al.*, 1979). Feeding adult flies on ketones or ketogenic substances (e.g., acetone or propan-2-ol) quickly leads to the conversion of all ADH molecules to the ADH-1 form (Schwartz and Sofer, 1976; Papel *et al.*, 1979). The interconversion is accompanied by a complete loss of ADH activity (ADH-1 is a dead-end ternary complex). The *in vivo* stability of the ADH molecules is enhanced after conversion (Anderson and McDonald, 1981a).

B. The Structure of ADH

There are several intriguing differences in properties between *Drosophila* ADH and mammalian ADHs, but since little is known about the structure of the *Drosophila* enzyme detailed functional comparisons cannot be made. The amino acid sequence of the enzyme has been fully determined (Thatcher, 1980) and confirmed by DNA sequencing (Goldberg, 1980; Benyajati *et al.*, 1980, 1981; Kreitman, 1983) with minor changes (see Ashburner, 1985). Amino acid substitutions have been described for several ADH variants (see Table 1). The ADH allozymes present are a unique "natural experiment" which can be exploited to probe structure–function relationships in this enzyme. Hypothetical secondary (Benyajati *et al.*, 1981; Thatcher and Sawyer, 1980) and tertiary (Thatcher and Retzios, 1980) structures have been proposed for the enzyme. The predicted structures contain an NAD^+-binding domain toward the N terminus of the molecule. The models account for the failure of ADH-N11 to bind NAD^+ caused by the substitution of aspartic acid for glycine at residue 14.

The other variants are not so directly informative, but nonetheless interesting. All have biochemical properties which, on preliminary investigation, were shown to resemble those of ADH-F or ADH-S very closely (Chambers *et al.*, 1984b). All of their described amino acid replacements (relative to ADH-S) occur at positions which are invariant in *Adh* genes from all other *Drosophila* species examined (see later). This includes position 192 which is substituted by threonine in ADH-F. The sole exception to this general picture is position 214. This residue is proline in *D. melanogaster* but threonine in all other species. The minor variant ADH-FCh.D. has serine at this sequence position (Chambers *et al.,* 1981a). The full biochemical consequences of amino acid substitutions at position 214 remain to be characterized but the role of residue 192 is becoming clearer. Enzyme inhibition studies using the dye Cibacron blue 3G-A, a cofactor analog, implicated this amino acid residue in coenzyme binding (Winberg *et al.,* 1982b) and kinetic studies (Winberg *et al.,* 1985) showed that it was also involved in hydride transfer. These findings serve to explain why the ADH-F and ADH-S allozymes differ in both substrate specificity and specific activity.

The locations of introns in the *Adh* gene are to some extent in accord with the predicted enzyme structure and also with the notion that intervening sequences separate exons which encode separate functional domains of proteins. The second and larger of the two introns divides the gene between the coenzyme binding domain and a hypothetical catalytic domain in the C-terminal part of the ADH molecule. The first intron follows lysyl residue 32 which lies in the junction between α-helical and β-sheet elements of the adenine mononucleotide binding section of the proposed coenzyme domain (Benjayati *et al.,* 1981). The division of mononucleotide binding folds into α-helix plus β-sheet exon units by introns has been reported in other proteins (Lonberg and Gilbert, 1985).

Further structural similarities with mammalian alcohol dehydrogenases are not apparent. The *Drosophila* enzyme is much smaller, is nonhomologous, lacks catalytic or structural zinc atoms (Chambers, 1984; Moxon *et al.,* 1985; Place *et al.,* 1980), and differs considerably in catalytic properties. There are, however, some paradoxical resemblances. Both types of enzyme are inhibited by pyrazole and metal-directed reagents (Winberg *et al.,* 1982b). Place (1985) attributed these similarities to evolutionary convergence in structure of their active site regions. It has also been suggested that the *Drosophila* enzyme may be structurally related to the ribitol dehydrogenase (EC 1.1.1.56) of *Klebsiella aerogenes* with which it shows restricted homology (Jörnvall *et al.,* 1981; and see later). Progress on structure–function analysis of *Drosophila* ADH is seriously hindered by the lack of a good three-dimensional model for the protein. This may be because crystals of *Drosophila* ADH suitable for X-ray crystallography are difficult to obtain. However, recent experiments have produced small ADH-F crystals which diffract X rays and show a threefold screw axis of symmetry (Chambers, Maeda, and Almo, unpublished results).

C. The Metabolic Role of ADH

The major function of *Drosophila* alcohol dehydrogenase is the utilization and detoxification of dietary alcohols. Certain other minor roles have been suggested from time to time, e.g., in juvenile hormone synthesis (Madhavan *et al.,* 1973) or pheromone metabolism (Winberg *et al.,* 1982a). However, these ideas have not been persued. Various interesting special functions of ADH-7IK in *Notch* mutant stocks have been proposed and are presently under investigation (Eisses, 1986; Eisses *et al.,* 1986). The relationship between alcohol metabolism and tolerance has been thoroughly investigated, and this work is described in detail later. It is worthwhile noting here that *D. melanogaster* does encounter high alcohol concentrations (commonly up to 5% v/v ethanol) at the sites where it lives and breeds (McKechnie and Morgan, 1982; McKenzie and McKechnie, 1979; Gibson *et al.,* 1981; Oakeshott *et al.,* 1982c).

Low concentrations of environmental alcohol are an exploitable resource. Above a certain optimum concentration, ethanol has a toxic effect and there is an upper threshold above which these costs outweigh the energetic benefits (Van Herrewege and David, 1980; Parsons, 1980, 1981). The nutritive value of alcohol is shown by several direct demonstrations that ethanol vapor can enhance the longevity of otherwise starving individuals (Starmer *et al.,* 1977; Van Herrewege and David, 1978). The direct product of the action of ADH on ethanol is acetaldehyde, and in all probability it is a buildup of this compound which is responsible for some of the toxic effects of ethanol (David *et al.,* 1978; Deltombe-Lietaert *et al.,* 1979; Depiereux *et al.,* 1985; Moxon *et al.,* 1985; Geer *et al.,* 1985; Gelfand and McDonald, 1980, 1983). Oxidation of the secondary alcohol propan-2-ol produces acetone, a ketone which is metabolically inert (Geer *et al.,* 1985) and excreted into the medium (Gonzàlez-Duarte and Vilageliu, 1985). Utilization of secondary alcohols quickly ceases due to the formation of a stable dead-end ternary complex ADH–NAD^+–acetone (Schwartz and Sofer, 1976; Schwartz *et al.,* 1979; see David *et al.,* 1981; Gonzàlez-Duarte and Vilageliu, 1985; Heinstra *et al.,* 1986a; Vilageliu-Arquès and Gonzàlez-Duarte, 1980, for recent discussions). It seems likely that the catalytic potential of ADH molecules which have been inactivated in this way is lost permanently. It is true that *in vivo* ADH activity levels recover slowly when propan-2-ol is removed (more than 96-hour recovery period; Gonzàlez-Duarte and Atrian, 1985; Papel *et al.*, 1979), but this is due entirely to *de novo* ADH synthesis and can be completely blocked by the administration of the protein synthesis inhibitor cycloheximide (Gonzàlez-Duarte and Atrian, 1986). However, enzyme "retroconversion" has been achieved *in vitro* by everyone who ever ran and stained an ADH gel; see Heinstra *et al.* (1986b) for a detailed explanation. To the existing array of catalytic activities must be added a newly discovered *Drosophila* enzyme; $NADP^+$-dependent aldo-keto reductase (AKR, EC 1.1.1.?). This activity has been reported from both *D. melanogaster* and *D. hydei*. The function of AKR seems to be in the regulation of internal acetaldehyde and acetone concentrations (Atrian and Gonzàlez-Duarte, 1985b; Gonzàlez-Duarte and Atrian, 1985).

Ingested ethanol ultimately produces an accumulation of triacylglycerol in the fat body (Geer *et al.*, 1985). Acetate is a presumed metabolic intermediate, but what is the pathway? For many years ADH existed in a state of limbo unattached to the rest of *Drosophila* intermediary metabolism. Any idea that aldehyde oxidase (AO, EC 1.2.3.1, encoded by *Aldox* at 3-56.7) was responsible for the conversion of acetaldehyde to acetate *in vivo* is now thoroughly discredited (David *et al.*, 1978; Garcin *et al.*, 1983a; Lietaert *et al.*, 1981, 1982). The void left by the departure of AO has been filled by a difficult choice of alternatives. First, aldehyde dehydrogenase activity (ALDH, EC 1.2.1.3) has been discovered in *D. melanogaster* (Chanteux *et al.*, 1985a,b; Garcin *et al.*, 1983b; Lietaert *et al.*, 1985). Second, ADH *itself* has been shown to be capable of oxidizing acetaldehyde to acetate (Brooks *et al.*, 1985; Eisses *et al.*, 1985a; Heinstra *et al.*, 1983; Moxon *et al.*, 1985). *Drosophila* aldehyde dehydrogenase activity was found to reside in a particulate heavy mitochondrial subcellular fraction. It cannot be attributed to contaminating ADH since ALDH activity is present in extracts of a $bAdh^{n4}$ strain of *D. melanogaster* which lacks both ADH and AO activities (Chanteux *et al.*, 1985a,b). The direct oxidation of acetaldehyde to acetate by ADH is unlikely to be significant *in vivo* because this activity has a high (>10) pH optimum (Heinstra *et al.*, 1983). However, ADH is capable of oxidizing ethanol to acetic acid under physiological conditions, provided that the acetaldehyde intermediate remains enzyme bound (Heinstra *et al.*, 1983; Moxon *et al.*, 1985).

A debate has been in progress for some time, albeit in a slightly muted fashion, concerning the mechanisms which control flux through the ethanol utilization/tolerance pathway(s). One hypothesis maintains that ADH activity is a limiting factor. Hence, the relative magnitudes of kinetic parameters (e.g., K_m for ethanol), turnover number, and number of enzyme molecules per cell (i.e., V_{max}) for ADH are the most important factors (Anderson *et al.*, 1981; Clarke and Allendorf, 1979; McDonald *et al.*, 1980). The alternative view is that the supply of NAD^+ produced from the reoxidation of NADH (possibly by the action of *sn*-glycerol-3-phosphate dehydrogenase; α-GPDH, EC 1.1.1.8) is rate limiting (McElfresh and McDonald, 1983). Despite numerous but not universal demonstrations that alcohol tolerance is correlated with overall ADH activity (see Schmitt *et al.*, 1986; Van Delden, 1982, 1984), net conversion of ethanol to carbon dioxide and triglyceride has been shown to be independent of ADH activity level (Middleton and Kacser, 1983). The NAD^+ supply hypothesis has fared slightly better. The intracellular $NADH/NAD^+$ ratio is elevated when ethanol is fed to *D. melanogaster*. However, resupply of NAD^+ is not controlled through α-GPDH (Geer *et al.*, 1983, 1985). Any metabolic interaction that exists between ADH and α-GPDH lies in their combined function in supplying precursors for lipid synthesis. Ethanol feeding also has other quite separate effects on lipid metabolism; for example, it leads to changes in membrane lipid composition, which has been implicated in the ethanol tolerance response (Geer *et al.*, 1986).

The relative contributions of all of these enzyme activities in the metabolism of alcohols, aldehydes, and ketones *in vivo* remain to be decided but it seems likely that the continued use of *Drosophila* structural gene mutants and radiotracers will play a prominent part in the elucidation of the various pathways and their attendant control mechanisms.

VII. Ecological Genetics of *Adh*

Ecology has for its central theme the study of adaptation. This work can take many forms: the observation of interactions of individual creatures one with another, the description of life histories, etc. Ecological geneticists have as their goal the understanding of the genetic bases of the observed adaptations. The task is often frustrating because of the utter complexity of most organisms and ecosystems. There are relatively few clear-cut genetic traits of direct adaptive significance which can be meaningfully analyzed with our present array of techniques. The adaptation of *Drosophila* species (notably *D. melanogaster*) to environmental alcohol is an exception and one in which the *Adh* gene itself occupies a central position.

The interaction of any organism with its environment has many facets. Thus, it is necessary to gather many different kinds of information to gain a full understanding of any adaptive process. As a consequence, ecologists need data provided by many different biological disciplines. In the same way, ecological genetics serves to unite fields such as behavioral genetics and population genetics which have traditionally been considered rather separate. A wide variety of experimental evidence is relevant to the adaptation of *Drosophila* species to alcohol-containing habitats. Some of the relevant facts relating to the biochemical properties of ADH and its function in alcohol metabolism have been described in the preceding section. The history of *Adh* research illustrates the way in which laboratory experiments can be used to complement the all-important field studies of particular model ecosystems. There are excellent general reviews of this area (Hoffman *et al.*, 1984; Parsons 1979, 1983) and the detailed reviews of *Adh* population genetics (David, 1977; Gibson and Oakeshott, 1981; Van Delden, 1982, 1984) provide much additional information. The 1982 Arizona Workshop proceedings volume (Barker and Starmer, 1982) contains papers dealing with some particularly well-studied cactus-based systems. This work is described later along with two other especially interesting field situations.

A. *Drosophila* AND ALCOHOL

Drosophila melanogaster lives and breeds among decomposing fruit and vegetables. The adults feed on microorganisms and are dependent on yeasts for some nutritional elements (see Clarke and Whitehead, 1984; Vacek, 1982). Conditions in these rotting piles of vegetation tend to be anaerobic and favor fermentation. This form of metabolic activity generates reduced excretion products such as ethanol and other alcohols. Ethanol levels measured at *D. melanogaster* breeding sites are very variable and sometimes quite high (Gibson *et al.,* 1981; McKechnie and Morgan, 1982; McKenzie and McKechnie, 1979; Oakeshott, *et al.,* 1982c). The extent to which such a niche can be exploited depends on the maximum alcohol concentration which the organism can utilize. *Drosophila melanogaster* is much more tolerant to environmental alcohol than other cosmopolitan *Drosophila* species, including its sibling species *D. simulans* (Daggard, 1981; Dickinson *et al.,* 1984; Parsons, 1980; Parsons and Spence, 1981a,b; Parsons and Stanley, 1981; Parsons *et al.,* 1979). Thus, *D. melanogaster* gains an edge over its potential rivals by being able to colonize and exploit resources unavailable to the others (Oakeshott *et al.,* 1982c). The adaptive trait, which enables *D. melanogaster* to breed where others cannot, is its high larval, and even higher adult, level of alcohol dehydrogenase activity compared to other species (Dickinson *et al.*, 1984). The poor performance of a *D. melanogaster Adh* null mutant relative to *D. simulans* in Parsons and Spence's (1981b) laboratory tolerance test is good evidence for the importance of ADH in detoxification.

Enhanced alcohol tolerance is one of several adaptations which have enabled the formerly tropical species, *D. melanogaster,* to occupy habitats successfully in temperate regions of the world. The accumulated evidence points to some of the physiological and biochemical adaptations enabling *D. melanogaster* to reproduce in alcohol-containing environments. For *D. melanogaster* populations, ethanol tolerence has been shown to increase with increasing distance from the equator in a number of different tests (Anderson, 1982; Anderson and Gibson, 1986; Cohan and Graf, 1985; David and Bocquet, 1975; Stanley and Parsons, 1981). A parallel increase in alcohol dehydrogenase activity has also been reported (Anderson, 1982; Anderson and Gibson, 1986). This pattern of ADH activity variation can be largely accounted for by the higher Adh^F frequencies observed in temperate *D. melanogaster* populations worldwide (Oakeshott *et al.*, 1982b). However, there is as yet no direct evidence for the existence of latitudinal ethanol concentration gradients in *Drosophila* food sources which would act as agents of natural selection. In *D. simulans* clinal patterns of ethanol tolerance and ADH activity are not observed (Anderson and Oakeshott, 1986). These facts are in full accord with the preceding discussion concerning the contrasting life-styles of these two sibling species.

Behavioral adaptations ensure that advantageous breeding sites will be exploited and their resources utilized. In *Drosophila* alcohol preference behavior has been shown to be species specific (Hoffman and Parsons, 1984; Parsons and Spence, 1980; David and Van Herrewege, 1983) and genetically programmed to some extent (Gelfand and McDonald, 1980, 1983; Hougouto *et al.*, 1982; Parsons, 1980). The answer to J. S. Jones' teasing question, Can genes choose habitats? (Jones, 1980), is clearly yes, in the case of the *Drosophila Adh* gene. Adult *D. melanogaster* are attracted to alcohols, especially ethanol, to a much greater extent than other species (Hoffman and Parsons, 1984); female *D. melanogaster* prefer to lay eggs on ethanol-containing medium (Hougouto *et al.*, 1982) and larvae will select media containing low ethanol concentrations and avoid those with high (Gelfand and McDonald, 1983; Parsons, 1977). The former study showed that larval behavior was dependent on the alcohol dehydrogenase activity level of the subjects and suggested that they were causally related by internal acetaldehyde pool size. These laboratory tests are fully in accord with the habitat preferences of *D. melanogaster* and other cosmopolitan *Drosophila* species reported in the field studies described earlier.

Within *D. melanogaster* the *Adh* polymorphism offers an opportunity to assess how large the contribution of a single gene might be to the complex adaptive and ecologically significant traits just described (see Ziolo and Parsons, 1982). Flies with Adh^F alleles have higher alcohol dehydrogenase protein and activity levels than flies with Adh^S alleles. Therefore, with identical genetic backgrounds (except for very close loci in linkage disequilibrium), Adh^F should have an advantage over Adh^S similar to that which *D. melanogaster* itself enjoys over its competitors. There is plenty of evidence that Adh^F and Adh^S homozygotes have different behavioral responses to ethanol (Fuyama, 1976, 1978; Gelfand and McDonald, 1980; Parsons, 1980). In many one generation tests Adh^F homozygotes survive better with ethanol as an energy source than Adh^S homozygotes. However, Adh^F has not always been found to outperform Adh^S in population cages and selection experiments (Gibson and Oakeshott, 1982; Oakeshott *et al.*, 1984a), nor do *Adh* allele frequencies always correspond with expectations in field situations (Gibson *et al.*, 1981; Hoffman *et al.*, 1984; Ziolo and Parsons, 1982). Thus, it is clear that *Adh* expression represents only a single factor in a complex series of phenotypic adaptations, but it is this part which *Adh* plays in these phenotypes which determines the dynamics of the *Adh* polymorphism and ultimately the evolution of the gene.

B. *Drosophila* Habitats of Special Importance

There are probably between two and three thousand living species of drosophilid flies in the world (Wheeler, 1981). We know next to nothing about most of them. Many cannot be kept alive long enough to permit laboratory study and it is often very difficult to discover where even well-characterized species live and breed. It is not surprising, then, that there are relatively few situations where detailed ecological research has been undertaken with this group of insects. There are two marked exceptions, both relevant to the *Adh* story and which extended our horizons beyond the rotting garbage piles to be found in orchards and city dumps. The development of these studies owes much to the inspirational field work of J. A. McKenzie and colleagues at the Chateau Tahbilk winery in Victoria, Australia (see McKenzie and McKechnie, 1981).

1. Vineyards and Winery Cellars

Wineries have proved to be popular study sites for *Drosophila* researchers in many parts of the world. The interior of a winery cellar and the neighborhood of wine fermentation vats are good places to find high ethanol environments and to test hypotheses developed from laboratory data and from studies on natural populations in orchards. The distribution of *Drosophila* species inside and outside wineries is in uniform accord with expectations; *D. melanogaster* is found at much higher frequencies inside wineries than is the closely related but relatively alcohol-sensitive species *D. simulans* (Marks *et al.,* 1980; Gibson *et al.,* 1981). The most extensive data collection is that of Monclus and Prevosti (1978–1979) who captured 24 *Drosophila* species in eastern Spain, 11 of which were found in association with human habitation and only 8 in winery cellars. Among those species exploiting the cellar habitat, *D. melanogaster* and *D. lebanonensis,* both species with high ADH acitivy (David *et al.,* 1979), made up almost 90% of the collection. *Drosophila funebris* was also quite numerous but not more common than is usual for domestic habitats in the region. This survey has prompted further work on the comparative structural and biochemical properties of the alcohol dehydrogenases from many of the *Drosophila* species found to be present in the Spanish wineries (Atrian-Ventura and Gonzàles-Duarte, 1981; Juan and Gonzàlez-Duarte, 1980, 1981; Vilageliu and Gonzàlez-Duarte, 1984; Vilageliu *et al.,* 1981; Winberg *et al.,* 1986).

When attention is focused on *D. melanogaster* alone, the situation is less clear. Three studies report Adh^F to be at higher frequency inside wineries than outside (in Spain, Alonzo-Moraga *et al.,* 1985, and Briscoe *et al.,* 1975; and Hickey and McLean, 1980, in Ontario, Canada). Other equally extensive studies (Gibson *et al.,* 1981; Marks *et al.,* 1980; McKenzie and McKechnie, 1978; McKenzie and Parsons, 1972, 1974a) found no differences in *Adh* allele frequencies between samples captured in corresponding locations. However, all of these workers, except for Hickey and McLean (1980), reported that flies from inside wineries had much greater tolerance of high ethanol concentrations. These findings once again illustrate the fact that ethanol tolerance is a complex character. Tolerance per se may be a coincidental property of relatively little ecological significance, because even in wineries environmental ethanol concentrations do not necessarily reach highly toxic levels (Gibson *et al.,* 1981). Resolution of the seeming paradox of winery cellars may lie in studies of the abilities of tolerant winery-derived strains to utilize alcohol, or else in the suggestion that many cellar habitats do not have higher alcohol concentrations than naturally fermenting fruit piles (Gibson *et al.*, 1981).

2. Succulent Cacti

During the last decade a coordinated multidisciplinary task force has initiated an ambitious research program into the *Drosophila* species and associated microorganisms that inhabit the decomposing parts of succulent cacti. There are many features of this system, including a detailed knowledge of the chemistry, biogeography, and systematics of the host cactus species, which serve to attract workers from many different fields. These advantages are described in detail by Barker and Starmer (1982). Most cacti of interest are endemic to Mexico, but many have been introduced to other parts of the world (see, for example, Murray, 1982). Indeed, even in New Zealand I pass an *Opuntia* (sp. unknown) with a single forlornly necrosing cladode on the way to my laboratory each day.

The ecosystem is complicated but exceptionally well characterized. The decaying cactus provides nutrients for the growth of microorganisms. These include fermenting and nonfermenting yeasts (Holzschu and Phaff, 1982) which provide the various *Drosophila* species with their food source (Starmer, 1982). Among the chemical components present are a number of volatile organic compounds (Fogleman, 1982; Starmer *et al.*, 1986) which may serve as larval attractants. A variety of alcohols, including propan-2-ol, have been detected often at quite high concentration. Individual members of the *mulleri* subgroup of the *Drosophila repleta* group occupy restricted yeast–cactus habitats (Starmer, 1982) and the alcohols present in each are utilized as nutrients (Batterham *et al.*, 1982; Starmer *et al.*, 1977).

Members of the *D. mulleri* subgroup have been shown to possess duplicate *Adh* genes by conventional genetic crossing and electrophoretic analysis (Batterham *et al.*, 1984; Barker, and Mulley, 1976; Oakeshott *et al.*, 1982a). Some species show varied and interesting developmental patterns of expression (Batterham *et al.*, 1983b) which may be related to the spatial relationship between their *Adh* genes (Fisher and Maniatis, 1985; Mills *et al.*, 1986). Alcohol dehydrogenase genes from several *mulleri* subgroup members have been cloned and sequenced. The structures of the *D. mulleri* genes have already been described and the other species examined so far appear to have a fairly similar arrangement. It is not clear at this stage from the Southern blotting data given in Mills *et al.* (1986) if 5′ pseudogenes are present in *D. mojavensis*, etc. It is probable that the DNA sequence of the entire *Adh-2* region of these species will need to be obtained in order to be certain. One interesting structural feature of the gene arrangements examined is that the two *Adh* genes of *D. mojavensis* and *D. arizonensis* are further apart than the *Adh* loci in the other *mulleri* species surveyed. This change coincides with the capacity to express *Adh-1* in ovaries which *D. mojavensis* and *D. arizonensis* alone possess (Batterham *et al.,* 1983b; Mills *et al.,* 1986).

The gene products from both *D. mojavensis Adh* loci have been purified and characterized biochemically (Batterham *et al.,* 1983a). The ADH isozymes of *D. mojavensis* are both small (24,000 molecular weight subunit) dimeric proteins with properties which closely resemble each other and *D. melanogaster* ADH. For example, they oxidize propan-2-ol more rapidly than ethanol. Similar results were obtained with crude extracts of *D. buzzatti* (Oakeshott *et al.,* 1982a). In all species examined the poplypeptide products of the two *Adh* loci combine to form an interlocus heterodimer, even though their amino acid sequences have diverged (by 10 amino acid substitutions in *D. mulleri*).

Some *mulleri* subgroup species have variant alleles at one or both loci, notably *D. mojavensis* and *D. buzzatti.* There is very little information concerning possible differences in properties between the allozymes, although the ADH-2^F and ADH-2^S variants of *D. mojavensis* have been shown to differ in thermal stability (Batterham *et al.,* 1982, 1983a; Starmer *et al.,* 1977). Despite limited biochemical knowledge, considerable effort has been directed toward the study of the *Adh* polymorphism in *D. buzzatti* (Fontdevila *et al.,* 1980; Oakeshott *et al.,* 1982). Biogeographical patterns of electrophoretic variation have been described (Barker and Mulley, 1976) and correlated with environmental variables (Barker, 1981; Mulley *et al.,* 1979). These studies suggested that the *Adh-1* polymorphism might be maintained by natural selection. Barker and East (1980) used this locus and two others (*Est-2* and *Pyr*) as subjects in a successful field trial to perturb allele frequencies in a wild population of *D. buzzatti.* Their experimental results provide some of the most convincing evidence for the differential action of natural selection on allozymes *in vivo.*

Work on the Australian *Opuntia–D. buzzatti* system is advancing quickly, with detailed studies of microorganisms and 2-propanol levels in habitats (Starmer *et al.,* 1986), effects of food components on *Adh-1* genotypes (Starmer and Barker, 1986), and laboratory population cage experiments (Cooke *et al.,* 1986).

3. *Hibiscus Flowers*

One further system deserves mention, if not for its advanced state of development, then surely for its charm and elegance (Holmes *et al.,* 1980; Parsons, 1984). Among the native Australian drosophilid fauna are species such as *D. hibisci* which breed in flowers. This species and others (e.g., *D. inornata*) are not attracted to fermenting fruit baits and are Adh^{null}. Other species such as *D. lativitta* and *D. nitidithorax* can be collected at fruit baits and when wild populations were examined electrophoretically they were found to be segregating for active and null *Adh* alleles. These two species are presently spreading into orchard and urban habitats although their ethanol resource-utilization threshold suggests that they do not presently pose a competitive threat to *D. melanogaster*. This example shows how rapidly evolution may proceed in some instances. A potentially simple event such as the putative reactivation of a silenced *Adh* gene may have permitted a pair of former floral specialists to adopt new life-styles as colonizing generalists.

VIII. Evolutionary Genetics of *Adh*

There are few scientific issues in evolution where we can find full accord among biologists. However, all are agreed on the central role and fundamental importance of evolutionary studies to all biological disciplines. Through the study of the preserved remains of once-living creatures evolutionists attempt to reconstruct history and to infer the nature of the processes by which those creatures now living came into being. All of the inferred processes involve some genetic basis, and it is the objective of evolutionary geneticists to describe and explain evolution in genotypic terms. Over the past two decades, these endeavors have been aided by the development of many sophisticated molecular techniques: protein electrophoresis, DNA–DNA hybridization, several immunological methods (e.g., microcomplement fixation), and protein sequencing. The application of these techniques has provided many challenges to conventional wisdom obtained through the classic methodologies of paleontology, morphological analysis, and cytology. Perhaps the most remarkable feature of the current molecular era is not the battles which have been fought between systematists ancient and modern but the broad extent of their overall agreement. Most recently, developments in molecular biology have enabled biologists to examine the genome of any organism directly and even to permit the molecular resurrection of extinct ones on occasion (Higushi *et al.,* 1984; Pääbo, 1985). These new methods, especially DNA sequencing, promise to create a real and far-reaching revolution in the study of the genetics of evolutionary change (see Lewontin, 1986). The *Drosophila Adh* gene is already at the forefront of this revolution.

The fossil record for *Drosophila* species is very sparse (Ashburner *et al.,* 1984; Beverley and Wilson, 1982, 1985; Throckmorton, 1975, 1982). Phylogenies of *Drosophila* species groups have been established based largely on anatomical (especially genital) and karyotypic features. Even dating the divergence of major species groups is difficult and relies, in the main, on the use of biogeographical data. Several modern methods have been applied to the problem: electrophoresis (Eisses *et al.,* 1979; Lakovara *et al.,* 1972; Triantaphyllidis *et al.,* 1980), microcomplement fixation (Collier and MacIntyre, 1977), and DNA–DNA hybridization (Zwiebel *et al.,* 1982). The results of these studies are reasonably consistent with conventional phylogenies. Zwiebel *et al.* (1982) further showed through hybridization experiments that a single gene, *Adh,* could provide an adequate representation of the whole genome. However, greater confidence should be placed in phylogenic reconstructions based on DNA sequences from several loci (Penny *et al.,* 1982). Alcohol dehydrogenase is one of the only genes for which information is available from sufficient *Drosophila* species to make even a single locus-based reconstruction possible.

A. The *melanogaster* Species Group

The species *D. melanogaster* is an atypical element of this species group because it possesses several alternative alleles at the *Adh* locus whereas the other members are invariant (Eisses *et al.,* 1979). The amino acid sequence and properties of the *D. melanogaster* variants have been described elsewhere. The DNA sequences of these variants have not been investigated extensively but that of $Adh^{FCh.D.}$ is known (C. Collet, personal communication). With one exception, all of the allozymes investigated differ from one or other of the two major allelic variants by a single amino acid substitution (see Table 1). The exceptional variant, ADH-UF, differs from ADH-F by two amino acid replacements. It seems likely that many silent-site base substitutions will be found when the DNA sequences of Adh^{UF}, etc., are determined. Such data may give an indication of how long each variant has persisted in natural populations (see Aquadro and Chambers, 1987). The $Adh^{FCh.D.}$ sequence does differ slightly from Adh^{F} in intron 1 and the 3′ noncoding region (C. Collet, personal communication). There is also considerable silent-site polymorphism among Adh^{F} and Adh^{S} alleles (Kreitman, 1983). This latter study serves as a timely and lasting reminder of how variable individual DNA sequences can be. The implication of this private DNA sequence variation is of great significance for evolutionary studies; i.e., no single individual DNA sequence for any gene may be taken as wholly representative of that gene in that species! Of course individual DNA gene sequences are not expected to differ much from the species consensus and, therefore, variability of this type should not cause too many major difficulties in phylogeny reconstruction. However, it does present pitfalls for the unwary when making hypotheses concerning control of gene expression from interspecific comparisons.

The sequences of *Adh* genes for several members of this species group have been reported (Ashburner *et al.,* 1984; Bodmer and Ashburner, 1984; Cohn *et al.,* 1984). The data show *D. mauritiana* (endemic to the island of Mauritius in the Indian Ocean) to be closely related to *D. simulans,* and that *D. orena* (from tropical West Africa) is more distantly related to the *D. melanogaster–D. simulans–D. mauritiana* cluster. This model is consistent with cytological data (Ashburner *et al.,* 1984) and DNA–DNA hybridization data (Zwiebel *et al.,* 1982). The Adh^S allele of *D. melanogaster* is revealed as the ancestral form since the Lys-192 AAG codon is present in *Adh* from each of the three other species group members. This finding confirms the earlier peptide mapping data of Juan and Gonzàlez-Duarte (1981). It is interesting to note that this lysyl residue is also preserved in even quite distantly related species, e.g., *D. immigrans* and *D. hydei* (Atrian and Gonzàlez-Duarte, 1985a; Vilageliu and Gonzàlez-Duarte, 1984).

B. Other *Drosophila* Species

The structures of alcohol dehydrogenase genes and proteins are being examined from an ever-increasing array of *Drosophila* species. There are now full DNA sequences for entire genes for members of several major species groups: *D. melanogaster* and relatives (see above, Bodmer and Ashburner, 1984), the *obscura* group (*D. pseudoobscura:* Schaeffer and Aquadro, 1986), the Hawaiian *Drosophila* (*D. affinisdisjuncta,* Rowan and Dickinson, 1987), and the *mulleri* subgroup (*D. mulleri,* Fischer and Maniatis, 1985; soon to be followed by *D. mojavensis,* Mills *et al.,* 1986). Descriptions of the major features of the structures and expression of these genes are given elsewhere in this article. It would be both premature and beyond the scope of this review to attempt a full analysis of patterns of evolution shown by these genes. However, a few interesting points can be made.

The gene as a whole seems to be evolving quite rapidly—not so quickly as mammalian fibrinopeptide sequences perhaps, but more quickly than is usual for some dehydrogenases (see Ballas *et al.,* 1984;

Wilson *et al.*, 1977). Silent-base substitutions have accumulated (at synonymous positions) within coding regions approximately 10 times faster than replacement changes. Even with this rapid rate of change at third codon positions, the coding region as a whole changes more slowly than introns or leader sequences. Introns are the most rapidly evolving component of the *Adh* gene and show size changes in addition to single base substitutions (see references above).

The problems of assigning divergence times and hence calculating evolutionary rates for *Adh* genes in *Drosophila* species is discussed by all of the above authors. Values for rates of sequences change have been calculated, based on historical and biogeographical data (Throckmorton, 1975) or based on a variety of general rate constants for molecular evolution, from electrophoretic genetic distance (Carson, 1976), DNA–DNA hybridization (Zwiebel *et al.*, 1982), or DNA sequence divergence (Li *et al.*, 1984, 1985; Miyata, 1982). Among molecular methods, the rate calibrations of Miyata and co-workers (Miyata, 1982; Hayashida and Miyata, 1983) seem most promising but should be applied with caution (Easteal and Oakeshott, 1985). As might be expected, there are considerable differences between the evolutionary rates calculated by strict application of the different methods. A reasonable consensus might be to speculate that the *Drosophila Adh* gene accepts one amino acid sequence change every two million years. This is equivalent to a unit evolutionary period (i.e., for 1% divergence between the sequences of any two species) of about 2.5 million years.

C. The Evolutionary Origin of *Drosophila* ADH

There remains one nagging question which must be addressed here. Where did the *Drosophila Adh* gene come from in the first place? The structural (Thatcher, 1980), catalytic (Hovik *et al.*, 1984; Winberg *et al.*, 1982a,b), and stereochemical (Benner *et al.*, 1985) differences between *Drosophila* and mammalian alcohol dehydrogenases argue strongly for a genuine case of evolutionary convergence. The absence of any zinc atoms in the *Drosophila* enzyme is particularly telling evidence. The only account presently advanced for the origin of this low molecular weight dehydrogenase is the model of H. Jörnvall and colleagues (Jeffrey and Jörnvall, 1983; Jörnvall *et al.*, 1981, 1984). Their scheme for the evolutionary origin of *Drosophila* ADH is part of a general attempt to explain the abundance and multiplicity of certain enzymes in higher organisms. An early evolutionary subdivision of dehydrogenases into "long" and "short-subunit" enzyme types is postulated to have preceded a lengthy later period of sequence divergence, but functional convergence, between members of the two type classes. The result of this process can be observed today in the diversity of alcohol and polyol dehydrogenases. On one hand, we have the long subunit metalloenzymes such as mammalian alcohol dehydrogenase and sorbitol dehydrogenase (EC 1.1.1.14), and on the other, the short subunit *Drosophila* alcohol dehydrogenase and *Klebsiella* ribitol dehydrogenase which do not contain metals. The best evidence in favor of this historical model is the rather limited amino acid sequence homology between these pairs of enzymes. There seems little doubt that the story will become clearer as more DNA sequence data appear. It seems amazing that no group has yet reported screening a *Drosophila* DNA library with a mammalian *Adh* probe or vice versa.

IX. Population Genetics of *Adh*

Just as evolutionary genetics reveals the path we have followed from the past to the present, so population genetics tells us where we are going in the future. A thorough description of the present biological state is required in terms of mutation rates, population sizes, and other population parameters before population genetics theory can be used as a reliable guide to new evolutionary destinations. Population genetics is unique among biological disciplines precisely because it possesses a large body of predictive theory. In fact, it might be said that we have amassed what almost amounts to an embarrassing wealth of such theory. What we currently lack in order to make adequate use of this resource is a sufficiently detailed description of the prevailing biological state. Experimental population geneticists are, therefore, busy measuring the extent of genetic variation in natural populations, trying to establish which perturbing forces are at work and to measure their intensities. Evolutionists struggle on, confident in the belief that those forces which we can detect in action today are the same ones which brought life hither from the past and which will direct it onward tomorrow.

We strive to comprehend the changes in developmental programming which have allowed so many differently designed creatures to evolve and populate the world in times past and present. Alas many, if not all, adaptive traits are so complex as to defy present methods of analysis. Therefore, geneticists must usually follow a reductionist approach and select a simple tractable system, such as a structural gene (like *Adh*) and its product. Occasionally, if one is lucky the chosen locus may be shown to play a greater or lesser role in some more complex traits (like alcohol tolerance). It should be clear from the information in this article and those other reviews mentioned earlier that the *Drosophila* alcohol dehydrogenase gene is well suited to such purposes.

The *Adh* gene in *Drosophila melanogaster* presents a two-allele allozyme polymorphism when examined by conventional single-condition electrophoresis (Kreitman, 1980). As such, the *Adh* locus is typical of many gene–enzyme systems found in *Drosophila* and other organisms (Lewontin, 1986). It has, however, figured more prominently than others in attempts to answer two fundamental questions concerning the population genetics of simple structural genes: (1) How may we best measure genetic variation at such loci? (2) What evolutionary forces are acting on this variation?

The answer to question 1 has been discussed at length by Lewontin (1986) together with the general patterns emerging from recent surveys of genetic variation. The problem requires no further comment from the present author except to describe work on *Adh* gene variation at the DNA level. Question 2 has remained vexingly unresolved for many years. Debate usually boils down to the issue of whether allozyme polymorphisms are maintained by natural selection or not—the so-called "selectionist" vs "neutralist" controversy. Random genetic drift as a process is undeniable and undenied, I think, even by the most ardent selectionists. Genetic drift is the inevitable consequence of sampling from a finite population each generation. Neutral theory (Kimura, 1982) simply proposes that this is the only force of any magnitude acting to alter allele frequencies at the majority of loci-encoding proteins. From this deceptively simple postulate there follows a series of definite and potentially testable predictions concerning the rates of protein evolution in terms of effective population size and mutation frequency. Many more tests of neutral theory have been proposed than can be described here (see Keith *et al.*, 1985; Lewontin, 1986; Mukai *et al.*, 1982), but much of the information that has been collected concerning *Adh* does bear directly on this issue.

A. Biogeography of *Adh* Variation

A random process, such as genetic drift, might be expected to produce random geographical patterns of genetic variation. It has long been realized that *Adh* alleles are distributed throughout natural populations in a far from random pattern (Johnson and Schaffer, 1973; Pipkin *et al.*, 1976; Vigue and Johnson, 1973). The frequency of Adh^F

increases with latitude both north and south of the equator, as revealed by large surveys on several continents (Malpica and Vassalo, 1980; Oakeshott *et al.,* 1982b, 1984b). Similar clinal patterns are also found for other *D. melanogaster* and *D. simulans* loci (Anderson and Oakeshott, 1984; Oakeshott *et al.,* 1984b). It is argued that these patterns reflect underlying latitudinal selection gradients.

The clines do not result from linkage disequilibrium of *Adh* allozymes with chromosomal inversions [e.g., In(2L)t, see Malpica *et al.,* 1987], which are themselves subject to natural selection (Lewontin *et al.,* 1981). Specific associations between *Adh* alleles (notably Adh^{S}) and inversions have been detected in some populations (data reviewed by Mukai *et al.,* 1982), but nonrandom associations between electrophoretically detectable alleles themselves are rarely, if ever, observed in *D. melanogaster* (e.g., Alahiotis *et al.,* 1976; Birley and Haley, 1987; Knibb, 1982, 1983; Laurie-Ahlberg and Weir, 1979). A second concern with electrophoretic survey data of this type is undetected (electrophoretically cryptic) variation. Such hidden variation, important though it is, creates no difficulties for interpretation of the *Adh* data. Cryptic *Adh* alleles are rare (see Ayala's 1982 analysis of Fletcher's 1980 data; Kreitman, 1980, 1983; and Lewis and Gibson, 1978), but a third heat-resistant allozyme Adh^{Fr} (Sampsell, 1977) is found at polymorphic frequencies in North America and a presumed Australasian counterpart $Adh^{FCh.D.}$ has been described (Gibson *et al.,* 1980). This minor *Adh* allele is widespread but at low frequency on both continents (Gibson *et al.,* 1982; Sampsell, 1977; Sampsell and Steward, 1983; Smith *et al.,* 1984; Wilks *et al.,* 1980). Possibly of greater concern are Adh^{null} alleles which have been reported to reach quite high (up to 3.9%) frequencies in some populations (Freeth and Gibson, 1985).

It is now certainly possible to predict the expected allozyme allele frequencies of *D. melanogaster* populations on other continents (Singh *et al.,* 1982; Singh and Rhomberg, 1987). Some independent survey data (e.g., David, 1982) are predicted with accuracy for more than one locus. With new data beginning to appear from other parts of the globe (e.g., Brazil, Charles-Palabost and Lehman, 1984; and Korea, Chung, 1977, 1984; Chung *et al.,* 1982; Chung and Yoon, 1980), a thorough test of such predictions will become possible. These large-scale nonrandom allele distributions must surely be more than coincidental products of genetic drift. Therefore, we can now turn to the examination of the possible nature of the hypothetical selective forces which may be responsible for their formation.

B. Environmental Correlates of *Adh* Allele Frequency

Do the latitudinal clines that we find for *Adh* and other *Drosophila* allozymes really reflect selection gradients imposed by the external environment on polymorphic genetic loci? Is it possible to identify climatic factors which may be responsible? These two important questions are prompted by the biogeographical data. The first approach to resolving the issue is to seek altitudinal data which would be expected to show parallel variation to the latitudinal data. Two studies of this type are to be found in the *Adh* literature and their results are consistent. They show that Adh^F increases in frequency with elevation on mountainsides in Russia (Grossman *et al.,* 1970) and Mexico (Pipkin *et al.,* 1976). All of these observations suggest temperature to be an important variable. The second approach is to correlate climatic data, such as average monthly maximum and minimum temperature at the collection site, with allele frequency. For the *D. melanogaster Adh* locus, the procedure has been regrettably rather uninformative (see Oakeshott *et al.,* 1982b, 1984b for discussions). This finding is not unexpected because the method employed is necessarily crude. Climatic conditions reported at a local weather station, however close to the field study site, may have little direct bearing on the microhabitat which the fly itself selects, although it must be admitted that climatic indices do in principle have the potential to direct our attention to relevant environmental parameters (see McKenzie and Parsons, 1974b; Barker, 1981).

A second environmental factor of recognized significance is alcohol. Both temperature and ethanol have figured prominently as variables in the *Adh* population cage experiments. Investigations of this type are fully described, explained, and discussed by Gibson and Oakeshott (1982), Kusakabe and Mukai (1984), Mukai *et al.* (1982), Oakeshott *et al.* (1984a, 1985), and Van Delden (1982, 1984). Cages have been run under almost every conceivable condition using a wide variety of alcohols. The following partial list of references bears testimony: Alahiotis (1982), Alahiotis and Kilas (1982), Birley (1984a), Birley and Haley (1987), Cavener and Clegg (1978, 1981a,b), Dorado and Barbancho (1984), Hayley and Birley (1983), Kerver and Van Delden (1984, 1985), Oakeshott *et al.* (1983), Sangiorgi *et al.* (1981), Van Delden and Kamping (1980, 1983), Wilson and Oakeshott (1986), Wilson *et al.* (1982), Yamazaki *et al.* (1983), Yoshimaru and Mukai (1979). From these data, we may tentatively draw the conclusion that Adh^F tends to be favored but not always fixed when ethanol is present.

Differences between the results of short- and long-term selection experiments of all types abound in the *Adh* literature. These problems are discussed at length by Gibson and Oakeshott (1982). There is considerable evidence to support their view that these differences in results are observed because *Drosophila* populations respond differently to selection regimes after they have been maintained in the laboratory for periods longer than 1 year. The ultimate goal of these experiments in their many and varied forms is to measure selection coefficients for the various *Adh* genotypes. For the present, it is probably fair to conclude that taken overall these experiments reveal the selection coefficients to be small and rather variable.

C. Biochemical Models

An alternative approach to explaining the maintenance of allozyme polymorphisms is the development of predictive biochemical models (Clarke, 1975; Koehn, 1978; Van Delden, 1984; Zera *et al.*, 1985). In short, the methodology is applied as follows. First, give due consideration to all of the biochemical properties of the allozymes in question. Next, select one property which differs between the variants and construct a model which explains the presence of the two (or more) competing enzyme forms in natural populations in terms of hypothetical selective advantages in different environments. Such models usually invoke some form of balancing selection but one novel frequency-dependent selection model based on K_m values has been described for the *Adh* system (Clarke and Allendorf, 1979). The assembled model should then account for the geographic distributions of alleles in natural populations, any seasonal fluctuations in allele frequency observed (Berger, 1971; Charles-Palabost *et al.*, 1985; Franklin, 1981; Girard and Palabost, 1976; Muñoz-Serrano *et al.*, 1986), and the dynamics of allele frequencies in captive populations when the environment is manipulated experimentally (i.e., the population cage results). Two criticisms can be made of the general approach: (1) When a pair of allozymes, such as ADH-F and ADH-S, differ in almost all of their biochemical properties, how should one select the single significant property, from among the many, for the basis of the model? (2) There is a strong temptation to utilize population survey data when constructing the model. A model so constructed then amounts to a biochemical rationalization a posteriori.

These criticisms should not be taken as an invitation to despair of ever finding the true molecular basis of the *Adh* (or any) polymorphism. However, they do suggest that we should scrutinize any models presented rather carefully. For the *Adh* polymorphism, the explanations proposed have been extensively described and discussed (Chambers, 1984; Chambers *et al.*, 1984b; Sampsell and Sims, 1982; Van Delden, 1982; Zera *et al.*, 1985). In brief, most authors suggest that the high level of ADH activity associated with the Adh^{F} allele gives its possessor a selective advantage in high-alcohol environments, whereas some biochemical property of the ADH-S allozyme (e.g., superior thermostability) conveys to its possessor a superior ability to survive elevated environmental temperatures (i.e., in the tropics). There seems little doubt that this overall picture is probably close to a true causal explanation. However, this view does lack detail and there are still a few unsolved riddles. For example, Sampsell and Barnette (1985) found that ADH levels increase in *D. melanogaster* cultured at high temperature (29°C) regardless of *Adh* genotype. Particularly troublesome is the third and minor member of the *Adh* polymorphism, Adh^{Fr}. The gene product of Adh^{Fr} (and $Adh^{FCh.D.}$) has catalytic properties which are indistinguishable from ADH-S (Chambers, 1984; Chambers *et al.*, 1984b; Gibson *et al.*, 1980) but the ADH-FCh.D. protein is much more heat stable than the enzyme produced by either of the two major alleles. The current biochemical model as presented cannot explain why Adh^{S} is not replaced by Adh^{Fr} in natural populations. A simplified biochemical model based on ADH activity levels alone (Chambers *et al.*, 1984b) provides a possible alternative hypothesis. It does predict the geographical distribution of $Adh^{FCh.D.}$ reasonably well (Gibson *et al.*, 1982) but problems arise when flux through the alcohol metabolism pathway as a whole is considered (Kacser and Burns, 1981; Middleton and Kacser, 1983). This alternative view is also unsatisfactory when one weighs the large body of evidence that directly implicates temperature in the maintenance of the polymorphism (Oakeshott *et al.*, 1985b; Sampsell and Sims, 1982). The role of environmental temperature variation may be more subtle than formerly porposed (Alahiotis, 1982).

D. Natural Selection and the *Adh* Polymorphism

Many of the publications presented here contain discussions of the possible ways in which natural selection might act to preserve the *Adh* polymorphism. The impression formed from reading them is that, among *Adh* workers, much of the former heat has been lost from the selectionist versus neutralist argument. Also there are no claims that any single universal mode of selection (such as heterozygote advantage) is at work at all times or in all populations. Neither party has emerged victorious, but instead, a new and amiable consensus seems to be abroad.

Theoretical population geneticists model simple one-locus, two-allele systems such as *Adh* in terms of genotypes, selection coefficients (s, t), and relative fitnesses:

FF	FS	SS	Genotype
homozygote	heterozygote	homozygote	
$1 - s$	1	$1 - t$	Fitness

Although selection coefficients may be considered to have global mean values ($\bar{s}$, $\bar{t}$), the actual values obtained in any one experiment or pertaining in any one natural situation may differ considerably. In other words, the selection coefficients like any other biological variables are quantities with means ($\bar{s}$, $\bar{t}$) and variances ($\sigma^2_{\bar{s}}$, $\sigma^2_{\bar{t}}$). From perusing the conflicting *Adh* population genetics data, it seems reasonable to say that the global mean values of the selection coefficients are small but that their variances are large. The numerous population cage experiments of Oakeshott, Wilson, and colleagues (Oakeshott *et al.*, 1983; Wilson and Oakeshott, 1986; Wilson *et al.*, 1982) provide an excellent illustration of this view.

The significance of these general observations about s and t values for *Adh* genotypes becomes apparent when they are presented as a plot in s, t space (see Fig. 4; after Oakeshott, Wilson, and colleagues but modified). It is true to say that a polymorphism, such as the one illustrated, is maintained by heterozygote advantage (stabilizing selection) taken over all populations and all environments. The global mean fitness set ($\bar{s}$, $\bar{t}$) falls in the quadrant I, where both s and t have positive values, and consequently the FS heterozygote has a higher global mean fitness than either homozygote. The stable equilibrium global allele frequencies will depend on the relative magnitudes of $\bar{s}$ and $\bar{t}$. However, all of this discussion tells us nothing about any individual population. Many local environmental factors (climate, food source), genetic factors (epistatic interactions, mutations), and random factors (effective population size) will cause particular s and t values to differ from $\bar{s}$ and $\bar{t}$. The actual populational values could lie anywhere within the dotted confidence limit.

If the population fitness set (s, t) falls in quadrant II then there is directional selection and Adh^S becomes fixed. Similarly, in quadrant IV fixation for Adh^F is expected. Quadrant III represents the region of heterozygote inferiority (disruptive selection), where any change from an unstable equilibrium allele frequency results in one allele or the other becoming fixed. Should the population reside in an area with an

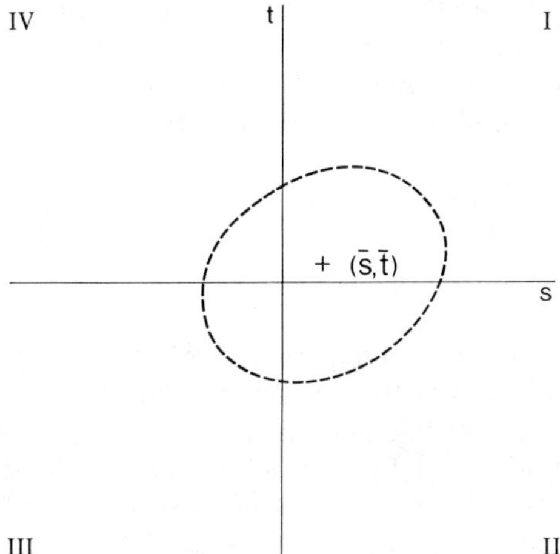

FIG. 4. Representation of a two-allele genetic polymorphism showing global overdominance (heterozygote advantage). The axes s and t are selection coefficients for the two homozygotes. The dotted envelope indicates hypothetical joint variances of mean values ($\bar{s}$ and $\bar{t}$). A full explanation is given in the text.

especially variable environment, the point (s, t) may be driven from one quadrant to another quite often. This process is called diversifying selection. Population cage experiments attempt either to measure s and t as estimators of $\bar{s}$ and $\bar{t}$, or to manipulate the environment in an attempt to discover factors capable of moving (s, t) to another position. For example, the inclusion of ethanol in the cage medium might be expected to move (s, t) for the experimental population toward or into quadrant IV (see Oakeshott *et al.*, 1983, 1984a; Van Delden, 1982, 1984).

If we accept this view of allozyme polymorphisms, there is little point ascribing their maintenance to any particular form of natural selection, because at any moment in time different populations of the same organism may be subject to quite different selective forces. Equally, there is little hope that we can decide form fitness estimates alone whether $(\bar{s}, \bar{t})$ is exactly coincident with (O;O), the condition of neutrality. With a truly neutral polymorphism, individual experimental procedures are likely to give nonzero estimates for s and t, and experimenters will differ wildly in their conclusions regarding the action of natural selection on the locus in question. Instead, one must look to other forms of evidence to decide if natural selection acts on any particular locus, and use population cages in an attempt to identify important environmental variables. The existence of the *Adh* clines, repeated over time and space, argues strongly that ($\bar{s}$, $\bar{t}$) for the *Adh* polymorphism resides in quadrant I, and many *Adh* population cage experiments do give *s* and *t* estimates which fall in this region (Oakeshott *et al.*, 1983, 1984a, 1985; Van Delden, 1982, 1984; Wilson *et al.*, 1982).

E. DNA Studies in Population Genetics

The development of recombinant DNA technology has presented population geneticists with many new opportunities. Not only is it possible to analyze genetic variation at the "ultimate" level, but it is also now possible to test hypotheses concerning the relative roles of processes such as selection, mutation, and historical accident in the evolutionary process. The initial impact of DNA studies on population genetics has been reviewed by Clegg and Epperson (1985) and by Lewontin (1986). The purpose of this section is to describe what has been learned about the *Adh* gene.

Two types of DNA study are currently in vogue, a coarse-grained analysis mapping restriction site polymorphisms and fine-grained analysis by DNA sequencing. The two methods have different applications. Restriction mapping surveys provide haplotypes in bulk which can be used to reconstruct phylogenies. DNA sequencing provides very detailed information for a relatively small number of subjects but which can be used to estimate levels of polymorphism in functionally differentiated segments of chromosomes (exons, introns, untranslated regions). Langley *et al.* (1982) first applied restriction mapping analysis to the *Adh* region of *D. melanogaster* and its close relatives. Their survey covered a 12-kbp region with *Adh* close to the middle of it. They demonstrated several polymorphic sites, including one case of a restriction site in strong linkage disequilibrium with an allozyme marker, and described a number of insertion/deletion polymorphisms, particularly in the region around 2 kbp away from the 3′ end of the coding region. Subsequent surveys (Aquadro *et al.*, 1986; Birley, 1984b; Golding *et al.*, 1986) supplied more data confirming these early findings and have given some new and interesting results. There is much more haplotype diversity among Adh^S lines than among Adh^F lines, which is in keeping with the idea that Adh^S is an ancestral form. Local linkage disequilibrium around *Adh* was shown to be quite marked and a quite surprising number of insertion/deletion polymorphisms were revealed, both 5′ and 3′ of the gene. The 3′ changes tended to involve bigger sections of DNA and many had arisen from the insertion or removal of transposable elements. A much smaller study has shown that the heat-resistant *Adh* alleles are derived from a limited number of *Adh*^{F} haplotypes (perhaps only one) and, thus, must be relatively recent in origin (C. F. Aquadro and G. K. Chambers, unpublished observations). However, it should be noted that *D. melanogaster* may not be typical of *Drosophila* species in general. In a restriction mapping survey of the *Adh* region of *D. pseudoobscura*, Schaeffer *et al.* (1986) found high haplotype diversity and low linkage disequilibrium. This study stands in strong contrast with the earlier work on *D. melanogaster*.

Kreitman's (1983) DNA sequencing analysis of 11 *Adh* alleles is already a landmark in population genetics. His data revealed that almost every individual fly possessed a unique DNA sequence for an *Adh* gene. Of even greater significance was the finding that all of the base substitutions occurred either in introns or at third positions in codons and only where they did not result in amino acid substitutions. The study confirmed that *Adh*^{F} alleles had less DNA sequence variation than *Adh*^{S} alleles. There was strong evidence of large-scale migration in *D. melanogaster* populations because identical *Adh*^{F} sequences were found in isolates from France, North America, and Africa. This strengthens the argument that the *Adh* clines (Oakeshott *et al.,* 1982b) are maintained by selection. The lack of single base substitutions resulting in coding changes in strong evidence for the purifying action of natural selection. This is supported by the observation that those allelic variants present in natural populations which have been shown to contain amino acid substitutions (ADH-FCh.D., ADH-F′, ADH-UF) have properties which resemble one or other of the two major alleles (Chambers *et al.,* 1984b). It can be argued that the rare allozymes represent neutral elements of selectively maintained classes. Newly arisen mutants with biochemical properties identical to either ADH-F or ADH-S could increase in frequency by genetic drift to reach polymorphic frequency, whereas newly arisen mutants with biochemical properties different from both ADH-F and ADH-S would be rapidly eliminated.

New techniques are being developed seemingly every minute in molecular biology and are sure to extend the analytical tool kit for population genetics. A combination of DNA sequencing and electrophoretic transfer now permits direct DNA sequence analysis of relatively crude genomic DNA (Church and Gilbert, 1984). A modification of this basic technique has been developed by Kreitman and Aguadé (1986) and uses four nucleotide-recognizing restriction enzymes. Their method enabled them to do a fine-grained (approximately 20% of all single base substitutions can be detected) and yet large-scale (87 lines from two North American *D. melanogaster* populations) survey. Their results are in agreement with the earlier findings concerning very limited recombination around the *Adh* gene. Further, they recorded no less than 50 distinct haplotypes for the 2.7-kbp DNA segment examined, but no significant differentiation between the two *D. melanogaster* populations from opposite coasts, although allozyme frequencies were very different between them. These data have been used to test the neutral theory prediction that quickly evolving regions of the genome will be characterized by high levels of polymorphism within species (Kimura, 1983). Comparison of Kreitman and Aguade's (1986) data with that of Coyne and Kreitman (1986) for *Adh*-flanking regions of *D. seychellia* led Hudson *et al.* (1986) to suggest that a balanced polymorphism was present at the *Adh* locus in *D. melanogaster.* However, it should be noted that Hudson and Kaplan (1986) came to a different conclusion from an analysis of Kreitman's (1983) from within *D. melanogaster* which is supported by Stephens and Nei's (1985) interspecific comparisons.

X. Conclusions

There is no further need to sing the praises of the *Drosophila* alcohol dehydrogenase gene–enzyme system. Its popularity with geneticists seems likely to continue for some time to come. The utility of Adh^{+} and Adh^{null} as selectable markers are beginning to promote the use of the *Adh* gene by workers primarily interested in other systems (Bonner *et al.,* 1984; Dudler and Travers, 1984; Keene and Elgin, 1984; Watts *et al.,* 1983) or even radically different problems such as control of insect pests (Robinson and Van Heemert, 1981), aging (Subrahmanyam *et al.,* 1984), or space exploration (Kloek *et al.,* 1976). The detailed molecular analysis of transcription at the *Adh* locus is developing rapidly and we can confidently expect to learn some of the principles which govern gene regulation in higher organisms from this work.

Interest in fine details of the mechanism of maintenance of the *Adh* polymorphism seems to be waning somewhat, bogged down by controversy. However, the *Adh* gene is enjoying a new lease on life as a subject of the current molecular biological approach to population genetics. It is also growing in significance and utility as a character in *Drosophila* phylogeny, in studies of enzyme evolution, and as a problem in biochemical evolution in its own right. A fuller appreciation of all three subjects will be gained if the present intensity of biochemical work on ADH is maintained and especially if a three-dimensional structure of the enzyme can be obtained.

It seems appropriate to close this essay by offering a small sociological insight gleaned from reading so many research papers on the *Drosophila Adh* gene. It seems to be that once any biologist (myself included) becomes entangled with the *Adh* locus, there can be no real hope of escape. One is forever tempted back to the laboratory bench to perform "just one more" critical experiment. With this thought in mind, good prospects for the future of the *Adh* gene–enzyme system seem assured.

Acknowledgments

I am much indebted to the large number of scientists who generously contributed so many helpful reprints and preprints. I am very grateful for the efforts of Ms. A. E. Chambers, Ms. A. P. Chambers, and Ms. I. J. Pomer who helped me organize them, and for the time spent conducting literature searches on my behalf by Mr. L. T. Diggins and Ms. E. Ng. I would like to express my appreciation of the expert work of Mrs. M. M. Howorth in preparing the manuscript, Mr. J. Casey for photography, and Ms. R. Mita for artwork. Dr. M. Ashburner, Dr. C. Benyajati, and Dr. J. S. McKinley-McKee made valuable suggestions which helped to improve the final manuscript.

References

Aaron, C. S. (1979). X-ray induced mutations affecting the level of the enzyme alcohol dehydrogenase E.C. 1.1.1.1 in *Drosophila melanogaster* progeny and genetic analysis of null-enzyme mutants. *Mutat. Res.* **63,** 137–138.

Aguadé, M., Cuello, J., and Prevosti, A. (1981). Correlated responses to selection for wing length in allozyme systems of *Drosophila melanogaster. Theor. Appl. Genet.* **60,** 317–327.

Alahiotis, S. N. (1982). Adaptation of *Drosophila* enzymes to temperature. IV. Natural selection at the alcohol dehydrogenase locus. *Genetica* **59,** 81–87.

Alahiotis, S. N., and Kilas, G. (1982). Mating propensity and variation in enzyme activity in long-term cage populations of *Drosophila melanogaster. J. Hered.* **73,** 53–58.

Alahiotis, S. N., Pelecanos, M., and Zacharopoulo, A. (1976). A contribution to the study of linkage disequilibrium in *Drosophila melanogaster. Can. J. Genet. Cytol.* **18,** 739–745.

Alonzo-Moraga, A., Muñoz-Serrano, A., and Rodero, A. (1985). Variation of allozyme frequencies in Spanish field and cellar populations of *D. melanogaster. Genet. Sel. Evol.* **17,** 435–442.

Anderson, D. G. (1982). Alcohol dehydrogenase activity and ethanol tolerance along the *Adh* cline in Australia. *In* "Advances in Genetics, Development and Evolution of *Drosophila*" (S. Lakovaara, ed.), pp. 263–272. Plenum, New York.

Anderson, D. G., and Gibson, J. B. (1986). Variation in alcohol dehydrogenase activity *in vitro* in flies from natural populations of *Drosophila melanogaster. Genetica* (in press).

Anderson, P. R., and Oakeshott, J. G. (1984). Parallel geographical patterns of allozyme variation in two sibling *Drosophila* species. *Nature (London)* **308,** 729–731.

Anderson, P. R., and Oakeshott, J. G. (1986). Ethanol tolerance and alcohol dehydrogenase activity in Australian populations of *Drosophila simulans*. *Heredity* **56,** 185–190.

Anderson, S. M., and McDonald, J. F. (1981a). A method for determining the *in vivo* stability of *Drosophila* alcohol dehydrogenase (E.C. 1.1.1.1). *Biochem. Genet.* **19,** 411–419.

Anderson, S. M., and McDonald, J. F. (1981b). Effects of environmental alcohol on *in vivo* properties of *Drosophila* alcohol dehydrogenase. *Biochem. Genet.* **19,** 421–430.

Anderson, S. M., and McDonald, J. F. (1983a). Biochemical and molecular analysis of naturally occurring *Adh* variants in *Drosophila melanogaster. Proc. Natl. Acad. Sci. U.S.A.* **80,** 4798–4802.

Anderson, S. M., and McDonald, J. F. (1983b). Changes in levels of alcohol dehydrogenase E.C. 1.1.1.1 during the development of *Drosophila melanogaster. Can. J. Genet. Cytol.* **23,** 305–314.

Anderson, S. M., Santos, M., and McDonald, J. F. (1980). Comparative study of the thermostability of crude and purified preparations of alcohol dehydrogenase (E.C. 1.1.1.1) from *Drosophila melanogaster. Dros. Inform. Serv.* **55,** 13–14.

Anderson, S. M., McDonald, J. F., and Santos, M. (1981). Selection at the *Adh* locus in *Drosophila melanogaster:* Adult survivorship–mortality in response to ethanol. *Experientia* **37,** 463–464.

Aquadro, C. F., Deese, S. F., Bland, M. M., Langley, C. H., and Laurie-Ahlberg, C. C. (1986). Molecular population genetics of the alcohol dehydrogenase gene region of *Drosophila melanogaster. Genetics* **114,** 1165–1190.

Ashburner, M. (1985). *Adh:* Alcohol dehydrogenase. *Dros. Inform. Serv.* **62,** 9–13.

Ashburner, M., Camfield, R., Clarke, B., Thatcher, D., and Woodruff, R. (1979). A genetic analysis of the locus coding for alcohol dehydrogenase and its adjacent chromosome region in *Drosophila melanogaster. In* "Eucaryotic Gene Regulation" (R. Axel, T. Maniatis, and C. F. Fox, eds.), pp. 95–106. Academic Press, New York.

Ashburner, M., Aaron, C. S., and Tsubota, S. (1982a). The genetics of a small autosomal region of *Drosophila melanogaster* including the structural gene for alcohol dehydrogenase. V. Characterisation of X-ray-induced *Adh* null mutations. *Genetics* **102,** 421–435.

Ashburner, M., Tsubota, S., and Woodruff, R. C. (1982b). The genetics of a small autosomal region of *Drosophila melanogaster,* including the structural gene for alcohol dehydrogenase. IV. *Scutoid,* an antimorphic mutation. *Genetics* **102,** 401–420.

Ashburner, M., Bodmer, M., and Lemunier, F. (1984). On the evolutionary relationships of *Drosophila melanogaster. Dev. Genet.* **4,** 295–312.

Atrian-Ventura, S., and Gonzàlez-Duarte, R. (1981). Comparison of some biochemical features of the enzyme alcohol dehydrogenase in sixteen species of *Drosophila. In* "Advances in Genetics, Development and Evolution of Drosophila" (S. Lakovaara, ed.), pp. 251–261. Plenum, New York.

Atrian, S., and Gonzàlez-Duarte, R. (1985a). Purification and molecular characterisation of alcohol dehydrogenase from *Drosophila hydei:* Conservation in the biochemical features of the enzyme in several species of *Drosophila. Biochem. Genet.* **23,** 891–911.

Atrian, S., and Gonzàlez-Duarte, R. (1985b). An aldo-keto reductase activity in *Drosophila melanogaster* and *Drosophila hydei:* A possible function in alcohol metabolism. *Comp. Biochem. Physiol.* **81B,** 949–952.

Auffret, A. D., Williams, D. H., and Thatcher, D. R. (1978). Identification of the blocked amino terminus of an alcohol dehydrogenase from *Drosophila melanogaster*. *FEBS Lett.* **90,** 324–326.

Ayala, F. J. (1982). Genetic variation in natural populations: Problem of electrophoretically cryptic alleles. *Proc. Natl. Acad. Sci. U.S.A.* **79,** 550–554.

Ballas, R. A., Garavelli, J. S., and White, H. B. (1984). Estimation of the rate of glycerol 3-phosphate dehydrogenase evolution in higher vertebrates. *Evolution* **38,** 658–664.

Barker, J. S. F. (1981). Selection at allozyme loci in cactophilic *Drosophila*. *In* "Genetic Studies of *Drosophila* Populations" (J. B. Gibson and J. G. Oakeshott, eds.), pp. 161–184. ANU Press, Canberra.

Barker, J. S. F., and East, P. D. (1980). Evidence for selection following perturbation of allozyme frequencies in a natural population of *Drosophila buzzatii*. *Nature (London)* **284,** 166–168.

Barker, J. S. F., and Mulley, J. C. (1976). Isozyme variation in natural populations of *Drosophila buzzatii*. *Evolution* **30,** 213–233.

Barker, J. S. F., and Starmer, W. T. (1982). "Ecological Genetics: The Cactus–Yeast–*Drosophila* Model System" Academic Press, Sydney.

Barnes, B. W., and Birley, A. J. (1978). Genetical variation for enzyme activity in a population of *Drosophila melanogaster*. IV. Analysis of alcohol dehydrogenase activity in chromosome substitution lines. *Heredity* **40,** 51–57.

Batterham, P., Starmer, W. T., and Sullivan, D. T. (1982). Biochemical genetics of the alcohol longevity response of *D. mojavensis*. *In* "Ecological Genetics and Evolution: The Cactus–Yeast–*Drosophila* Model System" (J. S. F. Barker and W. T. Starmer, eds.), pp. 307–322. Academic Press, Sydney.

Batterham, P., Gritz, E., Starmer, W. T., and Sullivan, D. T. (1983a). Biochemical characterisation of the products of the *Adh* loci of *Drosophila mojavensis*. *Biochem. Genet.* **21,** 871–883.

Batterham, P., Lovett, J. A., Starmer, W. T., and Sullivan, D. T. (1983b). Differential regulation of duplicate alcohol dehydrogenase genes in *Drosophila mojavensis*. *Dev. Biol.* **96,** 346–354.

Batterham, P., Chambers, G. K., Starmer, W. T., and Sullivan, D. T. (1984). Origin and expression of an alcohol dehydrogenase gene duplication in the genus *Drosophila*. *Evolution* **38,** 644–657.

Benner, S. A., Nambier, K. P., and Chambers, G. K. (1985). A stereochemical imperative in dehydrogenases: New data and criteria for evaluating function-based theories in bioorganic chemistry. *J. Am. Chem. Soc.* **107,** 5513–5517.

Benyajati, C. (1984). DNA sequences that are required for the alcohol dehydrogenase gene expression in *Drosophila melanogaster*. *Prog. Cancer Res. Ther.* **30,** 345–350.

Benyajati, C., and Dray, T. F. (1984). Cloned *Drosophila* alcohol dehydrogenase genes are correctly expressed after transfection into *Drosophila* cells in culture. *Proc. Natl. Acad. Sci. U.S.A.* **81,** 1701–1705.

Benyajati, C., Wang, N., Reddy, A., Weinberg, E., and Sofer, W. (1980). Alcohol dehydrogenase in *Drosophila:* Isolation and characterisation of messenger RNA and a cDNA clone. *Nucleic Acids Res.* **8,** 5649–5667.

Benyajati, C., Place, A. R., Powers, D. A., and Sofer, W. (1981). Alcohol dehydrogenase gene of *Drosophila melanogaster:* Relationship of intervening sequences to functional domains in the protein. *Proc. Natl. Acad. Sci. U.S.A.* **78,** 2717–2721.

Benyajati, C., Place, A. R., Wang, N., Pentz, E., and Sofer, W. (1982). Deletions at intervening sequence splice sites in the alcohol dehydrogenase gene of *Drosophila*. *Nucleic Acids Res.* **10,** 7261–7272.

Benyajati, C., Place, A. R., and Sofer, W. (1983a). Formaldehyde mutagenesis in *Drosophila:* Molecular analysis of *Adh*-negative mutants. *Mutat. Res.* **111,** 1–7.

Benyajati, C., Spoerel, N., Haymerle, H., and Ashburner, M. (1983b). The messenger RNA for alcohol dehydrogenase in *Drosophila melanogaster* differs in its 5′ end in different developmental stages. *Cell* **33,** 125–133.

Berger, E. M. (1971). A temporal survey of allelic variation in natural and laboratory populations of *Drosophila melanogaster*. *Genetics* **67,** 121–136.

Beverley, S. M., and Wilson, A. C. (1982). Molecular evolution of *Drosophila* and higher Diptera. I. Micro-complement fixation studies of a larval haemolymph protein. *J. Mol. Evol.* **18,** 251–264.

Beverley, S. M., and Wilson, A. C. (1985). Ancient origin for Hawaiian Drosophilidae inferred from protein comparisons. *Proc. Natl. Acad. Sci. U.S.A.* **82,** 4753–4757.

Birley, A. J. (1984a). Natural selection for the control of alcohol dehydrogenase activity in populations of *Drosophila melanogaster*. *Dev. Genet.* **4,** 407–424.

Birley, A. J. (1984b). Restriction endonuclease map variation and gene activity in the *Adh* region in a population of *Drosophila melanogaster*. *Heredity* **52,** 103–112.

Birley, A. J., and Marson, A. (1981). Genetic variation for enzyme activity in a population of *Drosophila melanogaster*. VII. Genotype–environment interaction for alcohol dehydrogenase activity. *Heredity* **46,** 427–442.

Bodmer, M., and Ashburner, M. (1984). Conservation and change in the DNA sequences coding for alcohol dehydrogenase in sibling species of *Drosophila*. *Nature* (*London*) **309,** 425–430.

Bonner, J. T., Parks, C., Parker-Thornberg, J., Mortin, M. A., and Pelhem, R. B. (1984). The use of promoter fusions in *Drosophila* genetics: Isolation of mutations affecting the heat shock response. *Cell* **37,** 979–992.

Brennan, M. D., Rowan, R. G., and Dickinson, W. J. (1984a). Introduction of a functional P-element into the germ line of *Drosophila hawaiiensis*. *Cell* **38,** 147–151.

Brennan, M. D., Rowan, R. G., Rabinow, L., and Dickinson, W. J. (1984b). Isolation and initial characterisation of the alcohol dehydrogenase gene from *Drosophila affinidisjuncta*. *J. Mol. Appl. Genet.* **2,** 436–446.

Briscoe, D. A., Robertson, A., and Malpica, J. M. (1975). Dominance at *Adh* locus in response of adult *Drosophila melanogaster* to environmental alcohol. *Nature* (*London*) **225,** 148–149.

Brooks, W. M., Moxon, L. N., Field, J., Irving, M. G., and Doddrell, D. M. (1985). *In vitro* metabolism of [2-^{13}C]ethanol by ^{1}H NMR spectroscopy using ^{13}C decoupling with the reverse DEPT polarisation–transfer pulse sequence. *Biochem. Biophys. Res. Commun.* **128,** 107–112.

Carson, H. L. (1976). Inference of the time of origin of some *Drosophila* species. *Nature* (*London*) **259,** 395–396.

Cavener, D. R., and Clegg, M. T. (1978). Dynamics of correlated genetic systems, IV. Multilocus effects of ethanol stress environments. *Genetics* **90,** 629–644.

Cavener, D. R., and Clegg, M. T. (1981a). Multigenic response to ethanol in *Drosophila melanogaster*. *Evolution* **35,** 1–10.

Cavener, D. R., and Clegg, M. T. (1981b). Temporal stability of allozyme frequencies in a natural population of Drosophila melanogaster. Genetics **98,** 613–623.

Chambers, G. K. (1984). The purification and biochemical properties of alcohol dehydrogenase-"Fast(Chateau Douglas)" from *Drosophila melanogaster*. *Biochem. Genet.* **22,** 529–549.

Chambers, G. K., Laver, W. G., Campbell, S., and Gibson, J. B. (1981a). Structural analysis of an electrophoretically cryptic alcohol dehydrogenase variant from an Australian population of *Drosophila melanogaster*. *Proc. Natl. Acad. Sci. U.S.A.* **78,** 3103–3107.

Chambers, G. K., Wilks, A. V., and Gibson, J. B. (1981b). An electrophoretically cryptic alcohol dehydrogenase variant in *Drosophila melanogaster*. III. Biochemical properties and comparison with common enzyme forms. *Aust. J. Biol. Sci.* **34,** 625–637.

Chambers, G. K., Fletcher, T. S., and Ayala, F. J. (1984a). Purification and partial characterisation of alcohol dehydrogenase, fructose 1,6-bisphosphate aldolase and the cytoplasmic form of malate dehydrogenase from *Drosophila melanogaster*. *Insect. Biochem.* **14,** 359–368.

Chambers, G. K., Wilks, A. V., and Gibson, J. B. (1984b). Variation in the biochemical properties of the *Drosophila* alcohol dehydrogenase allozymes. *Biochem. Genet.* **27,** 153–168.

Chanteux, B., Lechein, J., Dernoncourt-Sterpin, M., Libion-Mannaert, M., Wattiaux-De Coninck, S., and Elens, A. (1985a). Ethanol metabolising enzymes subcellular distribution in *D. melanogaster* flies homogenates. *Dros. Inform. Serv.* **61,** 47–48.

Chanteux, B., Libion-Mannaert, M., Dernoncourt-Sterpin, S., Wattiaux-De Coninck, S., and Elens, A. (1985b). Ethanol metabolising system in *Drosophila melanogaster:* Subcellular distribution of some main enzymes. *Experientia* **41,** 1543–1546.

Charles-Palabost, L., and Lehmann, M. (1984). The effect of the temperature upon the variability of the gene pool in a Brazilian population of *D. melanogaster*. *Dros. Inform. Serv.* **60,** 71–72.

Charles-Palabost, L., Lehmann, M., and Mercot, H. (1985). Allozyme variation in 14 natural populations of *Drosophila melanogaster* collected from different regions of France. *Genet. Sel. Evol.* **17,** 201–210.

Chia, W., Karp, R., McGill, S., and Ashburner, M. (1985a). Molecular analysis of the *Adh* region of the genome of *Drosophila melanogaster*. *J. Mol. Biol.* **186,** 689–706.

Chia, W., Savakis, C., Karp, R., Pelham, H., and Ashburner, M. (1985b). Mutation of the *Adh* gene of *Drosophila melanogaster* containing an internal tandem duplication. *J. Mol. Biol.* **186,** 679–688.

Chung, Y. J. (1977). A study of alcohol dehydrogenase alleles in Korean strains of *Drosophila melanogaster*. *Korean J. Breed.* **9,** 118–123.

Chung, Y. J. (1984). Biochemical genetic study of *Drosophila* populations in Korea. *Dros. Inform. Serv.* **60,** 77.

Chung, Y. J., and Yoon, Y.-S. (1980). A study on electrophoresis and activity of alcohol dehydrogenase of *Drosophila melanogaster*. *Korean J. Zool.* **23,** 125–136.

Chung, Y. J., Lee, Y. R., and Tai, H. J. (1982). Biochemical genetic study of *Drosophila* populations in Korea: Seasonal variations of alcohol dehydrogenase alleles and α-glycerophosphate dehydrogenase alleles. *J. Korean Res. Inst. Better Living* **30,** 73–81.

Church, G. M., and Gilbert, W. (1984). Genomic sequencing. *Proc. Natl. Acad. Sci. U.S.A.* **81,** 1991–1995.

Clarke, B. (1975). The contribution of ecological genetics to evolutionary theory: Detecting the direct effects of natural selection at particular polymorphic loci. *Genetics* **79,** 101–113.

Clarke, B., and Allendorf, F. W. (1979). Frequency-dependent selection due to kinetic differences between allozymes. *Nature (London)* **279,** 732–734.

Clarke, B., and Whitehead, D. L. (1984). Opportunities for natural selection on DNA and protein at the *Adh* locus in *Drosophila melanogaster*. *Dev. Genet.* **4,** 425–438.

Clarke, B., Camfield, R. G., Galvin, A. M., and Pitts, C. R. (1979). Environmental factors affecting the quantity of alcohol dehydrogenase in *Drosophila melanogaster*. *Nature (London)* **280,** 517–518.

Clegg, M. T., and Epperson, B. K. (1985). Recent developments in population genetics. *Adv. Genet.* **23,** 235–269.

Cohan, F. M., and Graf, J.-D. (1985). Latitudinal cline in *Drosophila melanogaster* for knockdown resistance to ethanol fumes and for rates of response to selection for further resistance. *Evolution* **39,** 278–293.

Cohan, F. M., and Hoffman, A. (1986). Genetic divergence under uniform selection II: Different responses to selection for knockdown resistence to ethanol among *Drosophila melanogaster* populations and their replicate lines. *Genetics* **114,** 145–163.

Cohn, V. H., Thompson, M. A., and Moore, G. P. (1984). Nucleotide sequence comparison of the *Adh* gene in three Drosophilids. *J. Mol. Evol.* **20,** 31–37.

Collier, G. E., and Macintyre, R. J. (1977). Microcomplement fixation studies on the evolution of α-glycerophosphate dehydrogenase within the genus *Drosophila*. *Proc. Natl. Acad. Sci. U.S.A.* **74,** 684–688.

Cooke, P. H., Barker, J. S. F., and East, P. D. (1986). Frequency and temperature dependent selection at the alcohol dehydrogenase-1 locus of *Drosophila buzzatii*. *Heredity* **57,** 47–51.

Courtright, J. B., Imberski, R. B., and Ursprung, H. (1966). The genetic control of alcohol dehydrogenase and octanol dehydrogenase isozymes in *Drosophila*. *Genetics* **51,** 1251–1260.

Daggard, G. E. (1981). Alcohol dehydrogenase, aldehyde oxidase, and alcohol utilisation in *Drosophila melanogaster, D. simulans, D. immigrans* and *D. busckii*. *In* "Genetic Studies of *Drosophila* Populations" (J. B. Gibson and J. G. Oakeshott, eds.), pp. 59–76. ANU Press, Canberra.

David, J. (1977). Significance of enzymatic polymorphism: Alcohol dehydrogenase in *Drosophila melanogaster*. *Ann. Biol.* **16,** 451–472.

David, J. R. (1982). Latitudinal variability of *Drosophila melanogaster:* Allozyme frequencies divergence between European and Afrotropical populations. *Biochem. Genet.* **20,** 747–761.

David, J. R., and Bocquet, C. (1975). Similarities and differences in latitudinal adaptation of two sibling *Drosophila* species. *Nature (London)* **257,** 588–590.

David, J. R., and Bocquet, C. (1977). Genetic tolerance to ethanol in *Drosophila melanogaster:* Increase by selection and analysis of correlated responses. *Genetica* **47,** 43–48.

David, J. R., and Van Herrewege, J. (1983). Adaptation to alcoholic fermentation in *Drosophila* species: Relationship between alcohol tolerance and larval habitat. *Comp. Biochem. Physiol.* **74A,** 283–288.

David, J. R., Bocquet, C., Van Herrewege, J., Fouillet, P., and Arens, M. F. (1978). Alcohol metabolism in *Drosophila melanogaster:* Uselessness of the most active aldehyde oxidase produced by the *Aldox* locus. *Biochem. Genet.* **16,** 203–210.

David, J. R., Van Herrewege, J., Monclus, M., and Prevosti, A. (1979). High ethanol tolerance in two distantly related *Drosophila* species: A probable case of recent convergent adaptation. *Comp. Biochem. Physiol.* **63C,** 53–56.

David, J. R., De Scheemaeker-Louis, M., and Pla, E. (1980). Evolution in the seven species of the *Drosophila melanogaster* subgroup: Comparison of the electrophoretic mobility of the enzymes produced by the *Adh* locus. *Dros. Inform. Serv.* **55,** 28–29.

David, J. R., Van Herrewege, J., De Scheemaeker-Louis, M., and Pla, E. (1981).

Drosophila alcohol dehydrogenase detoxification of isopropanol and acetone substrates not used in energy metabolism. *Heredity* **47,** 263–268.

Day, T. H., Hillier, P. C., and Clarke, B. (1974). The relative quantities and catalytic activities of enzymes produced by alleles at the alcohol dehydrogenase locus in *Drosophila melanogaster. Biochem. Genet.* **11,** 155–165.

Deltombe-Lietaert, M. C., Delcour, J., Lenelle-Montfort, N., and Elens, A. (1979). Ethanol metabolism in *Drosophila melanogaster. Experientia* **35,** 579–581.

Depiereux, E., Hougouto, N., Lechien, J., Libion-Mannaert, M., Lietaert, M. C., Feytmans, E., and Elens, A. (1985). Larval behavioural response to environmental ethanol in relation to alcohol dehydrogenase activity level in *Drosophila melanogaster. Behav. Genet.* **15,** 181–188.

Dickinson, W. J. (1980a). Compex cis-acting regulatory genes demonstrated in *Drosophila. Dev. Genet.* **1,** 229–240.

Dickinson, W. J. (1980b). Evolution of patterns of gene expression in Hawaiian picture-winged *Drosophila. J. Mol. Evol.* **16,** 73–94.

Dickinson, W. J. (1980c). Tissue specificity of enzyme expression regulated by diffusible factors: Evidence in *Drosophila* hybrids. *Science* **207,** 995–997.

Dickinson, W. J. (1983). Tissue specific allelic isozyme patterns and cis-acting developmental regulators. *Isozymes: Curr. Top. Biol. Med. Res.* **9,** 107–122.

Dickinson, W. J., and Carson, H. L. (1979). Regulation of the tissue specificities of an enzyme by a cis-acting genetic element: Evidence from interspecific *Drosophila* hybrids. *Proc. Natl. Acad. Sci. U.S.A.* **76,** 4559–4562.

Dickinson, W. J., Rowan, R. G., and Brennan, M. D. (1984). Regulatory gene evolution: Adaptive differences in expression of alcohol dehydrogenase in *Drosophila melanogaster* and *Drosophila simulans. Heredity* **52,** 215–225.

Dorado, G., and Barbancho, M. (1984). Differential responses in *Drosophila melanogaster* to environmental ethanol modification of fitness components at the *Adh* locus. *Heredity* **53,** 309–320.

Dudler, R., and Travers, A. A. (1984). Upstream elements necessary for optimal function of the hsp 70 promoter in transformed flies. *Cell* **38,** 391–398.

Easteal, S., and Oakeshott, J. G. (1985). Estimating divergence times of *Drosophila* species from DNA sequence comparisons. *Mol. Biol. Evol.* **2,** 87–91.

Eisses, K. Th. (1986). Interaction of *Notch* mutants and *Alcohol dehydrogenase* alleles in *Drosophila melanogaster.* Ph.D. thesis, Rijksuniversiteit at Utrecht.

Eisses, K. Th., Van Dijk, H., and Van Delden, W. (1979). Genetic differentiation within the *melanogaster* species group of the genus *Drosophila* (*sophophora*). *Evolution* **35,** 1063–1068.

Eisses, K. Th., Schoonen, G. E. J., Aben, W., Scharloo, W., and Thörig, G. E. W. (1985). Dual function of the alcohol dehydrogenase of *Drosophila melanogaster:* Ethanol and acetaldehyde oxidation by two allozymes, ADH-71K and ADH-F. *Mol. Gen. Genet.* **199,** 76–81.

Eisses, K. Th., Schoonen, W. G. E. J., Scharloo, W., and Thörig, G. E. W. (1986). Evidence for a multiple function of the alcohol dehydrogenase allozyme ADH^{71K} of *Drosophila melanogaster. Comp. Biochem. Physiol.* **85,** 759–766.

Fischer, J. A., and Maniatis, T. (1985). Structure and transcription of the *Drosophila mulleri* alcohol dehydrogenase genes. *Nucleic Acids Res.* **13,** 6899–6917.

Fletcher, T. S. (1980). Biochemical analysis of *Drosophila melanogaster Adh* alleles. Ph.D. thesis, University of California, Davis.

Fletcher, T. S., Ayala, F. J., Thatcher, D. R., and Chambers, G. K. (1978). Structural analysis of the Adh^S electromorph of *Drosophila melanogaster. Proc. Natl. Acad. Sci. U.S.A.* **75,** 5609–5611.

Fogleman, J. C. (1982). The role of volatiles in ecology of cactophilic *Drosophila*. *In* "Ecological Genetics and Evolution: The Cactus–Yeast–*Drosophila* Model System" (J. S. F. Barker and W. T. Starmer, eds.), pp. 191–208. Academic Press, Sydney.

Fontdevila, A., Santos, M., and Gonzàlez, R. (1980). Genotype–isopropanol interaction in the *Adh* locus of *Drosophila buzzatii*. *Experientia* **36,** 398–400.

Fox, A. S., Horikawa, M., and Ling, L.-N. L. (1967). The use of *Drosophila* cell cultures in studies of differentiation. *In Vitro* **3,** 65–84.

Franklin, I. R. (1981). An analysis of temporal variation of isozyme loci in *Drosophila melanogaster*. *In* "Genetic Studies of *Drosophila* Populations" (J. B. Gibson and J. G. Oakeshott, eds.), pp. 217–236. ANU Press, Canberra.

Freeth, A. L., and Gibson, J. B. (1985). Alcohol dehydrogenase and *sn*-glycerol-3-phosphate dehydrogenase null activity alleles in natural populations of *Drosophila melanogaster*. *Heredity* **55,** 369–374.

Fuyama, Y. (1976). Behavioural genetics of olfactory responses in *Drosophila*. I. Olfactometry and strain differences in *Drosophila melanogaster*. *Behav. Genet.* **6,** 407–420.

Fuyama, Y. (1978). Behavioural genetics of olfactory responses in *Drosophila*. II. An odorant specific variant in a natural population of *Drosophila melanogaster*. *Behav. Genet.* **8,** 406–414.

Garcin, F., Côté, J., Radouco-Thomas, S., Chawla, S., and Radouco-Thomas, C. (1983a). *Drosophila* ethanol metabolising system. Acetaldehyde oxidation in ALDOX-null mutants. *Experientia* **39,** 1122–1123.

Garcin, F., Côté, J., Radouco-Thomas, S., Kasienczuk, D., Chawla, S., and Radouco-Thomas, C. (1983b). Acetaldehyde oxidation in *Drosophila melanogaster* and *Drosophila simulans:* Evidence for the presence of an NAD^+-dependent dehydrogenase. *Comp. Biochem. Physiol.* **75B,** 205–210.

Gasser, S. M., and Laemmli, U. K. (1986). Cohabitation of scaffold binding regions with upstream enhancer elements of three developmentally regulated genes of *D. melanogaster*. *Cell* **46,** 521–530.

Geer, B. W., McKechnie, S. W., and Langevin, M. L. (1983). Regulation of *sn*-glycerol-3-phosphate dehydrogenase in *Drosophila melanogaster* larvae by dietary ethanol and sucrose. *J. Nutr.* **113,** 1632–1642.

Geer, B. W., Langevin, M. L., and McKechnie, S. W. (1985). Dietary ethanol and lipid synthesis in *Drosophila melanogaster*. *Biochem. Genet.* **23,** 607–622.

Geer, B. W., McKechnie, S. W., and Langevin, M. L. (1986). The effect of dietary ethanol on the composition of lipids of *Drosophila melanogaster* larvae. *Biochem. Genet.* **24,** 51–69.

Gelfand, L. J., and McDonald, J. F. (1980). Relationship between alcohol dehydrogenase E.C. 1.1.1.1 activity and behavioural response to environmental alcohol in *Drosophila melanogaster*. *Behav. Genet.* **10,** 237–250.

Gelfand, L. J., and McDonald, J. F. (1983). Relationship between alcohol dehydrogenase (Adh) activity and behavioural response to environmental ethanol in five *Drosophila* species. *Behav. Genet.* **13,** 281–293.

Gibson, J. B. (1970). Enzyme flexibility in *Drosophila melanogaster*. *Nature (London)* **227,** 959–960.

Gibson, J. B. (1972). Differences in the number of molecules produced by two allelic electrophoretic enzyme variants in *D. melanogaster*. *Experientia* **28,** 975–976.

Gibson, J. B., and Oakeshott, J. G. (1982). Tests of the adaptive significance of the alcohol dehydrogenase polymorphism in *Drosophila melanogaster:* Paths, pitfalls and prospects. *In* "Ecological Genetics and Evolution. The Cactus–Yeast–*Drosophila* Model System" (J. S. F. Barker and W. T. Starmer, eds.), pp. 291–306. Academic Press, Sydney.

Gibson, J. B., Lewis, N., Adena, M. A., and Wilson, S. R. (1979). Selection for ethanol tolerance in two populations of *Drosophila melanogaster* segregating alcohol dehydrogenase allozymes. *Aust. J. Biol. Sci.* **32,** 387–398.

Gibson, J. B., Chambers, G. K., Wilks, A. V., and Oakeshott, J. G. (1980). An electrophoretically cryptic alcohol dehydrogenase variant in *Drosophila melanogaster*. I. Activity ratios, thermostability, genetic localization and comparison with two other thermostable variants. *Aust. J. Biol. Sci.* **33,** 479–489.

Gibson, J. B., May, T. W., and Wilks, A. V. (1981). Genetic variation at the alcohol dehydrogenase locus in *Drosophila melanogaster* in relation to environmental variation: Ethanol levels in breeding sites and allozyme frequencies. *Oecologia* **51,** 191–198.

Gibson, J. B., Wilks, A. V., and Chambers, G. K. (1982). Genetic variation at the alcohol dehydrogenase locus in *Drosophila melanogaster:* A third ubiquitous allele. *Experientia* **38,** 653–654.

Girard, P., and Palabost, L. (1976). Étude du polymorphism enzymatique de 15 populations naturelles de *Drosophila melanogaster*. *Arch. Zool. Exp. Gen.* **117,** 41–57.

Goldberg, D. A. (1980). Isolation and partial characterisation of the *Drosophila* alcohol dehydrogenase gene. *Proc. Natl. Acad. Sci. U.S.A.* **77,** 5794–5798.

Goldberg, D. A., Posakony, J. W., and Maniatis, T. (1983). Correct developmental expression of a cloned alcohol dehydrogenase gene transduced into the *Drosophila* germ line. *Cell* **34,** 59–73.

Golding, G. B., Aquadro, C. F., and Langley, C. H. (1986). Sequence evolution within populations under multiple types of mutation. *Proc. Natl. Acad. Sci. U.S.A.* **83,** 427–431.

Gonzàlez-Duarte, R., and Atrian, S. (1986). Metabolic response to alcohol ingestion in *Drosophila hydei*. *Heredity* **56,** 123–128.

Gonzàlez-Duarte, R., and Vilageliu, L. I. (1985). Metabolic response to ethanol and isopropanol in *D. funebris* and *D. immigrans*. *Comp. Biochem. Physiol.* **80C,** 189–193.

Grell, E. H., Jacobson, K. B., and Murphy, J. B. (1965). Alcohol dehydrogenase in *Drosophila melanogaster* isozymes and genetic variants. *Science* **149,** 80–82.

Grell, E. H., Jacobson, K. B., and Murphy, J. B. (1968). Alterations of genetic material for analysis of alcohol dehydrogenase isoenzymes of *Drosophila melanogaster*. *Ann. N.Y. Acad. Sci.* **151,** 441–455.

Grossman, A. I., Koreneva, L. G., and Ulitskaya, L. E. (1970). Variability of alcohol dehydrogenase (Adh) locus in natural populations of *Drosophila melanogaster*. *Sov. Genet.* **6,** 211–214.

Hall, J. G., and Koehn, R. K. (1981). Letter to the editor on the biochemical differences between *Adh* allozymes in *Drosophila*. *Genetics* **98,** 669–672.

Hayashida, H., and Miyata, T. (1983). Unusual evolutionary conservation and frequent DNA segment exchange in class I genes of the major histocompatability complex. *Proc. Natl. Acad. Sci. U.S.A.* **80,** 2671–2675.

Hayley, C. H., and Birley, A. J. (1983). The genetical response to natural selection by varied environments. II. Observations on replicate populations in spatially varied laboratory environments. *Heredity* **51,** 581–606.

Heberlein, U., England, B., and Tjian, R. (1985). Characterisation of transcription factors that activate the tandem promoters of the alcohol dehydrogenase gene. *Cell* **41,** 965–977.

Hedrick, P. W., and McDonald, J. F. (1980). Regulatory gene adaptation: An evolutionary model. *Heredity* **45,** 83–97.

Heinstra, P. W. H., Eisses, K. Th., Schoonen, W. G. E. J., Aben, W., De Winter, A. J., Van der Horst, D. J., Marrewijk, W. J. H., Beenakkers, A. M. Th., Scharloo, W., and Thörig, G. E. W. (1983). A dual function of alcohol dehydrogenase in *Drosophila. Genetica* **60,** 129–137.

Heinstra, P. W. H., Eisses, K. Th., Scarloo, W., and Thörig, G. E. W. (1986a). Metabolism of secondary alcohols in *Drosophila melanogaster:* Effects on alcohol dehydrogenase. *Comp. Biochem. Physiol.* **83B,** 403–408.

Heinstra, P. W. H., Scharloo, W., and Thörig, G. E. W. (1986b). Alcohol dehydrogenase of *Drosophila:* Conversion and retroconversion of isozyme patterns. *Comp. Biochem. Physiol.* **83B,** 409–414.

Henikoff, S. (1983). Cloning genes for mapping transcription: Characterisation of *Drosophila melanogaster* alcohol dehydrogenase gene. *Nucleic Acids Res.* **11,** 4735–4752.

Henikoff, S., Keene, M. A., Fechtel, K., and Fristrom, J. W. (1986). Gene within a gene: Nested Drosophila melanogaster genes encode unrelated proteins on opposite DNA strands. *Cell* **44,** 33–42.

Hickey, D. A., and McLean, M. D. (1980). Seletion for ethanol tolerance and *Adh* allozymes in natural populations of *Drosophila melanogaster. Genet. Res.* **36,** 11–16.

Higuchi, R., Bowman, B., Freiberger, M., Ryder, O. A., and Wilson, A. C. (1984). DNA sequences from the quagga, an extinct member of the horse family. *Nature (London)* **312,** 282–284.

Hoffmann, A. A., and Parsons, P. A. (1984). Olfactory response and resource utilization in *Drosophila:* Interspecific comparisons. *Biol. J. Linnean Soc.* **22,** 43–53.

Hoffmann, A. A., Nielsen, K. M., and Parsons, P. A. (1984). Spatial variation of biochemical and ecological phenotypes in *Drosophila:* Electrophoretic and quantitative variation. *Dev. Genet.* **4,** 439–450.

Holmes, R. S., Moxon, L. N., and Parsons, P. A. (1980). Genetic variability of alcohol dehydrogenase E.C. 1.1.1.1 among Australian *Drosophila* species: Correlation of alcohol dehydrogenase biochemical phenotype with ethanol resource utilisation. *J. Exp. Zool.* **214,** 199–204.

Holzschu, D. L., and Phaff, H. J. (1982). Taxonomy and evolution of some ascomycetous cactophilic yeasts. *In* "Ecological Genetics and Evolution: The Cactus–Yeast–*Drosophila* Model System" (J. S. F. Barker and W. T. Starmer, eds.), pp. 127–142. Academic Press, Sydney.

Hougouto, N., Lietaert, M. C., Libion-Mannaert, M., Feytmans, E., and Elens, A. (1982). Oviposition-site preference and *Adh* activity in *Drosophila melanogaster. Genetica* **58,** 121–128.

Horikawa, M., Ling, L. L., and Fox, A. S. (1967). Effects of substrates on gene-controlled enzyme activities in cultured embryonic cells of *Drosophila. Genetics* **55,** 569–583.

Hovik, R., Winberg, J.-O., and McKinley-McKee, J. S. (1984). *Drosophila melanogaster* alcohol dehydrogenase: Substrate specificity of the Adh^F alleloenzyme. *Insect Biochem.* **14,** 345–351.

Hudson, R. R., and Kaplan, N. L. (1986). On the divergence of alleles in nested subsamples from finite populations. *Genetics* **113,** 1057–1076.

Hudson, R. R., Kreitman, M., and Aguadé, M. (1986). A test of neutral molecular evolution based on nucleotide data. *Genetics* **116,** 153–159.

Imberski, R. B., and Strömmen, C. (1972). Developmental changes in alcohol dehydrogenase activity in *Drosophila hydei*. *Dros. Inform. Serv.* **48,** 74.

Jeffrey, J., and Jörnvall, H. (1983). Enzyme relationships in a sorbitol pathway that bypasses glycolysis and pentose phosphates in glucose metabolism. *Proc. Natl. Acad. Sci. U.S.A.* **80,** 901–905.

Johnson, F. M., and Denniston, C. (1964). Genetic variation of alcohol dehydrogenase in *Drosophila melanogaster*. *Nature (London)* **204,** 906–907.

Johnson, F. M., and Powell, A. (1974). The alcohol dehydrogenase of *Drosophila melanogaster:* Frequency changes associated with heat and cold shock. *Proc. Natl. Acad. Sci. U.S.A.* **71,** 1783–1784.

Johnson, F. M., and Schaeffer, H. E. (1973). Isozyme variability in species of the genus *Drosophila*. VII. Genotype environment relationships in populations of *D. melanogaster* from the eastern United States. *Biochem. Genet.* **10,** 149–163.

Jones, J. S. (1980). Can genes choose habitats? *Nature (London)* **286,** 757–758.

Jörnvall, H., Persson, M., and Jeffrey, J. (1981). Alcohol and polyol dehydrogenases are both divided into two protein types and structural properties cross relate the different enzyme activities within each type. *Proc. Natl. Acad. Sci. U.S.A.* **78,** 4226–4230.

Jörnvall, H., Von Bahr-Lindström, H., Jany, K. D., Ulmer, W., and Fröschle, M. (1984). Extended superfamily of short alcohol-polyol-sugar dehydrogenases: Structural similarities between glucose and ribitol dehydrogenases. *FEBS Lett.* **165,** 190–196.

Juan, E., and Gonzàlez-Duarte, R. (1980). Purification and enzyme stability of alcohol dehydrogenase from *Drosophila simulans, Drosophila virilis* and *Drosophila melanogaster adh*^{s}. *Biochem. J.* **189,** 105–110.

Juan, E., and Gonzàlez-Duarte, R. (1981). Determination of some biochemical and structural features of alcohol dehydrogenases from *Drosophila simulans* and *Drosophila virilis*. *Biochem. J.* **195,** 61–69.

Kacser, H., and Burns, J. A. (1981). The molecular basis of dominance. *Genetics* **97,** 639–666.

Keene, M. A., and Elgin, S. C. R. (1984). Patterns of DNA structural polymorphism and their evolutionary implications. *Cell* **36,** 121–130.

Keith, T. P., Brooks, L. D., Lewontin, R. C., Martinez, J. C., and Rigby, D. L. (1985). Nearly identical allelic distribution of xanthine dehydrogenase in two populations of *Drosophila pseudoobscura*. *Mol. Biol. Evol.* **2,** 206–216.

Kelley, M. R., Mims, L. P., Farnet, C. M., Dicharry, S. A., and Lee, W. R. (1985). Molecular analysis of X-ray induced alcohol dehydrogenase (*Adh*) null mutations in *Drosophila melanogaster*. *Genetics* **109,** 365–377.

Kerver, J. W. M., and Van Delden, W. (1984). Adaptation of *Drosophila melanogaster* after long-term exposure to ethanol in relation to the alcohol dehydrogenase polymorphism. *Dros. Inform. Serv.* **60,** 130–131.

Kerver, J. W. M., and Van Delden, W. (1985). Development of tolerance to ethanol in relation to the alcohol dehydrogenase locus in *Drosophila melanogaster*. I. Adult and egg-to-adult survival in relation to *Adh* activity. *Heredity* **55,** 355–367.

Kimura, M. (1982). The neutral theory as a basis for understanding the mechanism of evolution and variation at the molecular level. *In* "Molecular Evolution, Protein Polymorphism and the Neutral Theory" (M. Kimura, ed.), pp. 3–56. Japan Scientific Societies Press, Tokyo, and Springer-Verlag, Berlin and New York.

Kimura, M. (1983). "The Neutral Theory of Molecular Evolution." Cambridge Univ. Press, London and New York.

Klemenz, R., Hultmark, D., and Gehring, W. J. (1985). Selective translation of heat shock mRNA in *Drosophila melanogaster* depends on sequence information in the leader. *EMBO J.* **4,** 2053–2060.

Kloek, G. P., Ralin, D. B., and Ridgel, G. C. (1976). Survivorship and gene frequencies of *Drosophila melanogaster* populations in abnormal oxygen atmospheres. *Aviat. Space Environ. Med.* **47,** 272–279.

Knibb, W. R. (1982). Chromosome inversion polymorphisms in *Drosophila melanogaster.* II. Geographic clines and climatic association in Australasia, North America and Asia. *Genetica* **58,** 213–221.

Knibb, W. R. (1983). Chromosome inversion polymorphisms in *Drosophila melanogaster.* III. Gametic disequilibria and the contributions of inversion clines to the *Adh* and *Gpdh* clines in Australasia. *Genetica* **61,** 139–146.

Koehn, R. K. (1978). Physiology and biochemistry of enzyme variation: The interface of ecology and population genetics. *In* "Ecological Genetics: The Interface" (P. Brussard, ed.), pp. 51–72. Springer-Verlag, Berlin and New York.

Korotchkin, L. I., Korotchkina, L. S., and Serov, O. L. (1972). Histochemical study of alcohol dehydrogenase in Malpighian tubules of *Drosophila melanogaster* larvae. *Folia Histochem. Cytochem.* **10,** 287–292.

Kreitman, M. (1980). Assessment of variability within electromorphs of alcohol dehydrogenase in *Drosophila melanogaster. Genetics* **95,** 467–475.

Kreitman, M. (1983). Nucleotide polymorphism at the alcohol dehydrogenase locus in *Drosophila melanogaster. Nature (London)* **304,** 412–417.

Kreitman, M. E., and Aguadé, M. (1986a). Excess polymorphism at the ADH locus in *Drosophila melanogaster. Genetics* **114,** 93–110.

Kreitman, M., and Aguadé, M. (1986b). Genetic uniformity in two populations of *Drosophila melanogaster* as revealed by filter hybridisation of four-nucleotide-recognizing restriction enzyme digests. *Proc. Natl. Acad. Sci. U.S.A.* **83,** 3562–3566.

Kubli, E. T., Schmidt, T., Martin, P. F., and Sofer, W. (1982). *In vitro* suppression of a nonsense mutant of *Drosophila melanogaster. Nucleic Acids Res.* **10,** 7145–7152.

Kusakabe, S., and Mukai, T. (1984). The genetic structure of natural populations of *Drosophila melanogaster.* XVIII. Clinal and uniform genetic variation over populations. *Genetics* **108,** 617–632.

Lakovaara, S., Saura, A., Lankinen, P., and Lokki, J. (1972). Evolution of enzymes and genetic distance in *Drosophila obscura* and *affinis* subgroups. *Congr. Int. Zool., 17th* **5,** 1–18.

Langley, C. H., Montgomery, E., and Quattlebaum, W. F. (1982). Restriction map variation in the *Adh* region of *Drosophila. Proc. Natl. Acad. Sci. U.S.A.* **79,** 5631–5635.

Laurie-Ahlberg, C. C. (1985). Genetic variation affecting the expression of enzyme-coding genes in *Drosophila*: An evolutionary perspective. *Isozymes Curr. Top. Biol. Med. Res.* **12,** 33–88.

Laurie-Ahlberg, C. C., and Stam, L. F. (1987). Use of P-element-mediated transformation to identify the molecular basis of naturally occurring variants affecting *Adh* expression in *Drosophila melanogaster. Genetics* **115,** 129–140.

Laurie-Ahlberg, C. C., and Weir, B. S. (1979). Allozyme variation and linkage disequilibrium in some laboratory populations of *Drosophila melanogaster. Genetics* **92,** 1295–1314.

Lee, C. Y. (1982). Alcohol dehydrogenase E. C. 1.1.1.1 from *Drosophila melanogaster. In* "Methods in Enzymology" (W. A. Wood, ed.), Vol. 89, pp. 445–450. Academic Press, New York.

Lefevre, G. (1976). A photographic representation and interpretation of the polytene chromosomes of the *Drosophila melanogaster* salivary gland. *In* "Genetics and

Biology of *Drosophila*" (M. Ashburner and E. Novitski, eds.), Vol. 1A, pp. 31–66. Academic Press, London.

Leibenguth, F. (1977). Selection against homozygotes in the *Adh* polymorphic system of *Drosophila melanogaster* associated with lethal factor *1(2)stm*. *Biochem. Genet.* **15,** 93–100.

Leibenguth, F., Rammo, E., and Dubiczky, R. (1979). A comparative study of embryonic gene expression in *Drosophila* and *Ephestia*. *Wilhelm Roux's Arch. Dev. Biol.* **187,** 81–88.

Leigh-Brown, A. J., and Lee, C. Y. (1979). Purification of alcohol dehydrogenase from *Drosophila* by general ligand affinity chromatography. *Biochem J.* **179,** 479–482.

Lewis, N., and Gibson, J. B. (1978). Variation in amount of enzyme protein in natural populations *Biochem. Genet.* **16,** 159–170.

Lewontin, R. C. (1986). Population genetics. *Annu. Rev. Genet.* **19,** 81–102.

Lewontin, R. C., Moore, J. A., Provine, W. B., and Wallace, B. (1981). "Dobzhansky's Genetics of Natural Populations, I–XLIII." Columbia Univ. Press, New York.

Li, W. H., Wu, C. I., and Luo, C. C. (1984). Nonrandomness of point mutations as reflected in nucleotides substitutions in pseudogenes and its evolutionary implications. *J. Mol. Evol.* **21,** 58–71.

Li, W. H., Wu, C. I., and Luo, C. C. (1985). A new method for estimating synonymous and nonsynonymous rates of nucleotide substitution considering the relative likelihood of nucleotide and codon changes. *Mol. Biol. Evol.* **2,** 150–174.

Libion-Mannaert, M., Delcour, J., Deltombe-Lietaert, M. C., Lenelle-Montfort, N., and Elens, A. (1976). Ethanol as a "food" for *Drosophila melanogaster*: Influence of the *ebony* gene. *Experientia* **32,** 22–23.

Lietaert, M. C., Libion-Mannaert, M., and Elens, A. (1981). Acetaldehyde metabolism in *Drosophila melanogaster*. Experientia **37,** 689–690.

Lietaert, M. C., Libion-Mannaert, M., Hougouto, N., and Elens, A. (1982). How can *Drosophila* flies without aldehyde oxidase detoxify acetaldehyde? *Experientia* **38,** 651–652.

Lietaert, M. C., Libion-Mannaert, M., Wattiaux-De Coninck, S., and Elens, A. (1985). *Drosophila melanogaster* aldehyde dehydrogenase. *Experientia* **41,** 57–58.

Lindsley, D. L., and Grell, E. H. (1968). "Genetic Variations of *Drosophila melanogaster*." Carnegie Institute of Washington, Publication No. 627.

Lindsley, D. L., and Zimm, G. (1985). The genome of *Drosophila melanogaster* progress report (genes A–K). *Dros. Inform. Serv.* **62.**

Lonberg, N., and Gilbert, W. (1985). Intron/exon structure of the chicken pyruvate kinase gene. *Cell* **40,** 81–90.

McDonald, J. F. (1983). The molecular basis of adaptation: A critical review of relevant ideas and observations. *Annu. Rev. Ecol. Syst.* **14,** 77–102.

McDonald, J. F., and Anderson, S. M. (1981). Letter to the editor—a reply. *Genetics* **98,** 673–675.

McDonald, J. F., and Ayala, F. J. (1978). Gene regulation in adaptive evolution. *Can. J. Genet. Cytol.* **20,** 159–175.

McDonald, J. F., Chambers, G. K., David, J., and Ayala, F. J. (1977). Adaptive response due to changes in gene regulation: A study with *Drosophila*. *Proc. Natl. Acad. Sci. U.S.A.* **74,** 4562–4566.

McDonald, J. F., Anderson, S. M, and Santos, M. (1980). Biochemical differences between products of the *Adh* locus in *Drosophila*. *Genetics* **95,** 1013–1022.

McElfresh, K. C., and McDonald, J. F. (1983). The effect of alcohol stress on nicotinamide adenine dinucleotide (NAD^+) levels in *Drosophila*. *Biochem. Genet.* **21,** 365–374.

McGill, S. (1985). Molecular studies of the *Adh* region of *Drosophila melanogaster*. Ph.D. thesis, University of Cambridge, Cambridge, England.

McKay, J. (1981). Variation in activity and thermostability of alcohol dehydrogenase in *Drosophila melanogaster*. *Genet. Res.* **37,** 227–238.

McKechnie, S. W., and Morgan, P. (1982). Alcohol dehydrogenase polymorphism of *Drosophila melanogaster*: Aspects of alcohol and temperature variation in the larval environment. *Aust. J. Biol. Sci.* **35,** 85–93.

McKenzie, J. A., and McKechnie, S. W. (1978). Ethanol tolerance and the *Adh* polymorphism in a natural population of *Drosophila melanogaster*. *Nature (London)* **272,** 75–76.

McKenzie, J. A., and McKechnie, S. W. (1978). A comparative study of resource utilisation in natural populations of *Drosophila melanogaster* and *D. simulans*. *Oecologia* **40,** 229–309.

McKenzie, J. A., and McKechnie, S. W. (1981). The alcohol dehydrogenase polymorphism in a vineyard cellar population of *Drosophila melanogaster*. *In* "Genetic Studies of *Drosophila* Populations" (J. B. Gibson and J. G. Oakeshott, eds.), pp. 201–216. ANU Press, Canberra.

McKenzie, J. A., and Parsons, P. A. (1972). Alcohol tolerance: An ecological parameter in the relative success of *Drosophila melanogaster* and *Drosophila simulans*. *Oecologia* **10,** 373–388.

McKenzie, J. A., and Parsons, P. A. (1974a). Microdifferentiation in a natural population of *Drosophila melanogaster* to alcohol in the environment. *Genetics* **77,** 385–394.

McKenzie, J. A., and Parsons, P. A. (1974b). Numerical changes and environmental utilisation in natural populations of *Drosophila*. *Aust. J. Zool.* **22,** 175–187.

Madhavan, K., Conscience-Egli, M., Sieber, F., and Ursprung, H. (1973). Farnesol metabolism in *Drosophila melanogaster*: Ontogeny and tissue distribution of octanol dehydrogenase and aldehyde oxidase. *J. Insect Physiol.* **19,** 235–241.

Malpica, J. M., and Vassallo, J. M. (1980). A test for the selective origin of environmentally correlated allozyme patterns. *Nature (London)* **286,** 407–408.

Malpica, J.-M., Vassallo, J. M., Frias, A., and Fuentes-Bol, F. (1987). On recombination among *In(2L)t, α-Gpdh,* and *Adh* in *Drosophila melanogaster*. *Genetics* **115,** 141–142.

Maniatis, T., Hardison, R. C., Lacy, E., Lauer, J., O'Connel, C., Quon, D., Sim, G. K., and Efstratiadis, A. (1978). The isolation of structural genes from libraries of eucaryotic DNA. *Cell* **15,** 687–701.

Marks, R. W., Brittnacher, J. G., McDonald, J. F., Prout, T., and Ayala, F. J. (1980). Wineries, *Drosophila,* alcohol and *Adh*. *Oecologia* **47,** 141–144.

Maroni, G. (1978). Genetic control of alcohol dehydrogenase levels in *Drosophila*. *Biochem. Genet.* **16,** 509–523.

Maroni, G. (1980). A duplication of *Adh* in association with *Sco*. *Dros. Inform. Serv.* **55,** 96–98.

Maroni, G., and Laurie-Ahlberg, C. C. (1983). Genetic control of *Adh* expression in *Drosophila melanogaster*. *Genetics* **105,** 921–923.

Maroni, G. G., and Stamey, S. C. (1983). Developmental profile and tissue distribution of alcohol dehydrogenase. *Dros. Inform. Serv.* **59,** 77–78.

Maroni, G., Laurie-Ahlberg, C. C., Adams, D. A., and Wilton, A. N. (1982). Genetic variation in the expression of *Adh* in *Drosophila melanogaster*. *Genetics* **101,** 431–446.

Martin, P. F., Place, A. R., Pentz, E., and Sofer, W. (1985). UGA nonsense codon in the alcohol dehydrogenase gene of *Drosophila melanogaster*. *J. Mol. Biol.* **184,** 221–229.

Middleton, R. J., and Kacser, H. (1983). Enzyme variation, metabolic flux and fitness: Alcohol dehydrogenase in *Drosophila melanogaster*. *Genetics* **105,** 633–650.

Mills, L. E., Batterham, P., Alegre, J., Starmer, W. T., and Sullivan, D. T. (1986). Molecular genetic characterisation of a locus which contains duplicate *Adh* genes in *Drosophila mojavensis* and related species. *Genetics* **112,** 295–310.

Miyata, T. (1982). Evolutionary changes and functional constraints in DNA sequences. *In* "Molecular Evolution, Protein Polymorphism and the Neutral Theory" (M. Kimura, ed.), pp. 238–266. Japan Scientific Societies Press, Tokyo, and Springer-Verlag, Berlin and New York.

Monclus, M., and Prevosti, A. (1978–1979). Cellar habitat and *Drosophila* populations. *Genet. Iber.* **30–31,** 189–201.

Moxon, L. N., Holmes, R. S., Parsons, P. A., Irving, M. G., and Doddrell, D. M. (1985). Purification and molecular properties of alcohol dehydrogenase from *Drosophila melanogaster*: Evidence from NMR and kinetic studies for function as an aldehyde dehydrogenase. *Comp. Biochem. Physiol.* **80B,** 525–535.

Mukai, T. Q., Yamaguchi, O., Kusakabe, S., Tachida, H., Matsuda, M., Ichinose, M., and Yoshimaru, M. (1982). Lack of balancing selection for protein polymorphisms. *In* "Molecular Evolution, Protein Polymorphism and the Neutral Theory" (M. Kimura, ed.), pp. 81–120. Japan Scientific Societies Press, Tokyo, and Springer-Verlag, Berlin and New York.

Mulley, J. C., James, J. W., and Barker, J. S. F. (1979). Allozyme genotype–environment relationships in natural populations of *Drosophila buzzattii. Biochem. Genet.* **17,** 105–126.

Munõz-Serrano, A., Alonso-Moraga, A., and Rodero, A. (1985). Annual variation of enzyme polymorphism in four natural populations of *Drosophila melanogaster* occupying different niches. *Genetica* **67,** 121–129.

Murray, N. D. (1982). Ecology and of the *Opuntia–Cactoblastis* ecosystem in Australia. *In* "Ecological Genetics and Evolution: The Cactus–Yeast–*Drosophila* Model System" (J. S. F. Barker and W. T. Starmer, eds.), pp. 17–32. Academic Press, Sydney.

Oakeshott, J. G., Gibson, J. B., Anderson, P. R, and Champ, A. (1980). Opposing modes of selection on the alcohol dehydrogenase locus in *Drosophila melanogaster. Aust. J. Biol. Sci.* **33,** 105–114.

Oakeshott, J. G., Chambers, G. K., East, P. D., Gibson, J. B., and Barker, J. S. F. (1982a). Evidence for a genetic duplication involving alcohol dehydrogenase genes in *Drosophila buzzatii* and related species. *Aust. J. Biol. Sci.* **35,** 73–84.

Oakeshott, J. G., Gibson, J. B., Anderson, P. R., Knibb, W. R., Anderson, D. G., and Chambers, G. K. (1982b). Alcohol dehydrogenase and glycerol-3-phosphate dehydrogenase clines in *Drosophila melanogaster* on different continents. *Evolution* **36,** 86–96.

Oakeshott, J. G., May, T. W., Gibson, J. B., and Willcocks, D. A. (1982c). Resource partitioning in five domestic *Drosophila* species and its relationship to ethanol metabolism. *Aust. J. Zool.* **30,** 547–556.

Oakeshott, J. G., Wilson, S. R., and Gibson, J. B. (1983). An attempt to measure selection coefficients affecting the alcohol dehydrogenase polymorphism in *Drosophila melanogaster* populations maintained on ethanol media. *Genetica* **61,** 151–159.

Oakeshott, J. G., Gibson, J. B., and Wilson, S. R. (1984a). Selective effects of the genetic background and ethanol on the alcohol dehydrogenase polymorphism in *Drosophila melanogaster. Heredity* **53,** 51–67.

Oakeshott, J. G., McKechnie, S. W., and Chambers, G. K. (1984b). Population genetics of the metabolically related *Adh, Gpdh* and *Tpi* polymorphisms in *Drosophila mela-*

nogaster. I. Geographic variation in *Gpdh* and *Tpi* allele frequencies in different continents. *Genetics* **63,** 21–29.

Oakeshott, J. G., Wilson, S. R., and Parnell, P. (1985b). Selective effects of temperature on some enzyme polymorphisms in laboratory populations of *Drosophila melanogaster*. *Heredity* **55,** 69–82.

Oakeshott, J. G., Cohan, F. M., and Gibson, J. G. (1985a). Ethanol tolerences of *Drosophila melanogaster* populations selected on different concentrations of ethanol supplemented media. *Theor. Appl. Gent.* **69,** 603–608.

O'Donnell, J., Gerace, L., Leister, F., and Sofer, W. (1975). Chemical selection of mutants that affect alcohol dehydrogenase in *Drosophila*. II. Use of 1-pentyne-3-ol. *Genetics* **79,** 73–83.

O'Donnell, J., Mandell, H. C., Krauss, M., and Sofer, W. (1977). Genetic and cytogenetic analysis of the *Adh* region in *Drosophila melanogaster*. *Genetics* **86,** 553–566.

Pääbo, S. (1985). Molecular cloning of ancient Egyptian mummy DNA. *Nature* (*London*) **314,** 644–645.

Papel, I., Henderson, M., Van Herrewege, J., David, J. R., and Sofer, W. (1979). *Drosophila* alcohol dehydrogenase activity *in vitro* and *in vivo*: Effects of acetone feeding. *Biochem. Genet.* **17,** 553–563.

Parsons, P. A. (1977). Larval reaction to alcohol as an indicator of resource utilisation difference between *Drosophila melanogaster* and *D. simulans*. *Oecologia* **30,** 141–146.

Parsons, P. A. (1979). Polygenic variation in natural populations of *Drosophila*. *In* "Quantitative Genetic Variation" (J. N. Thompson and J. M. Thoday, eds.), pp. 61–80. Academic Press, New York.

Parsons, P. A. (1980). Ethanol utilisation: Threshold differences among six closely related species of *Drosophila*. *Aust. J. Zool.* **28,** 535–541.

Parsons, P. A. (1981). Longevity of cosmopolitan and native Australian *Drosophila* in ethanol atmospheres. *Aust. J. Zool.* **29,** 33–39.

Parsons, P. A. (1983). Ecobehavioural genetics: Habitats and colonists. *Annu. Rev. Ecol. Syst.* **14,** 35–55.

Parsons, P. A. (1984). Colonizing species: A probe into evolution via the organism. *Endeavour* **8,** 108–112.

Parsons, P. A., and Spence, G. E. (1980). Larval responses to environmental ethanol in three *Drosophila* species: An indicator of habitat selection. *Aust. J. Zool.* **28,** 543–546.

Parsons, P. A., and Spence, G. E. (1981a). Ethanol utilisation: Threshold differences among three *Drosophila* species. *Am. Nat.* **117,** 568–571.

Parsons, P. A., and Spence, G. E. (1981b). Longevity, resource utilisation and larval preferences in *Drosophila*. Inter- and intraspecific variation. *Aust. J. Zool.* **29,** 671–678.

Parsons, P. A., and Stanley, S. M. (1981). Comparative effects of environmental ethanol on *D. melanogaster* and *D. simulans* adults, including geographic differences in *Drosophila melanogaster*. *In* "Genetic Studies of *Drosophila* Populations" (J. B. Gibson and J. G. Oakeshott, eds.), pp. 47–58. ANU Press, Canberra.

Parsons, P. A., Stanley, S. M., and Spence, G. E. (1979). Environmental ethanol at low concentrations: Longevity and development in the sibling species *Drosophila melanogaster* and *D. simulans*. *Aust. J. Zool.* **27,** 747–754.

Penny, D., Foulds, L. R., and Hendy, M. D. (1982). Testing the theory of evolution by comparing phylogenetic trees constructed from five different protein sequences. *Nature* (*London*) **297,** 197–200.

Pipkin, S. B., Franklin-Springer, E., Law, S., and Lubega, S. (1976). New studies of the alcohol dehydrogenase cline in *Drosophila melanogaster* from Mexico. *J. Hered.* **67,** 258–266.

Place, A. R. (1985). *Drosophila* alcohol dehydrogenase—a case for convergent evolution of a common active site. *Fed. Proc., Fed. Am. Soc. Exp. Biol.* **44,** 1060.

Place, A. R., Powers, D. A., and Sofer, W. (1980). *Drosophila melanogaster* alcohol dehydrogenase does not require metals for catalysis. *Fed. Proc., Fed. Am. Soc. Exp. Biol.* **39,** 1640.

Posakony, J. W., Fisher, J. A., and Maniatis, T. (1985). Identification of DNA sequences required for the regulation of *Drosophila* alcohol dehydrogenase gene expression. *Cold Spring Harbor Symp. Quant. Biol.* **50,** 515–520.

Retzios, A., and Thatcher, D. R. (1979). Chemical basis of the electrophoretic variation observed at the alcohol dehydrogenase locus of *Drosophila melanogaster. Biochemie* **61,** 701–704.

Robinson, A. S., and Van Heemert, K. (1981). Genetic sexing in *Drosophila melanogaster* using the alcohol dehydrogenase locus and a linked translocation. *Theor. Appl. Genet.* **59,** 23–24.

Rowan, R. G., and Dickinson, W. J. (1986). Two alternate transcripts coding for alcohol dehydrogenase accumulate with different developmental specificities in different species of *picture-winged* Drosophila. *Genetics* **114,** 435–452.

Rowan, R. G., and Dickinson, W. J. (1987). Nucleotide sequence of the genomic region coding for alcohol dehydrogenase in *Drosophila affinidisjuncta. Genetics* (submitted).

Rowan, R. G., Brennan, M. D., and Dickinson, W. J. (1986). Developmentally regulated RNA transcripts coding for alcohol dehydrogenase in *Drosophila affinisdisjuncta. Genetics* **114,** 405–433.

Rubin, G. M., and Spradling, A. C. (1982). Genetic transformation of *Drosophila* with transposable element vectors. *Science* **218,** 348–353.

Sampsell, B. (1977). Isolation and genetic characterisation of alcohol dehydrogenase thermostability variants occurring in natural populations of *Drosophila melanogaster. Biochem. Genet.* **15,** 971–988.

Sampsell, B. M., and Barnette, V. C. (1985). Effects of environmental temperatures on alcohol dehydrogenase activity levels in *Drosophila melanogaster. Biochem. Genet.* **23,** 53–59.

Sampsell, B., and Sims, S. (1982). Effects of *Adh* genotype and heat stress on alcohol tolerance in *Drosophila melanogaster. Nature (London)* **296,** 853–855.

Sampsell, B., and Steward, E. (1983). Alcohol dehydrogenase thermostability variants in *Drosophila melanogaster*: Comparison of activity ratios and enzyme levels. *Biochem. Genet.* **21,** 1071–1088.

Sánchez-Cañete, F. J. S., Dorado, G., and Barbancho, M. (1986). Ethanol and isopropanol detoxification associated with the *Adh* locus of *Drosophila melanogaster. Heredity* **56,** 167–175.

Sangiorgi, A., Pieragostini, E., Properi, I., Leggieri, P., and Cavicchi, S. (1981). Mimicry of isozyme adaptive advantage by gene association. II. *Adh* genotype frequency variations in *Drosophila* cage populations reared at different temperatures. *Genetica* **56,** 229–234.

Savakis, C., and Ashburner, M. (1985). A simple gene with a complex pattern of transcription: The alcohol dehydrogenase gene of *Drosophila. Cold Spring Harbor Symp. Qûant. Biol.* **50,** 505–514.

Savakis, C., Ashburner, M., and Willis, J. H. (1986). The expression of the gene coding for alcohopl dehydrogenase during the development of *Drosophila melanogaster*. *Dev. Biol.* **114,** 194–207.

Schaeffer, S. W., and Aquadro, C. F. (1987). Nucleotide Sequence of the *Adh* gene region of *Drosophila pseudo obscura* evolutionary change and evidence for an ancient gene duplication. *Genetics* **117,** 61–73.

Schaeffer, S. W., Aquadro, C. F., and Anderson, W. W. (1986). Restriction map variation in the alcohol dehydrogenase region of *Drosophila pseudoobscura*. *Mol. Biol. Evol.* **4,** 254–265.

Schmitt, L. H., McKechnie, S. W., and McKenzie, J. A. (1986). Associations between alcohol tolerance and the quantity of alcohol dehydrogenase in *Drosophila melanogaster* isolated from a winery population. *Aust. J. Biol. Sci.* **39,** 59–67.

Schwartz, M. F., and Jörnvall, H. (1976). Structural analysis of mutant and wild-type alcohol dehydrogenase from *Drosophila melanogaster*. *Eur. J. Biochem.* **68,** 159–168.

Schwartz, M., and Sofer, W. (1976). Diet-induced alterations in distribution of multiple forms of alcohol dehydrogenase in *Drosophila*. *Nature* (*London*) **263,** 129–130.

Schwartz, M., O'Donnell, J., and Sofer, W. (1979). Origin of the multiple forms of alcohol dehydrogenase E.C. 1.1.1.1. from *Drosophila melanogaster*. *Arch. Biochem. Biophys.* **4,** 365–378.

Singh, R. S., and Rhomberg, L. R. (1987). A comprehensive study of genic variation in natural populations of *Drosophila melanogaster*: I. Estimates of gene flow from rare alleles. *Genetics* **115,** 313–322.

Singh, R. S., Hickey, D. A., and David, J. (1982). Genetic differentiation between geographically distant populations of *Drosophila melanogaster*. *Genetics* **101,** 235–256.

Smith, M. R., Chambers, G. K., Brooks, L. D., Cohan, F. M., and Cohan, S. C. (1984). How many *Adh* clines on the west coast of North America? *Dros. Inform. Serv.* **60,** 188–190.

Sofer, W., and Hatkoff, M. A. (1972). Chemical selection of alcohol dehydrogenase-negative mutants in *Drosophila*. *Genetics* **72,** 545–549.

Spradling, A. C., and Rubin, G. M. (1982). Transposition of cloned P-elements into *Drosophila* germ line chromosomes. *Science* **218,** 341–347.

Stanley, S. M., and Parsons, P. A. (1981). The response of the cosmopolitan species *Drosophila melanogaster* to ecological gradients. *Proc. Ecol. Soc. Aust.* **11,** 121–130.

Starmer, W. T. (1982). Associations and interactions among yeasts, *Drosophila* and their habitats. *In* "Ecological Genetics and Evolution: The Cactus-Yeast-*Drosophila* Model System" (J. S. F. Barker and W. T. Starmer, eds.), pp. 159–174. Academic Press, Sydney.

Starmer, W. T., and Barker, J. S. F. (1986). Ecological genetics of the *Adh-1* locus of *Drosophila buzzatii*. *Biol. J. Linn. Soc.* **28,** 373–386.

Starmer, W. T., Heed, W. B., and Rockwood-Sluss, E. S. (1977). Extension of longevity in *Drosophila mojavensis* by environmental ethanol differences between subraces. *Proc. Natl. Acad. Sci. U.S.A.* **74,** 387–391.

Starmer, W. T., Barker, J. S. F., Phaff, H. J., and Fogleman, J. C. (1986). The adaptations of *Drosophila* and yeasts. Their interactions with the volatile, 2-propanol, in the cactus–microorganism–*Drosophila* model system. *Aust. J. Biol. Sci.* **39,** 69–77.

Stephens, J. C., and Nei, M. (1985). Phylogenetic analysis of polymorphic DNA sequences at the *Adh* locus in *Drosophila melanogaster* and its sibling species. *J. Mol. Evol.* **22,** 289–300.

Subrahmanyam, G., Kannan, K., and Reddy, A. R. (1984). Comparison of tryptic peptide profiles of alcohol dehydrogenase from *Drosophila melanogaster* at different ages: A rapid procedure using high performance liquid chromatography. *J. Biochem. Biophys. Methods* **10,** 153–162.

Thatcher, D. R. (1977). Enzyme instability and proteolysis during the purification of an alcohol dehydrogenase from *Drosophila melanogaster. Biochem. J.* **163,** 317–323.

Thatcher, D. R. (1980). The complete amino acid sequence of three alcohol dehydrogenase alleloenzymes ($Adh^{N\text{-}11}$, Adh^{S} and Adh^{UF}) from the fruitfly *Drosophila melanogaster. Biochem. J.* **187,** 875–886.

Thatcher, D. R., and Hodson, B. (1981). Denaturation of proteins and nucleic acids by thermal gradient electrophoresis. *Biochem J.* **197,** 104–110.

Thatcher, D. R., and Retzios, A. (1980). Mutations affecting the structure of the alcohol dehydrogenase from *Drosophila melanogaster. Protides Biol. Fluids* **28,** 157–160.

Thatcher, D. R., and Sawyer, L. (1980). Secondary-structure prediction from the sequence of *Drosophila melanogaster* (fruitfly) alcohol dehydrogenase. *Biochem. J.* **187,** 884–886.

Thatcher, D. R., and Sheikh, R. (1981). The relative conformational stability of alcohol dehydrogenase alleloenzymes purified from *Drosophila melanogaster. Biochem. J.* **197,** 111–117.

Thompson, J. N., Ashburner, M., and Woodruff, R. C. (1977). Presumptive control mutation for alcohol dehydrogenase in *Drosophila melanogaster. Nature (London)* **270,** 363.

Thörig, G. E. W., Schoone, A. A., and Scharloo, W. (1975). Variation between electrophoretically identical alleles at the alcohol dehydrogenase locus in *Drosophila melanogaster. Biochem. Genet.* **13,** 721–731.

Throckmorton, L. H. (1975). The phylogeny, ecology and geography of *Drosophila. In* "Handbook of Genetics" (R. C. Kind, ed.), pp. 421–469. Plenum, New York.

Throckmorton, L. H. (1982). Pathways of evolution in the genus *Drosophila* and the founding of the *repleta* group. *In* "Ecological Genetics and Evolution: The Cactus–Yeast–*Drosophila* Model System" (J. S. F. Barker and W. T. Starmer, eds.), pp. 33–47. Academic Press, Sydney.

Triantaphyllidis, C. D., Panourgias, J. N., Scouras, Z. G., and Ioannidis, G. C. (1980). A comparison of gene–enzyme variation between *Drosophila melanogaster* and *Drosophila simulans. Genetica* **51,** 227–231.

Ursprung, H., Sofer, W. H., and Burroughs, N. (1970). Ontogeny and tissue distribution of alcohol dehydrogenase in *Drosophila melanogaster. Wilhelm Roux Arch.* **164,** 201–208.

Vacek, D. C. (1982). Interactions between microorganisms and cactophilic *Drosophila* in Australia. *In* "Ecological Genetics and Evolution: The Cactus–Yeast–*Drosophila* Model System" (J. S. F. Barker and W. T. Starmer, eds.), pp. 175–190. Academic Press, Sydney.

Van Delden, W. (1982). The alcohol dehydrogenase polymorphism in *Drosophila melanogaster*: Selection at an enzyme locus. *Evol. Biol.* **15,** 187–222.

Van Delden, W. (1984). The alcohol dehydrogenase polymorphism in *Drosophila melanogaster* facts and problems. *In* "Population Biology and Evolution" (K. Wohrmann and V. Hoescheke, eds.), pp. 127–142. Springer-Verlag, Berlin and New York.

Van Delden, W., and Kamping, A. (1980). The alcohol dehydrogenase polymorphism in populations of *Drosophila melanogaster.* IV. Survival at high temperature. *Genetica* **51,** 179–185.

Van Delden, W., and Kamping, A. (1983). Adaptation to alcohols in relation to the alcohol dehydrogenase locus in *Drosophila melanogaster. Entomol. Exp. Appl.* **33,** 97–102.

Van Herrewege, J., and David, J. R. (1978). Feeding an insect through its respiration: Assimilation of alcohol vapours by *Drosophila melanogaster* adults. *Experientia* **34,** 163–164.

Van Herrewege, J., and David, J. R. (1980). Alcohol tolerance and alcohol utilisation in *Drosophila:* Partial independence of two adaptive traits. *Heredity* **44,** 229–235.

Van Herrewege, J., and David, J. R. (1984). Extension of life duration by dietary ethanol in *Drosophila melanogaster*: Response to selection in two strains of different origins. *Genetica* **63,** 61–70.

Vigue, C. L., and Johnson, F. M. (1973). Isoenzyme variability in species of the genus *Drosophila.* VI. Frequency–property–environment relationships of allelic alcohol dehydrogenase in *D. melanogaster. Biochem. Genet.* **9,** 213–226.

Vigue, C., and Sofer, W. (1976). Chemical selection of mutants that affect *Adh* activity in *Drosophila.* III. Effects of ethanol. *Biochem. Genet.* **14,** 127–135.

Vilageliu-Arqués, L., and Gonzàlez-Duarte, R. (1980). Effect of ethanol and isopropanol on the activity of alcohol dehydrogenase viability and life span in *Drosophila melanogaster* and *Drosophila funebris. Experientia* **36,** 828–830.

Vilageliu, L. I., and Gonzàlez-Duarte, R. (1984). Alcohol dehydrogenase from *Drosophila funebris* and *Drosophila immigrans*: Molecular and evolutionary aspects. *Biochem. Genet.* **22,** 797–815.

Vilageliu, L. L., Juan, E., and Gonzàlez-Duarte, R. (1981). Determination of some biochemical features of alcohol dehydrogenase from *Drosophila melanogaster, D. simulans, D. virilis, D. funebris, D. immigrans, D. lebanonensis,* comparison of their properties and estimation of the homology of the *Adh* enzyme of different species. *In* "Advances in Genetics, Development and Evolution of *Drosophila*" (S. Lakovaara, ed.), pp. 237–250. Plenum, New York.

Ward, R. D. (1974). Alcohol dehydrogenase in *Drosophila melanogaster* activity variation in natural populations. *Biochem. Genet.* **12,** 449–458.

Ward, R. D. (1975). Alcohol dehydrogenase activity in *Drosophila melanogaster*: A quantitative character. *Genet. Res.* **26,** 81–93.

Ward, R. D., and Hebert, P. D. N. (1972). Variability of alcohol dehydrogenase activity in a natural population of *Drosophila melanogaster. Nature (London)* **236,** 243–244.

Watts, F., Castle, C., and Beggs, J. (1983). Aberrant splicing of *Drosophilia* alcohol dehydrogenase transcripts in *Saccharomyces cerevisiae. EMBO J.* **2,** 2085–2091.

Wheeler, M. R. (1981). The Drosophilidae: A taxonomic overview. *In* "The Genetics and Biology of *Drosophila*" (M. Ashburner, H. L. Carson, and J. N. Thompson, eds.), Vol. 3A, pp. 1–97. Academic Press, London.

Wilks, A. V., Gibson, J. B., Oakeshott, J. G., and Chambers, G. K. (1980). An electrophetically cryptic alcohol dehydrogenase variant in *Drosophila melanogaster.* II. Post electrophoresis heat treatment screening of natural populations. *Aust. J. Biol. Sci.* **33,** 575–585.

Wilson, A. C., Carlson, S. S., and White, T. J. (1977). Biochemical evolution. *Annu. Rev. Biochem.* **46,** 573–639.

Wilson, S. R., and Oakeshott, J. G. (1986). Statistical analysis of data from gene frequency perturbation experiments. *Biometric. J.* (in press).

Wilson, S. R., Oakeshott, J. G., Gibson, J. B., and Anderson, P. R. (1982). Measuring selection coefficients affecting the alcohol dehydrogenase polymorphism in *Drosophila melanogaster. Genetics* **100,** 113–126.

Winberg, J.-O., Thatcher, D. R., and McKinley-McKee, J. S. (1982a). Alcohol dehydrogenase from the fruitfly *Drosophila melanogaster*: Substrate specificity of the alleleoenzymes Adh^S and Adh^{UF}. *Biochim. Biophys. Acta* **704,** 7–16.

Winberg, J.-O., Thatcher, D. R., and McKinley-McKee, J. S. (1982b). Alcohol dehydrogenase from the fruitfly *Drosophila melanogaster*: Inhibition studies of the alleleoenzymes Adh^S and Adh^{UF}. *Biochim. Biophys. Acta* **704,** 17–25.

Winberg, J.-O., Thatcher, D. R., and McKinley-McKee, J. S. (1983). *Drosophila melanogaster* alcohol dehydrogenase: An electrophoretic study of the Adh^S, Adh^F, and Adh^{UF} alleleoenzymes. *Biochim. Genet.* **21,** 63–80.

Winberg, J.-O., Hovik, R., and McKinley-McKee, J. S. (1985). The alcohol dehydrogenase alleleoenzymes Adh^S and Adh^F from the fruitfly *Drosophila melanogaster*: An enzymatic rate assay to determine the active-site concentration. *Biochem. Genet.* **23,** 205–216.

Winberg, J.-O., Hovik, R., McKinley-McKee, J. S., Juan, E., and Gonzàlez-Duarte, R. (1986). Biochemical properties of alcohol dehydrogenase from *Drosophila lebanonensis. Biochem. J.* **235,** 481–490.

Woodruff, R. C., and Ashburner, M. (1979a). The genetics of a small autosomal region of *Drosophila melanogaster* containing the structural gene for alcohol dehydrogenase. I. Characterisation of deficiencies and mapping of *Adh* and visible mutations. *Genetics* **92,** 117–132.

Woodruff, R. C., and Ashburner, M. (1979b). The genetics of a small autosomal region of *Drosophila melanogaster* containing the structural gene for alcohol dehydrogenase. II. Lethal mutations in the region. *Genetics* **92,** 133–149.

Yamazaki, T., Kusukabe, H., Tachida, M., Ichinose, H., Yoshimaru, H., Matsuo, Y., and Mukai, T. (1983). Reexamination of diversifying selection of polymorphic allozyme genes using population cages in *Drosophila melanogaster. Proc. Natl. Acad. Sci. U.S.A.* **80,** 5789–5792.

Yoshimaru, H., and Mukai, T. (1979). Lack of experimental evidence for frequency-dependent selection at the alcohol dehydrogenase locus in *Drosophila melanogaster. Proc. Natl. Acad. Sci. U.S.A.* **76,** 876–878.

Zera, A. J., Koehn, R. K., and Hall, J. G. (1985). Allozymes and biochemical adaptation. *In* "Comprehensive Insect Physiology, Biochemistry and Pharmacolgy" (G. A. Kerkut and L. I. Gilbert, eds.). Pergamon, Oxford.

Ziolo, L. K., and Parsons, P. A. (1982). Ethanol tolerance, alcohol dehydrogenase activity and *Adh* allozynes in *Drosophila melanogaster. Genetica* **57,** 231–237.

Zwiebel, L. J., Cohn, V. H., Wright, D. R., and Moore, G. P. (1982). Evolution of single-copy DNA and the *Adh* gene in seven drosophilids. *J. Mol. Evol.* **19,** 62–71.

GENETICS RESEARCH ON BRACONID WASPS

Daniel S. Grosch

Department of Genetics, North Carolina State University,
Raleigh, North Carolina 27695

I. Introduction

Over 20 years ago Anna R. Whiting (1961a) reviewed the research on the genetics of *Habrobracon* covering the years from 1917 to 1960. The purpose of the present review is to report on subsequent braconid research. In a few topics, omissions from the Whiting review are included as background information. Most of the investigators have used *Habrobracon juglandis* as the genus and species with which they worked. Taxonimists have revised this species to *Bracon hebetor* Say. Here *Bracon* and *Habrobracon* will be used synonymously. Some research with *Habrobracon serinopae* and a few related wasps will be discussed.

Much of the earlier research was devoted to sex determination and sex linkage in *Habrobracon* (Whiting, 1961a). The mapping experiments which established nine multiple alleles at the sex locus positioned 10 map units from that of the *fused* gene (antennal and tarsal segments) have never been refuted. However, since each allele is considered to be a chromosome segment they need to be analyzed by the modern techniques of molecular genetics. Heterozygous diploids are female; haploids and the rare homozygotes are male.

Bacci (1965) revised the Whiting symbolism in a discussion which appreciated that although primary and secondary sex characters act as a series of allelic factors they must be regarded as a polygenic group. Furthermore, a major problem arises from attempts to fit the braconid situation into a comprehensive whole consistent with the balance theory of *Drosophila*.

Although given a subordinate role, germ line cytology, linkage maps, and the description of nearly 100 visible mutations were accomplished before ionizing radiations and chemical mutagens preempted other interests. In this article, due attention is given to the effects of mutagenic agents, but some balance has been achieved by discussion of investigations of biochemical, developmental, behavioral, and aging phenomena.

With few exceptions among the 25,000 species the Braconidae are primary parasites useful for pest control. Their habits are those of a gregarious parasitoid that oviposits on the outside of a host it paralyzed. There are more than 8,000 titles in general biology and the taxonomy of braconids (Shenefelt, 1965, 1969), not to be reviewed here. Matthews (1974) recommended subdividing the Braconidae into only 20 subfamilies, although some entomologists have proposed as many as 31. Below the subfamilial level much remains to be worked out. The three braconid genera with the greatest number of species are *Bracon*,

Opius, and *Apantheles* (Matthews, 1974). Taxonomic debates are beyond the scope of this review and will not be presented.

II. Developmental Aspects

European authors provided an early background in wasp development. Seurat (1899) described morphological changes in the stages from the early larva to the adult form in several entomophagous wasps including braconids. Gross changes in the internal organs were depicted as well as the external features. A historical study of wasp metamorphosis appeared soon afterward (Anglas, 1901). Henschen (1928) worked out the time course of changes in the gonad anlagen from egg fertilization to the eclosion of the adult *Habrobracon*. Wing development was detailed by Schlüter (1933). Glover (1934) reported his detailed measurements of body size and parts for the development of *Bracon tachardiae.*

A. Embryogenesis

A wasp egg acquires its developmental potential during oogenesis. During the period before its release the developmental future of the egg is influenced by an associated mass of nurse cells and a surrounding layer of follicle cells. Nurse cell nuclei assume the major burden of synthesizing informational macromolecules and pump RNA along with cytoplasmic organelles through cytoplasmic bridges into growing occytes (Davidson, 1968). Ribosomes, mitochondria, membranes, and lipid droplets move through the ring canals to contribute to the egg cytoplasm (King and Cassidy, 1973). The follicle cells play a different role. Although they may synthesize some substances for oocyte storage, they act more as a selective barrier to mediate the transfer of materials from the insect's hemolymph into oocyte cytoplasm, especially vitellogenic protein synthesized in somatic tissues (Davidson, 1968). Thus an egg's volume receives gradients coming from several different directions. Much pours in from the end attached to the bundle of nurse cells, but also other materials flow in from the sides. Consequently oogenesis produces the cytoplasmic heterogeneity necessary for morphogenesis.

During the first 2 hours after oviposition most mitochondria are in the egg periplasm, typically arranged in clumps with their long axes parallel. At this time they are elongate with poorly developed cristae. At 3–4 hours in incipient blastoderm cells the mitocondria are 40%

lateral to the nuclei and 42% subnuclear. Most are spherical with well-developed cristae. These resemble the mitochondria seen in the oocytes (Cassidy and King, 1972). Presumably they are transmitted from the nurse cells in spherical condition, having assumed a simple morphology temporarily.

By 7–8 hours the fully formed blastodermal cells have about half the mitochondria in supranuclear position and many are no longer spherical. The same situation is seen in cells of the gastrula. No firm evidence has been obtained for differential distribution of mitochondria into presumptive germ layers. Other cellular constituents present in eggs include free ribosomes, lipid droplets, yolk spheres, and glycogen deposits similar to those found by Cassidy and King in oocytes (1972). Smooth and rough endoplasmic reticulum increase in quantity and structural complexity in the course of differentiation, becoming evident when the cells of the blastoderm form. At the same time, nucleoli appear in the nuclei (Amy, 1975).

Histological and cytological studies of developing embryos have not explained the differential mortality of male types reared in standard conditions. Sex determination by a series of multiple alleles was established over several decades (Whiting, 1961a). The biparental diploid females are heterozygous for two alleles, and the parthenogenetically produced normal males are haploids. Diploid homozygotes can be obtained only from two-allele crosses in *B. hebetor* but few such males survive to adult. In *H. serinopae* diploid males are much more viable.

Petters and Mettus (1980) confirmed a low hatchability in air of eggs containing diploid male embryos and demonstrated excellent subsequent survival to adulthood. However, when diploid male embryos were incubated in mineral oil their hatchability was equivalent to that of their haploid brothers. In other words the poor viability was traced to an aspect of normal rearing conditions. Perhaps diploid male embryos are more prone to desiccation than the characteristic haploid male type. There is a difference in cell size. Less likely seems the suggestion that something superficial on the host caterpillar promotes hatching which is simulated by mineral oil. Eggs deposited off the host can hatch but cannot develop unless the host is very near and can be reached quickly.

B. Genetic Mosaicism

Events at fertilization or at the initial mitotic divisions of embryogenesis have far-reaching consequences for an insect if they give rise to genetic differences between cleavage nuclei. Whether a gynandromorph is bilateral or anterior–posterior is one impressive example. The high degree of autonomy in insect development is an important contributing factor which results in sharp demarcation between tissues of different genotypes. Cells of the same type in patches of adult tissue represent a common origin by descent during development (Stern, 1968).

The patterns resulting in *Habrobracon* have been studied by use of appropriate gene markers in a cross which produces mosaics (Clark *et al.,* 1971). For example, females homozygous for the recessive determiners of ebony body color and orange eyes were crossed with haploid males carrying the nonallelic recessive gene for white eyes. In other crosses other genes for eye color, body color, and wing shape were used to identify the parental origin of mosaic patches in the offspring. A high proportion of mosaics showed more than one phenotype within the same eye, to which the term *intramosaic* applies. These can be used to estimate the number of cell clones making up the compound structure. In contrast, few ocelli showed intramosaicism. The incidence of intramosaicism was markedly different for structures derived from ectoderm: probabilities were, for eye, 0.21; wing, 0.05; leg, 0.02; ocellus, 0.005; and antenna, 0.004. Leg and wing correlation of genetic origin was high, 85%, for the appendages on the same side. Bilateral mosaicism is more common than anterior–posterior types. Mosaicism was more common in the thorax and abdomen than in the head.

The abdominal sternites develop independently from the external genitalia. Furthermore, the ectodermal phenotype is not necessarily indicative of the internal structures derived from mesoderm and endoderm. Also, in a mosaic, the sex of the internal organs of reproduction cannot be predicted from the external genitalia. In some cases both male and female organs were found inside the same abdomen (Egen, 1974). When derived from the same germ layer correlation between external and internal structures can occur.

Longitudinal striping of the appendages is characteristic of adult mosaics. Presumably stripes correspond to the original relative positions of the pronuclei, amplified by oriented cell divisions during development. A comparison of the appendages of braconids and of the mosaics in flies, moths, and bees suggests that longitudinal striped patterns are the rule for holometabolous insects (Postlethwait and Schneiderman, 1971). An exception is the striped pattern of mosaic wasp antennae, an appendage shape not available to *Drosophila.*

Insects employ oriented cell division in adult morphogenesis to a greater degree than cell migration. Cuticular surfaces other than the appendages also show elongate clonal stripes. In *Habrobracon* the mosaic patterns on body segments (Egen, 1974) are not as random as those in *Drosophila,* where the angle of first cleavage is variable (Parks, 1936) and the blastoderm nuclei have no set pattern of migration. Cleavage in *Habrobracon* gives rise to two pools of nuclei that maintain their relative positions (Speicher, 1936). The adult mosaic reflects the tendency for pools of nuclei of different origin to retain the anterior–posterior or left–right orientation of the original cleavage nuclei. These produce the broad gauged mosaicism of the abdomen except in the posterior.

Bilateral mosaic tergites and sternites provide evidence that each abdominal segment arises from two histoblasts. Each anlage is composed of about 12 cells at the time of determination (Egen, 1974). Longitudinal concordance values from pairwise comparison of successive half-segments are uniformly high, but these decrease sequentially as the distance from the reference segment increases. This indicates formation from different progenitor histoblasts rather than a proliferation of cells from a single progenitor. The haploid–diploid mosaicism results in a shift of the tergite midline toward the haploid side, presumably because of the smaller size of haploid cells. A sequence of several haploid tergites causes the abdomen to bend in that direction.

In the reproductive system development is completely autonomous with well-defined bilaterality. At the blastoderm stage two separate groups of cells are delineated for the genital disks (Egen, 1974). The direct apposition of the parts in the male and female genitalia and the occurrence of intramosaic genital ducts indicate the structures to be developmentally homologous in their origins from the genital disks.

Considerable insight into the nature of mosaicism was gained from the use of the examples produced by the *ebony* stock. However, one type reported in the early Whiting (Whiting *et al.,* 1934) observations has not been recovered by the Clark group. This was gynandroid, the sex intergrade which was a male mosaic with feminized genitalia. Two types of male tissue appeared to interact at their junction to change a clasper into a protuberance which resembled a sensory gonapophysis. Unfortunately there were no genetic markers in the parents producing these gynandroids. No feminization of haploid–haploid gynandromorphs has yet occurred in bilateral mosaics of visibly distinct tissues obtained by A. M. Clark's group (personal communication). A number of specimens with the penis composed of side-by-side haploid tissues of different phenotypes have occurred.

The filiform antennae of the genetic mosaics of *Habrobracon* are especially suitable for investigations of sexual dimorphism. They consist of 19 segments in the male but only 12 segments in the female. In mosaics, mixed male and female tissues occurred proximally, but female segments were never found distal to the twelfth segment (Clark *et al.,* 1973). In the transitional region of segments 10–13 a high frequency of incomplete or helical segmentation occurred. For example consider an antenna with honey-colored male tissue extending for three segments into the transitional zone on one side. The other half of the antenna was completed by unsegmented dark female cuticle little more than two segments long. Here there was a compensatory bend toward the shorter side. Defects in segmentation are associated with change in the polarity of antennal sensory placodes. In extreme cases the sensilla lay in a transverse direction.

Since no morphological abnormalities were found in the first seven segments of mixed antennae, these are considered to be homologous in the two sexes. Beyond this the two types of tissue could be out of phase, perhaps by a difference in growth rate prior to the event of segmentation. This appears to be a simultaneous event in braconids because the number of segments is complete as soon as segmentation becomes visible (Clark *et al.,* 1973).

During development eye disks do not undergo the cell rearrangements accompanying the unfolding and eversion required of appendages. As a result eye mosaics are characterized by patches that are more triangular than striped. In addition to explaining aspects of eye development, eye mosaics were used to investigate the morphogenetic relationship between the eyes and optic lobes of the brain (Hochberg, 1976). However, no brain mosaicism in association with mosaic eyes was demonstrated. The main experiment involved eye mosaics of the Z+A type obtained among the F_1 progeny of females with ebony (*e*) body color and white eyes (*wh*) crossed to males carrying a small eye trait (*se*) associated with cantaloupe (*c*) eye color and honey (*ho*) body color.

Small eye (*se*) causes the corneal lens to be malformed and reduced in thickness. Also the region normally occupied by a crystalline cone is replaced by an amorphous zone. In mosaic eyes the *small eye* ommatidia occur in isolated clusters. Even at the boundary between different patches of tissues, the facets are either normal or mutant, never of hybrid nature. This means a single ommatidial stem cell is responsible for the formation of each ommatidium. This is consistent with the ommatidium specific position effect pattern of variegation described for *Drosophila* by Shoup (1966) but differs from the Hotta and Benzer report (1973) of a mixture pigmented and unpigmented cells in single ommatidia in mosaics for the *Drosophila white eye* gene.

The number of cells primordial to the eye anlagen was estimated from the relative area of mutant tissue in mosaic eyes. The imaginal eye disk is formed from 17–20 primordial cells in either wild-type or *small eye* mutants. Therefore the deviation from wild type occurs after formation of the eye disk. Either the mutant cells develop slower than their wild-type counterparts, or mutant cells degenerate during larval development. The shape and size of the eye are altered if appreciable *se* tissue is present. The phenotypic modification occurring after formation of the eye disk is cell autonomous, and a more pronounced trait in diploids than in haploids demonstrates a gene dosage effect which decreases the size of the small eye.

C. Mosaic Production

Mosaic wasps appeared early in experimental samples of *Habrobracon*. Between 1921 and 1961 several hundred arose spontaneously. In the production of wasp gynandromorphs, chromosome loss or gene deletion does not apply to these haplo–diploid organisms. Instead the products of meiosis can escape disintegration in the egg interior so that two or more can take part in embryonic development. Also a sperm nucleus may begin dividing instead of participating in fertilization. The parthenogenetic production of the normal haploid males occurs because there is no block against the development of unfertilized eggs. In a mosaic, diploid tissues result from the union of a sperm and egg pronucleus. Haploid structures arise from an accessory pronucleus which may be of maternal or paternal origin.

Mosaicism appears more frequently in some stocks than in others, in which case it may occur in association with a single gene (Clark *et al.*, 1968). This argues for a genetic control of mosaic production which would influence the events that precede the union of egg and sperm nuclei, according to evidence obtained by the Clark group. In a survey of stocks the frequency of mosaicism was high only for crosses in which the mothers were homozygous for *ebony* (*e*/*e*) or for its allele *ebony*^{1} (e^1/e^1). Homozygosity for the other black body genes at other loci were not effective. Further experiments using females homozygous for *ebony* demonstrated about 5% mosaic and androgenetic progeny among fertilized eggs (Clark *et al.*, 1968), but no aberrant types were found from unmated females. Every aberrant type showed sperm involvement. Most frequent were haplo–diploid mosaics containing zygogenetic (Z) diploid parts. The haploid part was either gyno- (G) or androgenetic (A), or both. In addition, some other types were recovered: Z+Z, Z+Z+A, Z+Z+G, G+A. Some androgenetic males appeared in which only a sperm pronucleus had functioned.

The *ebony* mutant or another closely linked gene may delay the migration of the female pronucleus. This delay could provide an opportunity for cleavage of an egg or sperm pronucleus before a union of egg or sperm could occur. These events seemed vulnerable to environmental influences since they occur in laid eggs. Low temperature was considered to be the prime candidate. To test this hypothesis we placed samples of more than a thousand eggs of known age in a refrigerator at 5.5°C for 1 hour (Petters and Grosch, 1976). This cold shock applied to precleavage eggs nearly doubled the proportion of mosaics among the fertilized offspring. Postcleavage eggs were not influenced. No mosaics were obtained from a wild-type stock cold shocked during the sensitive period. Exposure to high temperature did not alter the proportion of mosaics obtained from the *ebony* stock.

A test cross experiment demonstrated that only one pronucleus is involved in producing the mosaics obtained from homozygous *ebony* females (Clark and Gould, 1972). In addition to *ebony*, small wing (*sw*), lemon (*le*) and honey (*ho*) body colors, and notched wing (*nw*) were used:

$$\frac{e}{e}\,\frac{sw}{sw}\,\frac{le\ +}{+\ ho}\,\textit{females} \times +\ +\ \underline{le\ ho}\ \text{or}\ +\ +\ \underline{le\ ho}\ nw\ \text{males}$$

Since this is a test cross in which a heterozygous linked pair of genes is crossed to a hemizygous recessive male, the genotype of the pronucleus can be determined for the zygogenetic phenotype. In the mosaic progeny showing crossover phenotypes all the gynogenetic parts showed the same phenotype as the zygogenetic parts for crossovers between the *le* and *ho* loci. Therefore, a female pronucleus cleaved before union of one of its daughter nuclei with a sperm nucleus. If two separate egg pronuclei were involved in the production of mosaics, the gynogenetic part could show any one of the three alternate genotypes. For example, when the zygogenetic $2n$ parts are lemon honey, the n parts arising from a second pronucleus could be wild type, or *lemon*, or *honey*. No such mosaic occurred.

D. Gynandromorph Behavior

Habrobracon is an excellent organism to use in the study of male and female behavior because the sex-related activities are clearly distinct for each sex and easy to recognize. Males respond to the presence of females by raising and vibrating their wings. Then they start on a search pattern. The male usually becomes aware of a female because of her odor (pheromones produced by the female). A male's long antennae are the most important receptors for this stimulus (Grosch, 1947), but sex mosaics can perceive female odor through short female antennae (Clark and Egen, 1975). The source of the pheromones is the anterior part of the female abdomen, as shown by experiments in which parts of females were crushed on bits of paper (Grosch, 1948). The squash preparation free of body fragments must be presented to males within 5 minutes because it loses its effectiveness quickly.

In addition to the pheromone-mediated long-range response, there is a contact response. Occasionally a male or gynandromorph who has shown no olfactory response may show the specific premating excitement and mount a female only after physical contact with her.

For many female insects there is no characteristic behavior except the absence of the male reaction. On the other hand, parasitic wasps have to find and cope with their specific prey. Female wasps find their host caterpillars by olfactory clues. Then they examine the lepidopteran hosts by antennal tapping and sting them in a characteristic manner. The distinctive female and male responses to the presence of caterpillars or potential mates have been used to determine nervous system phenotypes.

P. W. Whiting (1932) described the reactions of 62 gynandromorphs which had appeared in his cultures over a 5-year period. Depending upon the stock examined, these had occurred at a frequency of 1/1000 wasps. More recently, 276 sex mosaics were provided by the *ebony* stock which regularly produced 4 or 5 gynandromorphs per 100 offspring (Clark and Egen, 1975). In general, gynandromorphs with male heads behave like males, and those with female heads behave like females, which indicates good correlation between the phenotype of the external structures and the genotype of the brain. However, some with head mosaicism can behave like both sexes. That is, according to the stimulus provided the gynandromorph can respond either like a male or a female, giving the male response in the presence of a female and the female response to a caterpillar.

In determining the type of response, the external morphology, either sensors or genitalia, are unimportant. Most important is the genotype of the brain. In the 1975 study (Clark and Egen), 30 of the 276 gynandromorphs gave unexpected responses. Nineteen tried to mate with females and also to sting caterpillars. Whiting (1932) designated this behavior as bisexual. Four responded to neither females nor hosts. One tried to mate with caterpillars and with females, although it is possible that confusion resulted from a trace of female scent on the host. Another tried to sting females but was negative to caterpillars. Five others responded differently than would have been expected from the external morphology of their heads. Evidently, the insect brain operates primarily as a modifier on behavior patterns which are organized in the thorax and abdominal centers (Manning, 1967). In a gynandromorph the lower centers sometimes may break through.

E. Morphogenetic Fate Mapping

Insect mosaics have been powerful tools for placing the origin of adult structures on the cellular blastoderm. Petters (1977) constructed a fate map for 20 adult cuticular structures based on the examination of 1211 haplo–diploid mosaics. These were selected from 2,477 exceptional individuals among 49,055 offspring of the mosaic producing *ebony* stock. In addition to the ebony body color, white eyes, cantaloupe eye color, and honey body color were used as recessive nonallelic markers.

Distances on the fate map were estimated by assuming that the physical distance between determined sites on the blastoderm is related to the proportion of cases in which two adult structures differ in phenotype. As originally devised by Sturtevant (1929), the greater the distance between two blastodermal sites, the more often the mosaic dividing line will fall between them. His proportional units were called "sturts" by other investigators. In the sturtoid formulation now commonly used the maximum distance possible between two structures is taken to be 100. For example, in sturtoids the distance between antenna and compound eye locations is 11.7, while between antenna and lateral ocellus the distance is 14.5. In comparison there are the 77.3 sturtoids between antenna and external genitalia. Thus, the relative proximity is demonstrated for head structures and the relative distance from them to the genitalia. A list of external structures mapped is shown in Table 1.

The fate map obtained for 2422 sides resembles the adult metameric organization within a simpler outline to that of a braconid embryo (Fig. 1). Petters' (1977) fate map for a wasp agrees with maps obtained for other insects, *Drosophila* (Hotta and Benzer, 1972) and *Apis mellifera* (Milne, 1976). Of course, structures differ and dimensions vary but the same body plan emerges. In the wasp the blastoderm forms a single layer of cells encompassing the yolk and the cells that digest it already by 7–8 hours after fertilization. Organogenesis for the larva is completed by 29 hours but adult structures are only undifferentiated epidermal thickenings. These later become imaginal disks from which adult structures develop during metamorphosis.

At every molting period the division stages of larval epidermal cells makes possible a determination of chromosomal numbers. Male epidermis has the haploid number of 10 chromosomes throughout larval

TABLE 1
Abbreviations for External Structures Scored on *Habrobracon* Gynandromorphs[a]

Abbreviation	Structure	Abbreviation	Structure
A	Antenna	T4	Abdominal tergite 4
E	Compound eye	T5	Abdominal tergite 5
O	Lateral ocellus	T6	Abdominal tergite 6
W2	Mesothoracic wing	T7	Abdominal tergite 7
W3	Metathoracic wing	S3	Abdominal sternite 3
L1	Prothoracic leg	S4	Abdominal sternite 4
L2	Mesothoracic leg	S5	Abdominal sternite 5
L3	Metathoracic leg	S6	Abdominal sternite 6
T2	Abdominal tergite 2	S7	Abdominal sternite 7
T3	Abdominal tergite 3	G	External genitalia

[a] Numbers refer to the body region segment.

development (Grosch, 1950a). Nuclear size is small and remains small in the epidermis of appendages during the pupal stage. This has been verified again in whole mounts of developing antennal and leg buds (Abbott and Grosch, 1984).

F. Postembryonic Development

Bracon is a holometabolous insect with a larval stage showing no external evidence of the appendages characteristic of adults. These

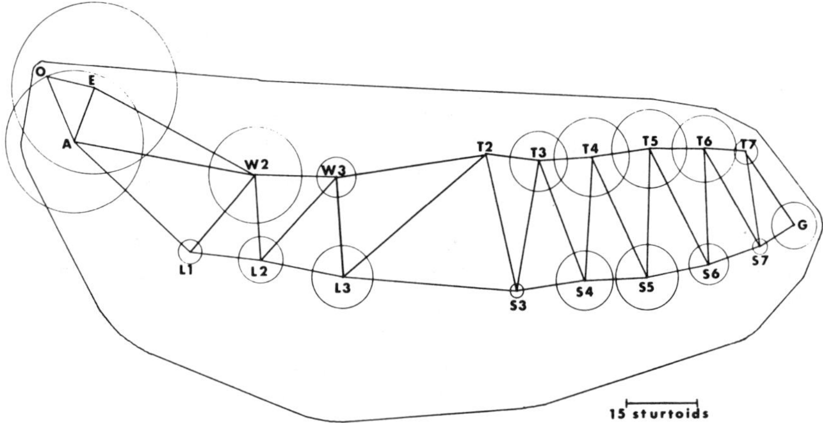

FIG. 1. *Habrobracon* fate map of blastoderm for external structures. Table 1 explains the abbreviations. The circles around each structure indicate the proportion of cases in which that structure is intramosaic. The outline around the map represents the dorsal and ventral midlines of the embryo. (From Petters, 1977.)

develop from imaginal disks. After the four instars of larval growth, the transformation into the adult requires two steps involving a molt from larva to pupa and another from pupa to imago.

Internally, the braconid larva is strikingly different from that of nonparasitic insects. Instead of converting ingested food into storage tissue, the wasp pumps fluid and tissue fragments from the host into an expansible midgut which occupies about 90% of the larval volume. It expands like a balloon because the larval midgut is not connected to the hindgut. Two large Malpighian tubules empty into the distal end of the hindgut. These structures as well as the silk glands, heart, and nerve cord are packed between the expanding gut wall and the body wall. The remaining space is taken up by a single layer of fat cells and a few urate bodies.

The cytological aspects of growth were studied in slides prepared from sections of impaternate male larve (Grosch, 1950a). Growth of the epidermis occurred chiefly by mitotic divisions with relatively little increase in cell and nuclear sizes. The haploid chromosome number was verified in division figures of the epidermal cells. In contrast, the larval midgut wall, Malpighian tubes, spinning glands, and fat cells grew entirely by increases in nuclear and cell size.

No mitotic cells were detected in the central nervous system on slides of 40 longitudinally sectioned and 180 transversely sectioned larvae, but a considerable increase in cytoplasmic material was evident (Grosch, 1950a). Subsequent examination of Feulgen-stained whole mounts of the chain of nerve ganglia revealed an occasional nerve cell in mid-mitosis. At any particular time only some ganglia showed a few dividing cells. Thus, the original conclusion holds that larval growth of the central nervous system is "chiefly" by increase in cytoplasmic materials (Grosch, 1950a).

The ventral nervous system of the braconid larva is formed of a double chain of 11 symmetrical ganglia (Genieys, 1925). Whole mounts show the chain of ganglia closely linked in a compact column for the first two instars. Initially, larval length barely exceeds the embryonic nerve cord. Successive ganglia are separated only by round openings which become oval when the longitudinal connectives begin their elongation. The long strands between ganglia become evident only in the last instar larvae and in pupae. This too is growth by increase in cytoplasm. Large nerves grow laterally into the body tissues as adult organs differentiate.

In the adult holometabolous insect, body ganglia are not present in the simple embryonic arrangement of one ganglion per segment posterior to the head. Instead, during metamorphosis ganglia unite in various combinations in different species.

Braconid wasps achieve an arrangement in which there are only seven ganglia behind the head. The prothoracic ganglion retains its identity but the large mesothoracic ganglion is a combination of two thoracic ganglia and one abdominal ganglion. The braconid's formation of an abdominal chain of four ganglia is similar to the arrangement in other Hymenoptera and more highly evolved than that of primitive orders of insects. However, the situation is far from the maximum fusion of units in Diptera, where up to eight pairs of ganglia unite to form one large mass in the anterior abdomen from which long nerves extend into the abdomen (Bodenstein, 1950).

The braconid abdominal ganglia are not of equal size in the adult. The ganglion found in the fourth segment is a composite of two larval ganglia. The single ganglia of the fifth and sixth segments are small, while the terminal ganglion positioned between the oviducts is a large, pear-shaped mass formed from at least two larval ganglia.

Whole mount preparations of other organs also provided instructive histological details not evident in mounted section. The cells of the functional spinning glands have four nuclei (Grosch, 1952) and other larval organs also are multinucleate. This explains why Genieys (1925) was puzzled by an apparent infrequency of cell walls between the nuclei of certain sectioned *Habrobracon* larval tissues. A multinucleate cell is achieved by amitotic divisions of enlarging nuclei in the sequence of 1,2,4, etc. Presumably this is the ultimate expression of nuclear lobations seen in the homologous organs of other insects (Fischer, 1935; Geitler, 1953). It is conceivable there is a physiological advantage in increasing the surface area of a giant nucleus by adopting an irregular shape; the braconids have gone a step further.

An even higher number of nuclei per cell occurs in the large, flat octonucleate cells comprising the wall of the larval midgut (Grosch, 1951). By the last larval instar the midgut wall is composed of large flat epithelial plates, each containing eight nuclei. Whole mounts of guts from earlier instars showed increase by even numbers of nuclei. The amitotic process starts with nuclear elongation which proceeds to a bilobate dumbbell shape. Finally, this separates at the thinned-down handle into two nuclei (Fig. 2E).

Uninucleate and binucleate cells are common in first instar tissues. In the absence of mitotic division rapid two-dimensional enlargement of each cell is necessary to contain the host caterpillar contents, which are being pumped into the blind-ended pouch of the midgut by the heavily muscled foregut. Cells of the Malpighian tubes and of the small glands of the head also have more than one nucleus.

The nuclear contents of highly specialized larval tissues are also

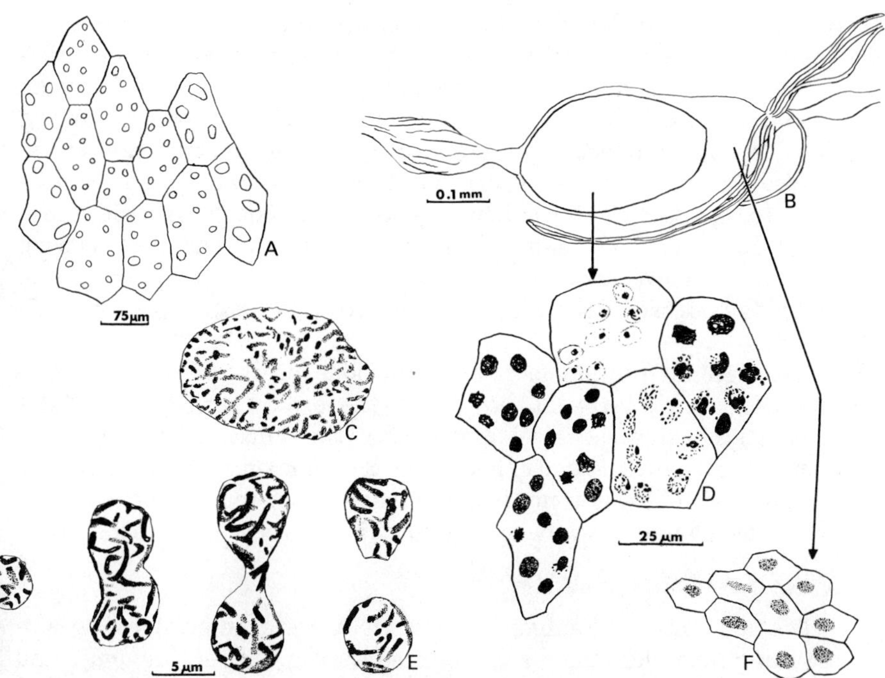

FIG. 2. Cytological aspects of braconid midgut development, Grosch (1951) observations. (A) Quadrinucleate and octonucleate epithelial plates of the gut wall of a second instar larva, derived from uninucleate and binucleate cells of the first instar. (B) The contracted remnant of the larval midgut undergoing histolysis within the lumen of the adult midgut during metamorphosis. Note that all eight Feulgen-stained nuclei in a specific unit are in a similar stage of degeneration (D), demonstrating intracellular physiological unity. (C) A magnified elliptical nucleus from (A) showing its endopolyploidy. (F) Conventional cells of the adult gut wall. Nuclear and cell area measurements of haploid and diploid adult midguts are available (Grosch and Clark, 1949).

consistent with the principle of an optimum presentation of increased surface area. A high degree of somatic polyploidy develops in each of the multiple nuclei in conjunction with a cyclic rhythm of change in Feulgen stainability. Replication of gene strings followed by their individualization in the nuclei of specialized tissues is much more common in the Insecta than the polytene ribbons of the Diptera (Geitler, 1953).

Typically, the larval tissues with polyploid multinucleate cells deteriorate and their debris is resorbed during metamorphosis. Grosch (1952) showed how the dispersed chromosomes of the larvae clump together in the nuclei of the silk glands after the cocoon has been spun. Then nuclei decrease in size and their chromatin condenses into a vacuolated amorphic mass. Subsequently, fragmentation of the gland is accompanied by the breakup of nuclei and the loss of stainability. Considerable evidence in a variety of insects of breakdown of larval tissues due to lysomal activity has become available. Phagocytosis helps to complete the process by ingestion of cell fragments (Sridhara, 1981). However, no research on the macromolecular changes has been performed on braconids to equal that on the Diptera and Lepidoptera, probably because of the reluctance of investigators to work on small insects.

The braconid, unlike most holometabolous insects, does not accumulate reserves in a large fat body during the feeding period. *Bracon's* fat body has its greatest development after the quiescent period in the cocoon has begun, instead of before the cocoon is spun or the pupal case is fashioned. The difference is in sequence rather than in the nature of events associated with the stored nutritional reserves (Grosch, 1949).

Nuclear Amounts of DNA

The amounts of Feulgen-stained DNA were determined with a scanning microdensometer for selected tissues from *H. juglandis* and *H. serinopae* males and females. The nuclei smeared out from syncytial preblastoderm embryos showed haploid amounts of DNA in parthogenetically produced males and the expected biparental diploid DNA content in females (Rasch *et al.,* 1977). However, cells from formalin-fixed, flattened tissues dissected out of adults contained additional amounts of DNA. Cells in the brain and cells in exuded hemolymph contained diploid and tetraploid amounts in both sexes. And in both sexes equivalent amounts of 8C, 16C, or 32C were found in nuclei of Malpighian tubules. The authors suggested an explanation of strandedness (Rasch *et al.,* 1977), but endopolyploidy is much more widespread in insects (Geitler, 1953). The condition is typical of specific types of specialized tissues concerned with the intensive production or excretion of materials. The supernumerary chromonemal replications which give rise to polytene structures in the Diptera individualize in specialized cells of the insects of other orders and provide polyploid nuclei. Grosch (1950a) reported that many of the endopolyploid cells of braconid larvae were dissembled during metamorphosis, but in all wasps investigated the blood–fat cell complex persisted into the adult (Schmieder, 1928). Other endopolyploid tissues in adult insects include the Malpighian tubes, gland cells, septa, and tracheal walls (Geitler, 1953). The latter two types present in organs can contribute nuclei to an organ's cell population which are not from constitutive cells.

The development of the gonads is most postembryonic. In the early embryo pole cells rest in a single caudal mass until after 20 hours at 30°C, when they assume a more anterior final position where they are divided into two lateral groups (Amy, 1961). A thin mesodermal sheath appears at about 26 hours around each mass of unchanged cells. During the larval period the ovarian mass elongates while each globular testis begins to enlarge with rosette contents (Henschen, 1928). A few hours before the prepupal stage each ovary anlage is divided into tubes by investing cortical cells. Connecting duct formation begins in the prepupal stage but the first oocyte does not appear until dark pigment is forming in the pupal eye. Enlarging oocyte–trophocyte masses enclosed in follicles become evident when mouth parts are pigmenting.

At adult emergence the polytrophic ovariole contains a distal germarium and a proximal vitellarium. Stem cell oogonia and mesodermal prefollicular cells fill the germarium while a single file of oocyte chambers of increasing size occupy the vitellarium. Inside each follicle are 32 fifth-generation descendants of a single cystoblast. These sister "nurse" cells are joined by canals, each of which is surrounded by a dilatable ring (Cassidy and King, 1972). The cell in optimum position for receiving molecules and cytoplasmic structures from its sister becomes the oocyte. Ultrastructural details of the transfer of cytoplasmic organelles to the oocyte along with the production of accessory nuclei and protein yolk spheres have been published (Cassidy and King, 1972). Also available is a study of the mitotic behavior of the follicle cells, the dynamics of cytoplasmic and nucleoplasmic growth of the endomitotic nurse cells, follicle cells, and oocytes, and their secretory activities (King and Cassidy, 1973). The oocyte's RNA is a product of nurse cell synthesis transported to the oocyte by cytoplasmic bridges, but vitellogenic protein reserves are mobilized from the fat body into the blood and thence to the oocyte (Davidson, 1968). Comparisons of the time spent in oogenesis and the rates of conversion of nutrients into ooplasm made between *Habrobracon* and *Drosophila* showed the wasp producing a batch of eggs equivalent to 12% of its body weight daily, while *Drosophila's* daily egg production amounts to 44% of its body weight (King and Cassidy, 1973). Because a drosophilid fly has 8–10 times more ovarioles than a braconid wasp, greater egg productivity is understandable.

In addition to the number of ovarioles the production of vitellogenic materials is important. In braconid females with one to five ovarioles in addition to the normal four, daily egg production did not increase significantly (Petters and Grosch, 1977). The rate of synthesis of vitellogenic proteins was a limiting factor. On the other hand, wasps with fewer than four ovarioles (one, two, or three) produced proportionately fewer eggs per day. In this case cell differentiation was the limiting factor.

Individuals with a decreased number of ovarioles were obtained by screening large samples of laboratory stocks to obtain the rare deviant types. Supernumerary ovarioles were induced by cold storage of fourth instar larvae (Grosch *et al.*, 1977). An initial indication that ovariole development could be modified by low temperature occurred in routine laboratory storage. When specific larval and pupal stages were stored for various lengths of time at 5 ± 1°C, the critical period was discovered to be between 76 and 96 hours and the optimum length of storage was found to be 2 weeks. Every number of ovarioles from 5 to 11 were induced, but females with 5 or 6 ovarioles were most frequent. Although the temperature used is cold enough to prevent the larval to pupal molt, it is not cold enough to halt growth, which suggests that a larger anlage at the time of ovariole establishment may be divided by investing sheath cells into more than the standard number of tubes.

III. Experiments with Adults

A. Venom and Apparatus

Stinging host caterpillars is a primary behavioral event which yields the fundamental source of food for both mother and offspring. Female braconids paralyze their prey by injecting their venom into the host's hemocoel. Although body movements stop, the host's heart continues to beat and visceral activity persists. Because *Bracon* deposits eggs outside the host, on or near its body, the host must be immobilized before the wasp oviposits. Characteristically, in the presence of a large supply of the preferred species, females sting more hosts than they parasitize. Subsequently, they relocate some of the paralyzed hosts for oviposition sites (Hagstrum and Smittle, 1977). Wandering hosts are attacked 10 times more frequently than experimentally confined hosts. Presence of a concealing layer of material makes no difference. Experiments at controlled densities demonstrated that an optimum ratio for biological control was one braconid female to 7.2 *Plodia* larvae (Reinert and King, 1971).

The venom has been studied by a number of investigators (Piek and Engels, 1969; Beard, 1972; Walther and Rathmayer, 1974; Visser *et al.*, 1976; Spanjer *et al.*, 1977). A high susceptibility to the venom's paralyzing action is restricted to members of the order Lepidoptera (Drenth, 1974). One part venom to 200×10^6 parts of host blood will halt movements within minutes. Very few species of the other higher orders of insects respond, and then only to very high doses of venom.

Bracon venom blocks the neuromuscular transmission at a presynaptic site (Piek and Engels, 1969). Visser *et al.* (1967) isolated and purified (28-fold) a paralyzing protein with a molecular weight of 61,000 (by gel chromatography) or 62,200 (by analytical ultracentrifugation). A modification of the Visser isolation procedure gave Spanjer *et al.* (1977) two products differing in the slopes of the dose–effect curves for paralysis, with molecular weights of 42,000 and 57,000.

Although the venom seems to have no direct effect on larvae of other parasitic wasps, the development of the endoparasitoid larvae of *Nemeritis* was arrested by host paralysis induced by venom injected by *Bracon* females. Young *Nemeritis* larvae were especially susceptible, but developmental inhibition also occurred for endoparasitoids that had grown for a week in host larvae before the braconid injected venom (Petters and Stefanelli, 1983). The inhibition of *Nemeritis* development is not caused by a direct effect on muscular function, but is due to an indirect effect related to changes in the host physiology. One type of change is a reduction in enzyme levels for cholinesterase (El-Sawof and Zohdy, 1976), and in the alkaline and acid phosphatases in paralyzed larvae (Zhody *et al.*, 1976). Another aspect is the oxygen content of host larvae, for which movement of the larvae is important (Fisher, 1961). However, the effect of venom on host respiration is not simply due to loss of locomotor function. A dose-related decline in respiration was demonstrated on motionless *Galleria* larvae by injection of a graded series of venom suspensions (Waller, 1964).

The venom apparatus of the Braconidae has been examined in 160 species (Edson and Vinson, 1979). Two types are recognized on the basis of their microscopic anatomy. Type 1, which includes the Braconinae and other subfamilies with ancestral characteristics, has a cone-shaped reservoir with a thick cuticular intima. Around it is a relatively thick muscular sheath which is innervated. It is supplied by a number of tubular glands. In contrast, the type 2 apparatus, found in subfamilies higher in the evolutionary scale, consists of a thin-walled reservoir with few muscles and filamentous glands, usually only two.

By resembling the apparatus in some Ichneumonidae (Togashi, 1976) this second type suggests an evolutionary continuity between ichneumonids and braconids.

More recently, the ultrastructures of the two types of venom apparatus have been compared (Edson *et al.,* 1982). No innervation of the type 2 musculature was detected. In both types a layer of squamous cells is adjacent to the cuticle with a single layer of cuboidal secretory cells behind it. The arrangement is similar to the class 3 described by Noirot and Quennedey (1974). The cytoplasm is dominated by endoplasmic reticulum, but also contains free ribosomes, Golgi bodies, and mitochondria. In contrast to earlier postulations of negligible metabolic activity in cells of the reservoir, electron miscroscopy disclosed an abundance of mitochondria, rough endoplasmic reticulum, free ribosomes, microtubules, coalescing vesicles, and pinocytotic structures. Evidently the reservoir does not function solely as a storage facility (Edson *et al.,* 1982).

An ultrastructural analysis of the paralyzing action of braconid venom has been performed using the femoral retractor unguis muscle from *Schistocerca gregoria* hind legs (Walther and Reinecke, 1983). Paralyzed and unparalyzed preparations, either resting or stimulated, were compared. No structural damage occurred at the nerve terminal but paralysis prevented the depletion of vesicles which occurred upon nerve stimulation of controls. However, an increase in the number and size of vesicles resulted soon after paralysis and continued in experiments run up to 6 days. Mitochondria swelling occurred when either controls or experimentals were stimulated. This resulted from a rise in interterminal calcium ions but did not occur when divalent calcium was replaced by magnesium ions. Thus, *Habrobracon* venom does not block the depolarization-dependent Ca^{2+} influx into the nerve terminal. Implied is a lack of interference with a transmitter-related process. Instead the venom seems to block vesicle exocytosis. This conclusion is consistent with the view that exocytosis is the mechanism whereby transmitter quanta are released in insect neuromuscular synapses (Walther and Reinecke, 1983).

Genetic differences have been demonstrated by biological assays which employed larvae from several species of Lepidoptera. Venom from different species of wasp vary in their effectiveness on a particular lepidopteran larva. Conversely, a specific venom from a specific wasp varies in its effectiveness on different Lepidoptera. In fact, *Anagasta,* a commonly used host, is less sensitive to *B. hebetor* venom than are *Galleria* or *Plodia* larvae (Beard, 1972). Nevertheless, discrimination between host species as ovipositional sites is not necessarily correlated with the effectiveness of induced paralysis. Undoubtedly, the biochemical development of venom and the morphological development of the injection apparatus played a role in the evolutionary adaptation to simultaneous changes in host species. This is another aspect not yet investigated.

B. Behavioral Studies

1. Standard Conditions

The first instar larva must be near or on a host caterpillar when it emerges from the egg. Therefore, the organism dies if the mother has erred in choosing a site for oviposition, or the egg has been accidently moved from where she laid it. The young larva must quickly attach to a suitable host and begin to pump up its midgut by sucking in host tissue. Ordinarily, it does not move from its feeding site until it is a fully fed fourth instar, or its host has been depleted of soft tissue by other feeding larvae. Even then it does not move far before spinning a cocoon.

Varied behavior is not evident until adulthood. Most obvious are courtship and mating behaviors which involve a complex series of interrelated activities. In the presence of a female a braconid male raises and vibrates his wings, an almost universal element of parasitoid mating behavior (Matthews, 1975). Then he engages in excited searching movements holding his wings raised. A strong response is a sustained reaction. When he comes in contact with a female, he orients himself and tries to mount her. She may crouch in a receptive position or may flip him off, rejecting him with her wings and hind legs.

The normal male sexual response was examined by Grosch (1947) who demonstrated that a male's long-range perception of the female was mediated primarily by antennal sensory structure, responding to some substance present on the anterior female abdomen (Grosch, 1948). The experimental approach was to amputate antennae or to impede the reception of stimuli by coating antennae or eyes with celloidin or asphaltum black, respectively. Antennaless males were unable to recognize the presence of females unless a male bumped into a female. Then mounting was attempted. Body fragments and bits of filter paper on which fragments were crushed served to identify the site of the stimulus which elicited sexual excitement in males. Subsequently, widespread evidence for pheromonal sex attractants has been obtained in various parasitic wasps (Matthews, 1975).

Grooming behavior in *Bracon* is sexually dimorphic. The females of wild-type stock 1 and different mutant stocks spent more time grooming their head structures, while males spent more time grooming their wings (Thelen and Farish, 1977). The slower grooming movements of braconids are a distinct advantage for analytical quantitative analysis, in contrast to the rapid movements of flies. Fifteen distinct motor patterns in *Bracon* stocks correspond to the fixed action patterns described for the family Braconidae in a study of 115 species, representing 33 families of wasps (Farish, 1972). Qualitatively all the mutant strains of the 1977 Thelen and Farish report performed the 12 most frequent grooming movements. However, some morphological mutants such as wasps with long antennae and notched wings showed deviant frequencies in the grooming pattern. From the standpoint of proprioceptive feedback, altered morphology might provide the grooming leg with the wrong message. Little variation in behavior is associated with changes in eye or body color.

No novel movements were observed. Furthermore, grooming behavior was entirely normal in wasps with vestigial wings or antennectomized wasps. These braconids groomed where the structure should have been. Normal patterns of grooming movements occurring in the absence of the proper recipient structure suggest a central nervous system control, and not merely a reflex response of the groomed structure (Thelen and Farish, 1977). In short, there is a remarkable qualitative homogeneity in braconid grooming movements. Neither major gene differences nor accumulated genetic changes resulting from strain isolation have influenced the fixed action patterns of grooming behavior. Evidently, the complex behavioral sequences are genetically well buffered against perturbation.

Females are recognized behaviorally by their response to a suitable host caterpillar. They can sense a host's presence from a distance, and in warehouses their sensing the presence of prey causes their penetration into bins of grain. Upon approaching a smooth-skinned lepidopteran larva, the wasp taps it with her antennae. If the host seems suitable she climbs up on the caterpillar, bends her abdomen, and inserts her sting. Antennaless females will sting caterpillars kept with them in a small vial (Clark and Egen, 1975), but take longer to do so.

The searching patterns which precede a wasp's contact with a host were traced by Murr (1930), who used braconids of both American and German origin. The last 5.1 cm of a searching pattern always led directly to the host. In containers lacking a host but filled with caterpillar-scented air, long convoluted searching patterns occurred. When one antenna was amputated a female made simple short explorations in a scent-free container. When host scent was provided, a pronounced tendency to veer toward the side with the antenna occurred in the course of long meandering search patterns.

Typically the duration of *Habrobracon* flight is brief, less than a minute. After a short flight, exploration normally continues on foot. Flightless laboratory strains have evolved in stocks maintained in shell vials for decades. Members of these strains cannot be stimulated to fly although the nonflight wing movements of grooming and courtship are normal.

Crosses between normal and flightless wasps demonstrated that the flightless trait is recessive (Petters *et al.,* 1978). In diploid–haploid gynandromorphs, morphogenetic fate mapping localized the defect in the anterior ventral region, which implicates either muscle or nervous tissue. Innervation was probably normal because it is necessary for the degree of muscle development observed. On the other hand, indirect flight muscle fixed in Bouin's solution showed anomalous nodal protuberances along their lengths, which sectioning and staining corroborated histologically. However, an electron microscopy study did not reveal any obvious departure from normal cell ultrastructure. Also, no differences were demonstrated in determinations of α-glycerophosphate dehydrogenase activity in flightless and wild-type adults.

In addition to the examples discussed above, there are a number of known braconid behavioral mutants not yet adequately investigated. Quivering types of wasps have occurred in both the Whiting and Petters laboratories. In an ethyl methanesulfonate experiment Brad Loucas found an antennapedia mutant (*ap*) in which the terminal segments of antennae develop into a leglike structure. The perception of females is hindered in males, and the recognition of hosts by females is less efficient. Differences in female performance of the six or seven components of a female behavior during her paralyzing a host caterpillar have not been analyzed quantitatively.

2. *Exceptional Conditions*

Whiting (1961a) reported a behavioral difference in *B. hebetor* stock 17-o^{i} that did not occur in wild-type stock 33. Mechanical stimulus evoked a collapse into immobility, after which the mutant types flew erratically. This appears to be an example of death feigning similar to those found in nearly all orders of insects (Frost, 1959).

Recently, the group led by Petters found stress sensitivity in 11 different *B. hebetor* stocks. Wild-type stock 33 has been reared in the laboratory for many years, two other stocks were derived from wild wasps trapped in Lumberton, North Carolina, and five more were from Pennsylvania collections. Three mutant laboratory stocks were also tested. Although the frequency varied, all stocks tested carried the trait.

In response to a sharp tap of the container, a typical "shake-down" laboratory procedure, positive individuals retract their legs and fold their antennae back over the thorax. After this they remain motionless for 10–30 seconds. In some cases motionless periods in excess of a minute have been observed. Habituation occurs in which wasps fail to show a second response for several hours.

Crosses between positive and negative parents indicate that there is a single dominant major gene involved. A great majority of offspring were positive when one of the parents was positive. Furthermore, a 1:1 ratio occurs among the parthenogenetically produced sons of heterozygous virgin females (Mettus and Petters, 1981). After establishing positive and negative lines, an attempt was made to alter the proportion of individuals showing the trait. However, selection for and against the trait neither increased nor decreased the proportion of stress-sensitive individuals in 21 inbred generations of the positive line and 12 generations of the negative line. The failure to produce 100% positive individuals may be explained by a 95% penetrance in females and a lower value in males. The genetic basis of death feigning in braconid wasps differs from that in *Tribolium* beetles, where there is additive action of heritable factors and an opportunity for progressive selection (Prohammer and Wade, 1981).

A somewhat different immobility response was discovered in ground-based experiments ancillary to the United States biosatellite space flights (Grosch, 1972). Although there is no basis for expecting animal tissues to respond in the fashion of an orthogonic plant organ, rotation was suggested by botanists as a universal method for counteracting the unilateral stimulation from gravity.

Rotation of groups of males in gelatin capsules on a horizontal disk subsequently increased the rate of mating in direct correlation with the rate of rotation used. In contrast, when the capsules of wasps were carried on vertical disks, no significant shortening of mating time was observed. Neither type of experiment explained the prolonged period between unpacking the space capsule and successful mating of the orbited males.

Another difference after rotation on a vertical versus a horizontal disk was the positioning of wasps. Individuals distributed themselves at random on the walls of capsules spun horizontally at all rates of rotation. In contrast, when disks were rotated in a vertical plane a new behavioral reaction was observed. Then wasps associated in small groups or clumps oriented with their legs toward the interior of the group and holding it together. The antennae were folded down flat along the dorsal body surface. Thus the group presented a smooth streamlined aspect with its long axis parallel to that of the capsule containing it (Fig. 3). Heads pointed in either direction. Usually, half the heads were at one end of the clump and the other half at the opposite end. During the rotation the clump floated freely up and down, up on the downstroke and down on the upstroke of the attachment site of the rotating disk.

After rotation, if the mass was gently eased from the capsule and left undisturbed the clump remained motionless for several minutes. In that aspect it appears to involve the death feigning response.

The original purpose of rotation tests was to throw light on differences in the time required to consummate mating. The mechanical approach failed, but during the tests an overriding influence of barometric pressure emerged. Particularly, the males were reluctant to mate at low pressure, and were overexcited and ineffectual at high pressure. A medium range of barometric pressure proved optimum. At the time Biosatellite II was unpacked, the barometer at Cape Kennedy was rising and the ground control males accomplished 243 observed matings within 3 hours. In contrast, fewer than 33% of the 254 males recovered from the space flight mated in 13 hours at Honolulu, where low pressure from an approaching storm dominated the weather pattern. Thus, ambient conditions on earth rather than influences encountered in space flight explain the male eagerness to mate or their apathy.

Although not applicable to the biosatellite situation, another laboratory has investigated the influence of genetic relatedness and the time to accomplish mating in one *B. hebetor* stock (Petters *et al.*, 1982). F_1 virgin females were assigned to individual vials for each of three mating tests: female–son, female–brother, and female–unrelated male. The males were obtained directly from cocoons and had no prior contact with other wasps of either sex. The number of male courtship performances and the time elapsed between introducing the two sexes and successful mating were recorded. Under standardized conditions

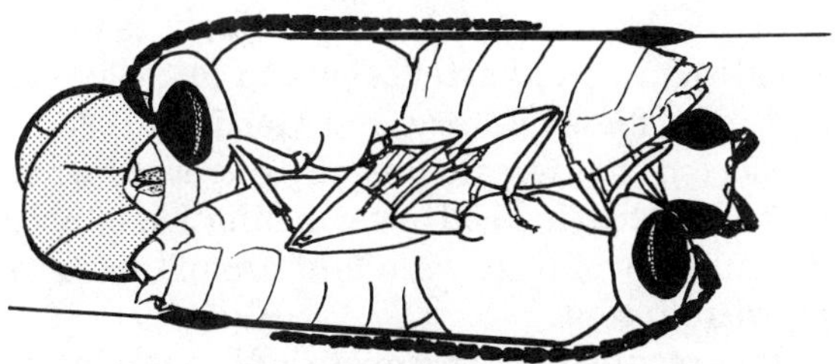

FIG. 3. A clump of three wasps of identical sex that floated in a rotating capsule. Presumably larger clumps float in the hypogravity of the space craft.

in experiments repeated thrice the average time until mating was shortest for the mother–son matings. Further experiments with other stocks of braconids need to be performed. In several other species of parasitic wasps considered to be highly inbred, sperm from the first mating usually have utilization precedence (Walker, 1980).

3. *Sex Ratio*

In wasps where normal males develop from unfertilized eggs, mating is important to produce biparental females. A mathematical model proposes that highly inbred wasps should produce highly female-biased sex ratios (Hamilton, 1967). Also, the selective advantage of a precise sex ratio has been calculated. Some laboratory studies of wasps from a different family showed sex ratios of only one male per small clutches of offspring and more males in larger clutches (Green *et al.*, 1982). This does not happen in braconids.

Bracon hebetor wasps reared in isolated vials consistently have a two-thirds female to one-third male ratio in our laboratory, but this does not hold when oviposition is disturbed by jostling in crowded conditions or when aged males supply an inadequate amount of sperm. Large numbers of offspring are predominantly male in *Bracon.*

Fifty *B. hebetor* females from the fields of Israel allowed to oviposit on successive series of hosts throughout their lifetimes produced 36 ± 9% females in their progeny, but only male progeny were produced by elderly females. During the period when both sexes occurred, 41% of them were female. Decreased numbers of offspring were explained when dissections of old females revealed their ovarioles depleted of oocytes.

Delays up to 3 weeks between mating and oviposition did not increase the proportion of female offspring, but a delay before mating did increase the female progeny by up to 53.6%. Other influences also play their roles. Differential mortality favors an increase in adult males when males survive high density better than females (Rotary and Gerling, 1973). Also, males are more likely to act as a cannibal than be cannibalized in circumstances when first-day eggs are unfertilized, haploid, and male (Rotary and Gerling, 1973). In the same paper it was reported that female progeny increased when host larvae were the only food source, instead of a combination of hosts and honey. The maximum number of eggs were laid around hosts when some of their webbing was present.

External influences undoubtedly operate through internal morphology and physiology. Although semen transport often is an active process, muscular contractions of various ducts can also be a factor

(Davey, 1965). Female insects commonly modulate the movements of sperm in their reproductive tracts (Walker, 1980). Braconids have spheroid spermathecae which is correlated in Hymenoptera with either monogamy or with the precedence use of first insemination for fertilization. The curved deferent spermatic canal receives a duct from a spermathecal gland halfway along it. The lumen of the canal is so narrow that only one sperm can pass along it. A bundle of muscles extends from the spermatheca along the curve of the tube (King, 1962).

Sperm movement is believed to be activated by a secretion of the accessory gland, because King (1962) caused sperm tails to change from gentle undulation to lashing movements by transferring them from one pH to another. Also spermathecae removed from Ringer's solution to buffer solutions at pH 5.59–8.04 showed a twitching of capsule and duct. Immediately after twitching started, the duct in the region of its bend tended to straighten by muscular action. *Nasonia* (*Mormoniella*) organs very similar to those of braconids were used in these tests.

Fertilization occurs as eggs pass down the oviduct. Orientation of the egg in the oviduct facilitates sperm entry if an egg is oriented in such a way that the single micropyle apposes the spermathecal duct. This may not always occur in braconids because eggs have an elongated curved shape. If some of the eggs regularly move past the duct opening with the micropyle pointing away from this source of sperm, unfertilized eggs result despite an adequate supply of sperm dispensed with regularity. The interrelations of the complex physiological and morphological aspects leading to wasp sex ratios is still not integrated, especially because males result from unfertilized eggs.

After reaching the base of the stylet an egg must enter the space between the first and second valvulae through which both the eggs and the venom pass to the outside. The passage's diameter is only a fifth of the width of an egg. Only the flexibility of the chorion and the fluid contents of the egg make is possible for a egg to move down the bore to be squeezed out near the tip.

C. Aging Studies

The first investigation of braconid life-span showed that starving females lived much longer than starving males (Grosch, 1950b). An obvious difference was the greater amount of stored nutritional reserves in oocytes. When the weights of unfed living wasps were obtained, females proved to be heavier. Furthermore, the weight loss over time was greater for males than for females (Griggs, 1959).

Determinations on homogenates showed that females contain more total nitrogen than males. In Section III,D, *Biochemical Investigations,* additional sex differences in enzyme activity are described. Radiation-induced life shortening is covered in Section IV,A.

The influence of diet on adult life-span has been studied by Clark's group. What the females ingest can make a significant difference in their longevity. *Habrobracon serinopae* females from a wild type and from a mutant stock lived twice as long if fed honey instead of host caterpillars (Clark, 1963), but the ovaries atrophied. Nevertheless, the rate at which oocytes are resorbed is slower on a honey diet than that occurring during starvation (Benson, 1973). For stock 1 *B. hebetor,* however, there was no correlation between longevity and diet-induced female sterility. No significant difference in life-span occurred between females fed the two diets. A small difference was found in stock 33, and a large difference appeared for stock 25 heterozygous for cantaloupe eye. Evidently, the genetic constitution of the organism influences the response.

A related aspect is the accumulation of urates in the fat body. This storage excretion characteristic of many kinds of insects takes the form of white opaque spheres scattered among the throphocytes, which are full of lipids and glycogen. When *Bracon* wasps are fed water-diluted honey or aqueous sucrose solutions they accumulate less urate than caterpillar-fed wasps (Clark and Smith, 1967b). This quantitative difference is expressed as the size of the spherical accumulations attained by a particular day in adult life. Genetic constitution as well as diet plays an important role. *Habrobracon serinopae* has obviously smaller urate spheres than *H. juglandis* at equivalent stages of the life cycle. More subtle differences among different strains of *H. juglandis* can be demonstrated by calculating the means and standard deviations of urate cell diameters.

Rearing temperature is an additional influence. Life is prolonged and urate deposition is decreased by maintaining *Bracon* wasps at lower temperatures. On the other hand, incubation temperatures of less than 30°C did not increase the adult life-span of *Mormoniella* (Clark and Kidwell, 1967). A comprehensive study of the influence of temperature on survivorship of sucrose-fed *Mormoniella vitripennis* used three rearing temperatures: 18, 25, and 30°C. Respectively, 4, 8, and 10 generations were studied.

Another approach to studying adult life-span involved rearing wasps at one temperature and holding them as adults at another temperature (Clark and Kidwell, 1967). For each culture subjected to this regimen there was an increase in adult life-span when the adult temperature was decreased. For example, consider these rearing/adult temperatures. The 30/30 group (temperature in degrees Centigrade) had a mean life-span of 35.8 ± 0.6 days, the 30/25 group's mean was 44.7 ± 0.6 days, and the 30/18 group's mean was 63.6 ± 0.9 days. These kinds of experiments are reminiscent of the permissive and restrictive temperature investigations of the products of specific qualitative genes. Thus, aging in insects can be approached from the viewpoint that adult life-span is a gene-controlled trait with variable expression amenable to methods used in developmental genetics.

However, the major activity of the females appears to have no significant effects on longevity. Clark and Smith (1967a) found no correlation between age at death and number of eggs laid by individual females of *H. serinopae* and *H. juglandis*. This supports the conclusion that fecundity is not a causal factor in the aging process. Indeed, *H. serinopae* lives twices as long as *H. juglandis* and the former's egg production is nearly thrice that of *H. juglandis*. The survival curves for the two species are parallel. In the Grosch laboratory, where a great deal of data on egg production has been collected while assaying the effects of radiations and chemical agents, the senile decline in egg production is regarded as an indication rather than a cause of old age. Additional evidence was supplied by F_1 females from fathers receiving 3000 R of X rays (Heidenthal *et al.*, 1972). The females were separated into two classes, high and low, on the basis of the hatchability of their eggs. Regardless of the variablity in hatchability and hence the heterozygosity of mothers for detrimental genes, the F_1 wasps that laid more eggs also lived longer. Thus, high fecundity give evidence of the biological fitness of females.

D. Biochemical Investigations

1. *Eye Pigments*

Fourteen eye color mutants were identified, described, and characterized for *H. juglandis* (Van Pelt, 1969). These are positioned at five nonlinked loci: *cantaloupe, garnet, orange, red,* and *white*. Each locus has an allelic series and each mutation is recessive to wild type. Paper chromatograms were obtained of wild-type and mutant adults using *n*-propanol : NH_3 and *n*-butanol : acetic acid : water as solvents for two sets of tests. Simultaneously, the same procedures were used for Oregon R and 7, common mutant strains of *Drosophila melanogaster*. The drosopterins, xanthopterins, and isoxanthopterins of Drosophilidae are not present in *Habrobracon*. However, the related compounds biopterin, isosepiaterin, and 2-amino-4-hydroxypteridin were identified in the braconid by R_f values and the colors of spots revealed under untraviolet light. These same three compounds are also present in the *Ephestia* host (Ziegler, 1961). The ommochromes characteristic of arthropods are produced as the single system for eye color in *Habrobracon*, with certain pteridines available only in a very subordinate role. The metabolic pathway to the ommochromes leads from tryptophan through kynurenine. A block just past kynurenine is characteristic of kynurenine hydroxylase and associated with the *cinnabar* gene in *Drosophila* and with the *orange* alleles in Habrobracon. Nonautonomy was shown in gynandromorphs and by eye disk implants in early experiments (Whiting, 1961a). In other orders of insects the same type of block often results in yellow or white eyes (Ziegler, 1961).

In *Habrobracon* the majority of the eye colors exhibit dosage compensation in which males resemble females, but one case differs. In *garnet* and in *garnet-2* stocks, haploid males have dark red eyes. The biparental females and rare diploid males have medium red eyes. To further establish the eye color difference due to a doubled gene condition, the *garnet* gene was transferred into haploid gynoid males that show the secondary sex characters of females. The eyes were dark red.

When wild-type braconids are raised at temperatures from 17 to 37°C there is no change in eye color. This is in contrast to normal body color, which is darker in wasps reared at lower temperatures. However, some eye color mutants are modified by the rearing temperature. Cantaloupe eyes that are light at 30°C are dark orange when reared at 37°C. Red eyes (*rd*) are lighter at 17°C while barolo eyes (rd^b) are lighter at 37°C. The *garnet* alleles are especially variable. *Garnet* (*g*) and *garnet-2* (g^2) eyes are light red at 17°C, very dark at 25°C, dark red at 30°C, decreasing in pigment at 32°C, and almost colorless at 37°C. In contrast, plum eyes (g^p) show only a slight increase in pigment with increase in temperature, and rose eyes (g^r) are unchanged.

The genes concerned with eye pigment formation seem to have begun acting during the molting and shedding of the larval cuticule but color becomes visible about 20 hours later. Higher temperatures prior to shedding do not influence pigment development. Wasps transferred to a higher temperature after the larval cuticle is shed show variability in eye color.

Leigenguth (1967) made quantitative determinations of the intermediate products of the tryptophan to ommochrome pathway, using extracts of wild-type wasps from stock 33 and mutants of two alleles at the *orange* locus. The extracts were chromatographed in an *n*-propanol–sodium citrate mixture which separated kynurenine, kynurenic acid, 3-hydroxykynurenine, and xanthurenic acid. The compounds were identified as fluorescent spots at 365 nm.

Wasp eggs had kynurenine predetermined by the maternal genes. The higher concentration of kynurenine in orange-eyed flies than in wild type occurred in all subsequent stages of development. No 3-hydroxykynurenine was synthesized in wasps homozygous or hemizygous for either of the *orange* alleles. The same rate of accumulating kynurenine was found for both mutant stocks except for an unexplained difference in 3-day-old male pupae. High levels of kynurenine in both mutant and wild-type stocks resulted when they were reared on host larvae injected with tryptophan. Nearly all kynurenic and xanthurenic acids were excreted during the molt to prepupal stage. Low constant amounts were present in pupae along with high concentrations of kynurenine and 3-hydroxykynurenine.

Enzyme assays showed high transaminase activity in young feeding larvae and low activity in old larvae. High activity appeared again in pupal and adult stages. Stage-specific high and low concentrations of the transaminase cofactor pyridoxal 5-phosphate and α-ketoglutarate occur in association with the amounts of substrate. In braconids, enzyme activities appear to be regulated by the amount of essential cofactors and specific inhibitors rather than the rate of enzyme formation (Leibenguth, 1967).

Almost no tryptophan pyrrolase occurs in 2-day-old pupae but freshly emerged adults have high pyrrolase levels. Either an inhibitor is excreted with the meconium or it is inhibited somehow. Unlike in *Drosophila* tissues or rat liver, neither injected substrate nor the resulting high amount of kynurenine caused an adaptive increase in tryptophan pyrrolase activity (Leibenguth, 1967).

In white-eyed mutants the amount of kynurenine is the same as that of the braconid wild type but there is much less hydroxykynurenine, evidently due to a lower kynurenine hydroxylase activity in white-eyed wasps (Leibenguth, 1971). A failure of nonautonomous eye colors to darken to wild type when their possessors were reared on wild-type hosts was a biochemical enigma. Host hemolymph and fat body contained 3-hydroxykynurenine which feeding transferred into the wasp gut and in turn appeared in the braconid's hemolymph. A relatively rapid transamination to xanthurenic acid subsequently excreted during the molt to the prepupal stage explained the enigma (Leibenguth, 1970). This is considered to be a mechanism evolved to reduce high larval chromagen contents when ommochrome synthesis is occurring only in pupal eyes. The mechanism is not required in insects of other orders where various internal tissues develop coloration.

2. *Enzymes*

In 1965 Lin completed a thesis for which he investigated cytological and biochemical differences between two species of braconids which are difficult to distinguish morphologically. Although originating on two different continents and reproductively isolated, *H. juglandis* and *H. serinopae* have identical wing venation, body shapes, and range in the number of antennal segments. The latter is a key factor of braconid taxonomy.

Both species have 10 pairs of chromosomes but the smallest pair are dots in *H. juglandis* and V's in *H. serinopae*. Also, the three larger pair of *H. juglandis* are almost twice the length of the largest pairs of *H. serinopae* (Lin and Grosch, 1982).

Polyacrylamide gel electrophoresis of the general protein patterns and of three dehydrogenases (lactic, malic, and xanthine) appeared to be identical for *H. juglandis* and *H. serinopae* (Lin and Grosch, 1982). A discontinuous buffer system of starch gel electrophoresis was used to characterize the esterase banding of the same two species. However, in addition to a California strain of *H. juglandis,* a Raleigh, North Carolina, laboratory strain, a wild-type strain collected in a Lumberton, North Carolina, warehouse (Lum.), and a stock collected in Natchitoches, Louisiana, by Lin (Lin) were investigated (E. Anderson, unpublished). She too found the *H. serinopae* pattern of bands quite similar to that of a laboratory stock (no. 33). The first band appeared identical in all the stocks tested while five other groups, II through VI, differed slightly in position and widths of bands. Some of the groups had two or three bands which might resolve into the extra three groups, which brings the total up to nine for the Louisiana Lin stock.

The lack of variability revealed by electrophoresis of strains of *Bracon* is consistent with the situation typical for the entire order of Hymenoptera. Strikingly low levels of genetic heterozygosity were found for five species of wasps and two species of bees in which isozymes were sought at 12–19 loci per species. The variation in mean heterozygosity for these seven Hymenoptera species representing two superfamilies and seven genera was less than that for seven species within a single *Drosophila* genus (Metcalf *et al.,* 1975). In another study using electrophoresis, one of the seven species tested was a braconid, *Opius juglandis*. Nine of its 13 loci were invariant (Lester and Selander, 1979). A general conclusion is that the mean heterozygosity of 0.043 for Hymenoptera is lower than the values for all other insect species yet surveyed electrophoretically. Snyder (1974) found allozymic variability lacking in three bee species in a survey of 22 enzymatic and 3 general protein loci. Pamila *et al.* (1978) analyzed 15 hymenopteran species for up to 18 loci. Other authors checked Hymenoptera including ants at one or a few loci (see Lester and Selander, 1979). In haplo–diploid organisms selection against deleterious alleles can be more effective and the time for fixation of new mutants is shorter than in diploid populations of the same size.

The succinoxidase systems of *B. hebetor* and *H. serinopae* were determined to see if they differed (Mizianty, 1967). Following the lead of Griggs (1959) tetrazolium salts were used as metabolic indicators. Data obtained by spectrophotometry were recorded as micromoles of reduced dye (formazan) produced per hour per organism, or as per unit dry weight (Mizianty, 1967). When triphenyltetrazolium chloride (TTC) was used as the indicator *B. hebetor* showed higher activity than *H. serinopae*. Also, as Griggs (1959) discovered, the males of each species had higher activity than the females. From the standpoint of the rate of living hypothesis this is expected because of the longer lifespan of *H. serinopae*. The addition of cytochrome *c* to the system increased activity to levels that reduced the difference between species. Furthermore, NADP addition increased activity enough to eliminate species differences. In contrast, when 2-*p*-iodophenyl-3-nitrophenyl-5-phenyltetrazolium chloride (INT) was the indicator no significant differences between sexes or species were observed. A difference in properties can be explained by the fact that the two tetrazolium salts accept electrons by different donors. In TTC assays electrons pass through cytochromes *c* and *a* and cytochrome oxidase, while INT accepts electrons from cytochrome *b*. When succinic dehydrogenase activity from the head, thorax, and abdomen was compared, virtually all activity was localized in the thorax (Griggs, 1959).

Not all types of enzymes show a higher activity in males. Studies on the phosphomonoesterases of *Bracon* demonstrated that females had a marked increase in acid phosphatase on emergence from cocoons and a continued increase throughout adulthood (Herr, 1953). In males the increase at emergence was modest and enzyme activity remained constant with age. Diploid male assays demonstrated the difference due to sex, not ploidy. The sex difference characterized adult types. No sex differences were evident for pupae.

Xanthine dehydrogenase (XDH) activity was determined in parental and F_1 braconids following the Biosatellite II flight. A fluorometric method of assay was used. The enzyme activity levels were decreased in all of the orbited adult groups in comparison with their appropriate ground control groups. Furthermore, for wasps positioned around the γ-radiation source, there was a dose-related XDH difference between experimental and control groups. These results agree with those of *Drosophila* from the same flight (Keller, 1970).

Depressed XDH levels in F_1 progeny were obtained from orbited males mated to nonorbited females, a situation which ruled out maternal or cytoplasmically transmitted influences. The changes seemed to be of such low magnitude that the usual genetic indicator systems (such as recessive lethals and visible mutations) were not sensitive enough to detect alterations unless mutagenic induction was increased by radiation exposure. However, the more sensitive quantitative XDH control system was capable of detecting changes (Keller, 1970).

3. *Uric Acid*

The accumulation of nitrogenous waste in the form of uric acid is common among terrestrial insects. In addition to excretion via Malpighian tubules, specialized cells of the insect fat body can sequester urates. Using a photometric method, Lin (1965) demonstrated the change in uric acid content during different stages of development for both *H. juglandis* and *H. serinopae*. From zero in eggs the amount increased to above 20 μg per wasp, a peak reached in the white pupae stage. From this peak there was a decline down to 4–8 μg per wasp at eclosion. The values for *H. serinopae* were consistently lower than those of the same sex of *H. juglandis*. In each species male values were lower than female values (Lin and Grosch, 1982).

After eclosion the amount of uric acid present in adults depends on diet. Braconids feeding on 10% sucrose remained at the eclosion level. Females fed caterpillars or both sexes fed 5% bovine serum albumen showed an increase during the first week but never achieved the peak found in early pupae. Body burdens of urates are large enough to make their sites of accumulation visible through the abdomen's body wall. The difference between the two species is enough to help explain the longer life-span and high fecundity of *H. serinopae* (see Clark and Smith, 1967b).

Quantitative determinations of xanthine dehydrogenase activity demonstrated interspecies differences in enzyme activity in the same manner as excretory product levels. Again females had higher activity than males. This is a sex difference and not influenced by ploidy, as shown by *H. serinopae* diploid and haploid males possessing the same level of activity.

In an autoradiographic study of the disposition of tracer after a feeding of [^{14}C]hypoxanthine most of the radioactive material was incorporated into the urate cells. Besides material still in the gut the only other tissues acquiring the label were the nurse cells and oocytes. The radioactivity of the eggs laid demonstrated that ingested hypoxanthine was used continuously for increasing the egg cytoplasm (Lin, 1965). Decreasing the food supply caused more radioactive material to be incorporated into eggs. Evidently, urate cells can serve as a reservoir for a component of oocyte growth.

4. *Lipid Nutrition*

The fatty acid composition of nearly all parasitoid wasps is qualitatively the same as those of the individual hosts on which the insects were reared. Quantitatively compositions had varying degrees of differences from those of the hosts. Thirty species from five families of Hymenoptera including *Bracon* and 11 other genera of Braconidae were analyzed for their 12- to 18-carbon fatty acid content (Thompson and Barlow, 1974). Total lipids extracted with chloroform–methanol (2:1), saponified and esterified, provided methyl esters for analysis by gas liquid chromatography and flame ionization. Usually when wasps were shifted to alternate hosts they took on the characteristic fatty acid content of the new host rather than retaining the phenotype conditioned by the standard host. Such adaptability is advantageous when a variety of lepidopteran larvae are present, but no single type exists in abundance.

For subsequent experiments, some wasps were reared on chemically defined diets. Both sets of experiments indicated that polyphagous parasites seem to lack the metabolic regulation over fatty acid levels that exists in most insects. This is advantageous because it enables a parasite to develop on numerous host species. The mere presence of free fatty acids at levels found in hosts has little influence on the parasites survival except under special conditions such as host starvation (Thompson, 1977).

E. Population Genetics

There have been relatively few studies of mass populations of *Bracon*. Supplying large number of hosts repeatedly is part of the problem. Sanitary conditions in a large population cage is another, although occasionally we have used a Wallace-type *Drosophila* cage for one or two generations of massive outcrossing to improve fitness of a laboratory stock. For a 20-generation study of natural selection, Dyson (1965) used a wide-mouth glass jar of cubical shape, 12 cm in each of three dimensions. A fine-meshed wire screen covered the mouth, held in place by a gasket and a perforated metal jar lid. For sanitation a jar change can be made every generation. Benson (1973) used plastic boxes 17.5 × 11.5 × 6.0 cm for his determination of life tables.

Ordinary laboratory glassware was used to obtain an inbred population from a *small wing, white eye* mutant stock by mating the same male to daughters from subsequent generations (Scossiroli and von Borstel, 1962). The life of the male was prolonged by storage in a refrigerator between matings. The mating scheme should reduce heterozygosity by one-half each generation. At the third generation a brother of the females used was substituted for the original male. After the sixth generation selection was started in two lines, using only those wasps beyond 1.47 standard deviations from the mean for the main wing length. Although isogenicity of the population was considered to approach 97%, selection increased the ratio of the main wing length to the total length of head plus thorax.

Two working hypotheses were proposed: (1) Inbreeding is not effective in producing isogenic stocks and there may have been undetected selection for heterozygosity. (2) Inbreeding is effective but gene mutability may be fairly high in *Habrobracon*.

Selection under identical conditions was repeated in a second experiment, and again long- and short-winged substrains were produced (Scossiroli and von Borstel, 1963). However, when the backcrossing was extended for 12 generations the attempt to select wing length substrains from three replicates failed. Therefore, homozygosity can be obtained for quantitative characters effecting wing length. In other words the hypothesis of high spontaneous mutation rate was invalidated.

Crozier (1977) maintained that hymenopteran population theory is sex-linked theory because of the deviations expected to result from the absence of interactions between autosomal and sex-linked loci in the haploid males. Also, oscillations in gene frequency occur in populations with discrete nonoverlapping generations. Crozier cited Dyson's (1965) report of fluctuations in the frequencies of two genes in *B. hebetor*. Dyson (1965) used three stocks: no. 76, an inbred wild type; honey body color (*ho*); and orange eyes (*o*).

An initial reciprocal cross made between the honey and orange stocks produced *ho*/+ *o*/+ virgin females which provided four types of haploid male sons. The wild-type males were discarded and the *honey, orange,* and *honey–orange* males were mated to wild-type stock 76 virgin females. These matings of one female by one male took place in individual vials. Virgin daughters from the crosses were mated as follows:

$$ho/+ \ +/+ \times ho\ + \qquad +/+ \ o/+ \times +\ o \qquad ho/+ \ o/+ \times ho\ o$$

After mating these wasps were transferred to jars to serve as population zero. Two replicates were set for each of the three crosses. Each jar received 150 females and 200–250 males. Host caterpillars were spread on wide-mesh nylon nets layered at the bottom of the jars. Parent wasps were cleared from the jars by suction after 3 days. Upon emergence, offspring were counted according to phenotype and a random sample of mated females was transferred to a fresh jar for the next generation. These procedures assured nonoverlapping generations.

The data from 20 generations show the mutant *honey* allele at a selective disadvantage in both *honey* and *honey–orange* populations. Average adaptive values showed a greater selective differential in *honey* versus wild-type females than in males. Male mating propensity was decreased, female fecundity was lower, and fewer adult progeny were obtained. In contrast, the proportion of *orange* alleles showed little change over 20 generations and none of the fitness tests differed consistently from wild type. To further investigate this, wasps from the fourteenth generation of orange populations were used to initiate new populations with high (0.9) and low (0.36) proportions of *o*. If true selective equilibrium occurred the proportion of *orange* alleles would converge toward the equilibrium value. Instead, the frequencies of the *orange* alleles fell in both populations. Dyson (1965) suggested an influence of the genetic background of chromosomal areas close to the *orange* locus. Adaptive values showed no heterozygote superiority. If anything they were slightly inferior to homozygotes. Also, no gene interaction was demonstrable.

Benson (1973) studied intraspecific competition in the population dynamics of *B. hebetor* using *Cadra cautella,* a tropical warehouse moth, as the host. Key factor analysis identified the searching ability of the adult female as a most important feature in a continuous host–parasite interaction. A mathematical model was presented which predicts the distribution of eggs by the parasite. The model considers the different numbers of eggs laid at each attack for different proportions of the total number of attacks. Egg sterility was small and not related to the number of eggs per host. However, there is a direct density-dependent mortality factor in that eggs and larvae can be killed by older larvae feeding on the same host. The larvae of later instars can die from a scramble competition reaction to high larval density. Premature movement from the host can result in larvae which succeed in spinning a cocoon and enter metamorphosis but fail to emerge as adults. Benson (1973) observed differential mortality acting more strongly against females to alter sex ratio. As indicated elsewhere in this review, this is not the only type of alteration of sex ratios.

IV. Experiments with Radiations

A. Effects on the Organism

In 1919 Davey discovered an apparent prolongation of life form exposing *Tribolium* beetles to X rays. Subsequently, Cork (1957) sought and found a favorable dosage for this type of *Tribolium* response after Sullivan and Grosch (1953) discovered that massive doses of X rays prolonged the life of adult braconids. However, the data for wasps were obtained under starvation conditions, and even if food is available irradiated adults may not feed (Grosch, 1956a). Evidently, there is a decreased expenditure of energy by insects made lethargic by irradiation, and they live off their stored reserves. As a result females that can resorb their oocytes in addition to their fat body live longer than the n or $2n$ males that possess only fat body.

The 100% acute lethal dose for wasps is about 250 kR. At doses between 6 and 150 kR adult life-span decreases in dose-related fashion for feeding adults (Clark, 1961). Above 150 kR wasps do not feed, even on honey water. In addition to these observations, Clark (1961) found that adult life-span is decreased for braconids irradiated as larvae or pupae at doses that allow most of the wasps to survive to maturity. Decreases in adult life-span appeared also in Erdman's (1961) study which was primarily concerned with the destruction of ovaries in developing females. Five kR for larvae and 15 kR for pupae are near the limiting doses for acute exposures. However, irradiating early embryos does not alter the life-span of survivors to adulthood (Clark and Osmun, 1967).

Table 2 shows that radiation-induced decrease in life-span is markedly influenced by genome number rather than sex, in contrast to the normal aging process. A 60-kR exposure to X rays reinforced the evidence that irradiated diploid males live longer than comparable haploid males (Clark *et al.*, 1963). None of the haploids survived 20 days as adults, but diploid males lived nearly 50 days.

A higher degree of polyploidy was provided by triploid *Mormoniella* (a chalcidoid wasp) where diploid males again proved more radiore-

TABLE 2

The Median Life-Span of *Habrobracon serinopae* Adults after Exposure to X Rays[a]

X-Ray dose (R)	Males						Females	
	Haploid				Diploid		Diploid	
	vl		+		+/*vl*		+/*vl*	
	Days	Percentage of controls	Days	Percentage of controls	Days	Percentage of controls	Days	Percentage of controls
0	62	100	62	100	62	100	92	100
10,000	39	63	41	66	52	84	91	99
20,000	29	47	31	50	51	82	87	95
30,000	23	35	23	35	44	71	65	71
50,000	18	29	18	29	37	60	46	50

[a] The diet was honey. From Clark and Rubin (1961).

sistant than haploid males when life-span was determined (Clark and Cole, 1967). Irradiated diploid and triploid females showed similar life-spans whether 10% sucrose or *Sarcophaga* larvae were fed. Again the evidence points to genetic damage, probably induced lethal mutations acting in the haploid males. Again the pattern differed from that of naturally aging adults.

The lack of proliferating cells in adult somatic tissues confers a relative radiotolerance. However, an obvious morphological change with age is an increase in urates accumulated in the braconid abdomen. The rate of accumulation depends upon diet and rearing temperature, but at equivalent culturing conditions X rays increased the rate of urate deposit (Clark and Smith, 1967b) in a dose-related pattern. A steep slope for the urate quantities plotted against age of wasps exposed to 50 kR corresponds with an abrupt drop in the survivorship in the same wasps.

Because of the susceptibility of haploids to radiation-induced life shortening, recessive gene mutations for life-span were sought and found (Baird and Clark, 1971). These were isolated from an *H. serinopae* stock by screening the haploid grandsons of males receiving 4000 rad of X rays. Two mutations attracted particular attention: in days, mean life-span was 22.1 ± 0.3 for *a171*, 27.6 ± 0.8 for *a160*, and 32.4 ± 0.6 for controls. Survivorship curves were shifted forward in time and the mutations differed from each other in time of onset of mortality and the range over which deaths occurred. In addition, unlike the depleted, flattened condition characteristic of control specimens at death, *a171* caused an appreciable proportion of its group to die with full abdomens. Also, *a160* mutants developed a feeble condition with a predisposition to drown in the droplet of sucrose solution provided daily as food. These results demonstrate that the termination of adulthood for haploids can be the consequence of single gene mutations. As such there is an analogy to the phase specificity of lethal action in preadult insect development which gives support to Clark's (1964) "developmental model" for the control of the duration of life.

A different approach to induced genetic load employed the progeny testing of female offspring after fathers were exposed to 3000 R of X rays (Heidenthal *et al.*, 1972). F_1 virgin females were separated into two groups on the basis of the hatchability of their eggs. Low hatchability, 0–69%, was assumed to indicate heterozygotes, and high hatchability, 69–100%, identified homozygotes. Thus identified, records were kept of the number of eggs laid by individuals on the 3 days of early peak production. Their life-spans were also recorded. F_1

individuals that showed high hatchability of their eggs averaged significantly more eggs laid and tended to live longer than the "heterozygotes." In other words there were positive correlations between the components of fitness studied. Instead of egg production wearing out the female, high fecundity indicates a well-functioning organism most likely to experience longevity.

A quantitation of the genetic variability induced in body weights by γ rays was carried out using both sexes (Dalebroux and Kojima, 1967), diploid females, and haploid males. During the preceding decade ionizing radiations were shown to be effective in inducing new genetic variability in quantitative traits of other insects and cultivated crop plants. In *Bracon* an inbred wild type, line 76, and another inbred line homozygous for honey body color was used. Each line was maintained in 15 replicates started from five males and five females.

The experiments were started with two virgin sisters. One was irradiated and the other was a nonexposed control. Their sons were mated to females from the two different strains. Matings to virgins from strain 76 provided a homozygous genetic background, or mating to *honey* virgins resulted in a heterozygous genetic background. An initial phase of the investigation was directed to dose response to a series of 1.0, 1.5, 2.5, and 4.0 kR of ^{60}Co γ rays. A ratio between variances among family means in irradiated and control material showed the maximum response was obtained from 1.5 kR. The second phase was a detailed study using the 1.5-kR dose.

A striking increase of genetic variance was found for induced mutation-heterozygous females in either homozygous or heterozygous genetic background. A considerable increase of genetic variance also occurred among males segregating for induced mutations in an otherwise homozygous line. This refers to males produced by mutation-heterozygote females with the heterozygous genetic background.

The range of doses chosen for the initial experiments was based on the assumption that the radiosensitivity of loci in the polygenic systems would be similar to that of genes responsible for visible mutations. The yield of mutations obtained implies that the choice has feasibility, although the higher F ratio in the second experiment cast some doubt upon the definitive nature of the initial dosage–yield relationship. Nevertheless, it was demonstrated that induced mutations can increase or decrease body weight. Also induced mutations affecting body weight are not necessarily all recessive to the wild-type alleles. Little or no nonallelic interactions occurred between induced mutations and the rest of the genes affecting body weight.

B. Effects on Sperm

Male sterility may be due to (1) absence of sperm, (2) inactivation of sperm, or (3) sperm carrying dominant lethality (Whiting, 1961a). The first condition is improbable for adult wasps in tests with any mutagen because of the large number of successive matings required to exhaust the sperm supply. Fourteen matings by individual males in less than 2 hours were observed in the Whiting laboratory (Whiting, 1961a). Hase (1922) reported 11 successful copulations per hour per male, and up to 30 attempts to copulate in a half-hour. Over a 2-week period male attempts to copulate decreased from multiple attempts down to single attempts per hour. In my laboratory we have noted life-spans decreasing to an average of 1 week when males prefer mating to eating. If no sperm were supplied to females during mating, egg hatchability should be that of virgin females who produce only haploid males.

The inactivation of sperm is more often encountered in experiments with chemical toxicants. In radiation experiments sperm inactivation seems to be demonstrable only at doses which induce dominant lethal events in the majority of sperm.

Induced dominant lethality has been investigated more in sperm than in any other cell type. The fully differentiated tissue of braconid testes at adult emergence presents an extremely uniform populations of cells. The classic braconid papers published before 1961 were summarized in the review by Whiting (1961a). Subsequently it became possible to compare results from different orders of insects. In species of Diptera and Hymenoptera there is a striking similarity of the radiation damage to sperm evident in the dose–hatachability curves plotted over doses up to 8 kR (von Borstel, 1963). Most induced dominant lethals are expressed in the embryo.

One of the important experiments on the flight of Biosatellite II employed 287 males carried in five polypropylene packages designed specifically for them. Multilayered modules carried tiers of wasps positioned in relation to a ^{85}Sr source so that groups would receive four different doses during space flight: 2400, 1200, 700, and 360 R. Earth-based controls included an exact mock-up of the space craft with an identical radiation source. The holders were designed to shield the ^{85}Sr before and after the exposure period. Thus, after the craft was in orbit radiation was delivered only during the extremely low gravity when the living system was removed from the influence of Earth's gravitational field.

We had chosen the most homogeneous of cell types, static mature sperm, for this test. As a result there were no significant differences between the dominant lethals in sperm of the flight and ground-based control males. Furthermore, the differences among the modules at the same package site were small, verifying the essential similarity of reciprocate modules at each exposure. Also, there were no effect of space flight on the radiation-induced recessive lethal mutation frequencies, or on the inherited partial sterility in sperm (von Borstel *et al.*, 1971).

C. Effects on the Oogenetic Series

1. *Hatchability*

Although numerous experiments showed a linear decline in hatchability with an increased exposure of metaphase I ova (Whiting, 1961a), some aspects of the contrast between the metaphase and prophase responses to radiation remained to be studied. Low temperature during a 1.5- to 2-minute X-ray exposure did not significantly alter subsequent egg hatchability. On the other hand, low temperature before and after an exposure decreased the damage from 1263 R (Whiting, 1961b). Presumably low temperature prevented lateral fusion of the broken ends of dyad chromatids, thereby permitting restitution on return to incubator temperatures (30°C). Before irradiation, females were kept without food for 8 hours in the freezing compartment of an ordinary refrigerator (about 5°C). After irradiation, they were stored similarly for 40 minutes. Irradiated females stored 12 hours at 30°C laid first-day eggs with significantly higher hatchability than a simultaneously exposed group allowed to oviposit immediately after irradiation.

A comparison of the mutation rates induced in metaphase and prophase cells was reported in 1963 (Whiting). In braconids, normal haploid males develop from unfertilized eggs and each chromosome has "uncovered" loci like those of a *Drosophila* X. Accordingly, visible mutations are detectable at all points on 10 chromosomes in the F_1 sons of irradiated females. However, a mutation induced in an egg is represented only by the phenotype of a single male. Scoring with more assurance is possible when grandsons are used where a mutation appears in half of the males. This is accomplished by mating treated females to untreated males and setting virgin female offspring on hosts in separate containers.

By this procedure a dose-related yield of visible mutations was obtained for prophase oocytes receiving a series of 11 doses of X rays ranging from 1,100 to 43,260 R. Over this range the proportion of visible mutations increased from 0.63 to 12.00%. A dose of 1,100 R delivered to first metaphase ova induced 5.32% visible mutations. Therefore, a particular dose given to a female yielded many more mutations recovered from exposed metaphases than from prophases. The diakinesis stage is in the same range of radiation sensitivity as the first meiotic metaphase (Smith and Whiting, 1966). This contrast between two oogenetic cell types was further demonstrated in a comparison of induced dominant and recessive lethal mutations. Table 3 shows rates for these along with that of visible mutations.

After most of the investigations of eggs laid during the first few days had been performed, stages earlier and later than first meiotic prophase and metaphase were examined (von Borstel and St. Amand, 1963). When the dominant lethals induced in oogonia were studied, a new type of lethality was discovered which was designated type V (von Borstel, 1961). The embryos died during late embryogeny about the time the embryo's genes must take over mRNA production. This suggests a destruction of gene function, but some depressed hatchability occurs at this same time in eggs from unirradiated females. An enhancement of normal aging processes must also be considered. After females received a 3-day chronic exposure to γ rays, a large number of embryonic deaths occurred in late development, for four total doses (Grosch, 1968). Particularly clear was a correlation of the increase in unhatched eggs with the mother's age at the time of deposit.

The transitional cells situated between the oogonia and differentiating trophocyte–oocyte nests are more radiosensitive than oogonia when judged by their low egg production. However, this might be

TABLE 3
A Comparison of Rates for Three Types of Mutations Induced in First Meiotic Prophase or Metaphase of Oocytes in *Bracon hebetor* Females[a]

	Rate of mutation (%)		
	Prophase		Metaphase
Type of mutation	1,100 R	25,000 R	1,100 R
Dominant lethals	2.06	78.70	85.22
Recessive lethals	0.66	15.00	15.93
Visibles	0.64	7.79	5.32

[a] From Whiting (1963).

spurious because nests of differentiating cells cannot be replaced by repopulation as can occur in a mass of stem cells.

A study of postovipositional eggs employed 500 R, a dose which approximates the 50% hatchability dose for metaphase ova still in the female. Eggs irradiated within a minute after deposit showed exactly the same response outside the females as they did inside sister females (von Borstel and St. Amand, 1963). Therefore, the response is truly one of stage in division and not a conditioning by maternal environment. The 50% lethal dose continues to be about 500 R through anaphase I and II, but for a short time afterward the nucleus becomes very sensitive to radiation; 500 R reduces viability below 10%. Presumably this is the time for replicating a new set of chromosomes. Afterward the pronucleus gradually becomes more radioresistant, until in prophase of the first mitotic division 500 R allows 85% of the exposed eggs to hatch. Again at metaphase of mitosis, cells seem to have sensitivity similar to those in first meiotic metaphase.

As summarized above, the 1 : 20 radiosensitivity ratio of prophase I to metaphase is not restricted to dominant lethality, but also exists for recessive lethals and visible mutations. In contrast, a later section reports experiments with alkylators which shows that the ratio does not hold for recessive lethals and visible mutations.

2. *Eggs Laid*

Radiation alters the pattern of egg production by braconid females. The Feulgen staining of whole mounts renders visible the induced damage to ovarian cell types *in situ.* In turn cell destruction is expressed by decreases or absence of eggs laid. Nuclear pyknosis typically precedes cell dissolution which results in empty spaces or gaps in the single series of oogenetic cells in each ovariole (Grosch, 1974). At eclosion the two ovarioles that comprise each braconid ovary are mature and radiation does not reduce their number. In other orders of insects, where ovaries complete development in the adult, radiation can decrease the ovariole number (Grosch, 1974).

On the basis of severity of effect, radiation dose levels fall into three categories: those casuing (1) permanent infecundity, (2) temporary infecundity, or (3) decreased egg production especially during the period corresponding to (2). Depending upon dose rate and linear energy transfer (LET), permanent infecundity is reached at a dose between 5 and 10 kR when all oogonia are irrevocably damaged. Cell deterioration results first in amorphous debris, followed by its voiding as excreta. Then only a collapsed empty tube remains. After an acute dose, permanent infecundity begins on the seventh day (Fig. 4).

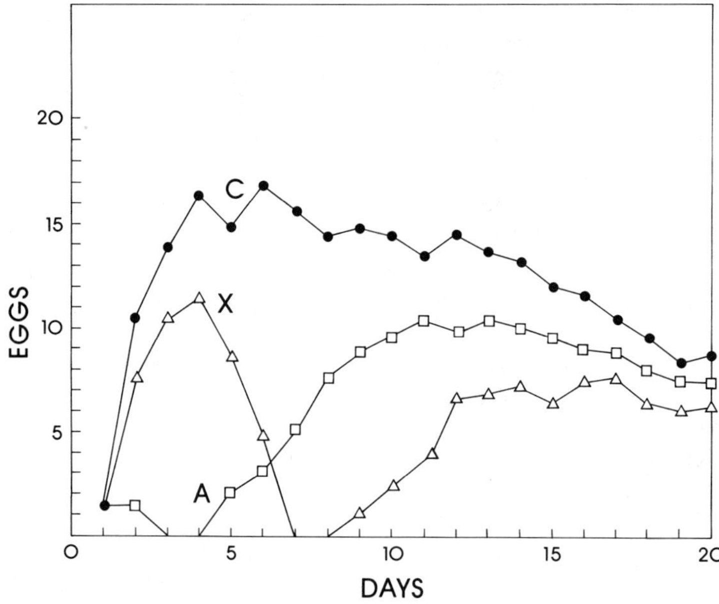

FIG. 4. The contrast between the pattern of daily egg production by females exposed to X rays (X) or fed an antivitellogenic agent (A). Irradiation damages mitotic cells in the ovariole's germarium. Shown in the response to 3000 R; higher doses cause longer periods of infecundity. Control values are represented by C.

The other two categories result from proportionately lower doses which give rise to a family of curves characterized by a trough which broadens and deepens with increased dose. The nadir point typically occurring on day 7 or 8 posttreatment reflects transitional cells in mitotic division (Grosch, 1974). The last oocytes differentiated before the female's exposure to radiation gives rise to eggs laid on the fifth day. When infecundity does not persist the ovarioles are repopulated from stem cells in the blind end of the tubes.

Even after high doses, infecundity does not necessarily persist if the radiation dose is protracted or fractioned. However, a chronic dose of γ rays from ^{85}Sr spread over 2 and 3 days shifted the nadir from the seventh day characteristic of an acute dose to day 8 or 9. Also, for an equivalent total dose the valley was not as broad or deep (Grosch, 1968). In an earlier experiment a low-level feeding of ^{32}P shifted the nadir to day 12 (Grosch and Sullivan, 1954).

More attention was given to experiments using fractioned doses, because altering the conditions during the interfraction period gives information about the recovery potential. Because an acute dose of 5000 R causes permanent infecundity, two 2500 R exposures of X rays were used to determine the effects of increasing the length of the interfraction period. Different doses and dose rate do not alter the numbers of eggs laid during the first 5 days because they are derived from oocytes already differentiated before exposure. Subsequently, the number of eggs laid per day is greatly influenced by the time between fractions. Very few eggs appear after the sixth day if the second exposure occurs in less than 2 hours. Each increase of an hour up to a 6-hour interfraction period raises egg production by about 10%. Above 6 hours and up to 24-hour interfraction periods a plateau is maintained at 60% of control values. Evidently the maximum amount of recovery is obtained within 6 hours (Grosch, 1962b). A 4-hour interfraction period which provides a recovery phase nearly as good as much longer periods was used for most of the tests in a series. These tests showed that the temperature must be above 20°C to promote good recovery. No recovery occurs when the interfraction temperature is below 15°C. Therefore, separated fractions of ionizing radiation are additive only at temperatures low enough to inhibit metabolism. On the other hand temperatures slightly above 30°C can enhance recovery somewhat (Grosch, 1956b). Conceivably, a number of enzyme-controlled steps are influenced in the recovery process. Attempts to interfere by using chemical inhibitors administered via the digestive or respiratory tracts have not been uniformly successful. A feeding of $NiSO_4$, or a juvenile hormone analog, or antibiotics that inhibit protein synthesis can prolong the period of induced fecundity (Grosch, 1975). However, these act by interfering with vitellogenesis rather than enhancing nuclear damage or interfering with DNA repair.

D. Modifying Agents

The chemicals used for the radiosensitization of certain vertebrate tissues have not been considered applicable to insects. Although the anoxic regions of mammalian tumors are especially vulnerable to radiation sensitization by electron-affinic compounds, the tracheal system supplying air to insect tissues is structurally strong and delivers enough air to render cells as sensitive as possible by affinic agents.

A more feasible demonstration of related matters is the comparison of results when braconids are exposed to X rays in the presence or absence of oxygen. Eclosion was decreased for pupae irradiated in an oxygen-filled chamber (Clark and Herr, 1955). The Whiting (1961a) review included initial experiments which established that when adult braconids were irradiated in nitrogen, all types of mutations were reduced both for eggs and in sperm. Also all types were reduced at the same relative frequency. In one test, helium instead of nitrogen sufficed to decrease the effects of X-ray doses. When *Habrobracon* were exposed during postembryonic stages in the presence of air, nitrogen, carbon dioxide, or hydrogen they were more tolerant to radiation when another gas had been substituted for air in the exposure chamber (Clark and Herr, 1955). Because nitrogen seems to be the most inert gas of the series, we used it in X-ray experiments in which fecundity and fertility were the criteria of damage. A nitrogen atmosphere afforded nearly complete protection from an acute dose of 2500 R (Grosch and Clark, 1961). Twice as much radiation had to be used on young virgin females as is used to obtain permanent infecundity in wasps exposed in air. The hatchability of the eggs laid was also improved greatly by using a nitrogen-flushed chamber to hold the mothers during exposure. In this type of experiment high dose rates ensure that the adults lack air for only a few minutes.

In addition to nitrogen protection against radiation damage expressed as mutations, cytoxicity, and arrested development, it has been applied to experiments on radiation's influence upon aging. Adult life-span was decreased less after exposure to ionizing radiation in the presence of nitrogen than in the presence of air (Clark and Smith, 1968). An additional inference is that insect life-span can be modified by radiation in the absence of oxygen.

Vapors, aerosols, and fumigants are equally suitable for exploiting the tracheal route of entry. Triethylamine (TEA), the active ingredient of a commercially available insect anesthetic, gives off a vapor which if present during a radiation exposure decreases egg production below that of radiation alone for every subsequent day from youth to senility. This case features cell destruction in the ovarioles of females. On the other hand, mortality of embryos, which reflects induced genetic lethals, is not increased by the presence of TEA vapors. The decrease in hatchability is similar with or without TEA vapor at the time of exposure (Grosch, 1982). A different response occurred in males given an anesthetizing dose of TEA. Most of them failed to respond to the presence of females, and very few biparental offspring resulted.

Nongaseous chemical protection has featured aminothiol compounds, which provide a universality of action (Bacq and Alexander, 1961). However, until Grosch (1960) studied their influence on the reproductive performance of *Bracon* females, insects were thought to be exceptions. Attempts to demonstrate radioprotection in adult male insects failed because cell division is rare in somatic tissues and the sperm are differentiated static cells.

In females, protection of oogenesis and of egg hatchability was shown for both cysteine and glutathione when either was fed to females before exposure to X rays. The effect was most evident in the eggs laid between 5 and 12 days after treatment. This is, it applied to the cells in transition from oogonia to trophocyte units which provides a population of proliferating cells. Egg production is altered by immediate chromosomal damage while embryonic death in the eggs is due to delayed effects.

A class of compounds which do not afford universality of radioprotective action are the chelating agents. Although effective in prolonging life-span if present in irradiated mice (Bacq and Alexander, 1961), EDTA shortened the life-span of irradiated wasps (LaChance, 1958). Also, in combination with modest doses of radiation EDTA caused more dominant lethals in the zygotes than found in either irradiated or EDTA-fed groups. Subsequently, three other metal-chelating agents in common usage were tested. One, diethyl dithiocarbamate (DIECA), combined with X rays decreased egg production below levels obtained from either used alone (Grosch, 1975). Neither 8-hydroxyquinoline nor *o*-phenanthroline were effective, possibly because of limitations in solubility and palatability.

The chelated removal of divalent cations which are cofactors required for enzyme action would explain the decrease in egg production characterized by a deficiency of eggs from vitellogenic oocytes. Consistent with this explanation are similar patterns of decrease obtained with enzyme inhibitors such as sodium azide, or with toxic metals, or with juvenile hormone analogs. These agents do not increase the action of radiation. Instead their type of damage reduces egg production of the first week, followed by radiation damage to the cells destined to produce the second week's eggs.

A different type of attack upon egg production caused a drop to a nadir point on the thirteenth day after treatment. This resulted when hydroxyurea interfered with recovery from radiation-induced damage by 2500 R of ^{137}Cs γ rays before a second dose of 2500 R was delivered 4 hours later. This seems to be during the S phase of DNA for the five mitotic divisions needed to produce the 32-cell nest in each developing follicle (Grosch, 1971).

The influences of the mechanical forces of rocket launching, orbital flight, and vehicle recovery had to be assessed in the United States Biosatellite experiments. These included acceleration, vibration in three dimensions, reentry deceleration, and buffeting during an aircraft midflight air snatch. However, the simulated hard-environments experiments revealed only a slight decline in eggs produced. Particularly vulnerable were oocytes dependent on their cytoplasmic connections with the cells of the trophocyte mass, a delicate situation that can be disturbed by mechanical stress.

The γ source was unshielded only during the orbital flights. Most surprising was the cancellation of the characteristic pattern of radiation destruction of mitotic cells present in ovarioles between the oogonial and oocyte regions. Furthermore, excepting metaphase I results the only evidence of low hatchability was in third-day eggs derived from cell nests with important cytoplasmic interconnections. Thus despite chronic irradiation, weightlessness contributed to the result of unimpaired cell types which permitted a surprisingly high fecundity, plus the adequacy of these eggs to support embryonic development. Possibly, the inactivity of packaged wasps floating in a state of weightlessness provides an opportunity for cell metabolism to repair presumptive radiation lesions.

In addition to the flight simulation experiments performed at Cape Kennedy and at Ames Research Center, we used vibration tables, centrifuges, clinostats, and sources of γ radiation in a series of simulation experiments. A portable power source assured immediate transitions between sequential treatments (Buchanan, 1970). However, no combinations of treatments performed on earth succeeded in eliminating the vulnerability to radiation damage of the series of cells transitional between oogonia and the follicles containing oocyte–trophocyte units (Grosch, 1972). Characteristically this damage is revealed during the second week after exposure as a trough or valley in the pattern of oviposition per day. Furthermore, the eggs laid during this period ordinarily have poor hatchability. All flight simulation experiments lacked a way to mimic weightlessness for insects and failed to protect against radiation-induced destruction of gametogenetic cells and developing embryos.

The final experiment combined the selection of treatments considered optimum in this sequence:

		Rotate 60 rpm				
Vibrate	Centrifuge	Before	During	After	Centrifuge	Vibrate
10—10,000 Hz	600 rpm at 8 *g*		^{60}Co 2300R		600 rpm at 8 *g*	10—10,000 Hz
1 minute	2.5 minutes	8 minutes	2 days	8 minutes	2.5 minutes	1 minute

In this experiment, a number of groups, 20 wasps each, were treated in the following:

Group 1. Complete sequence. The length of each vibration period was 15 minutes.
Group 2. Complete sequence. The length of each vibration period was 1 minute.
Group 3. No vibration. Treatment otherwise identical to Group 2.
Group 4. No rotation during irradiation. Treatment otherwise identical to Group 2.
Group 5. Eight-minute pause before and after rotation and irradiation. Treatment otherwise identical to Group 2.
Group 6. Control. No radiation. No mechanical forces.
Group 7. Radiation only.
Group 8. One minute vibration—2 days irradiation—1 minute vibration.
Group 9. No rotation during irradiation. Treatment otherwise identical to Group 2.
Group 10. Neither rotation nor pause before starting rotation and irradiation. Treatment otherwise identical to Group 2.
Group 11. Treatment with Group 2 except that females were refrigerated at 5°C for 6 hours after metaphase and prophase classes of eggs were laid.

Mean life-spans equal to or higher than the untreated control value attested to the somatic fitness of all the experimental groups. Oviposition plotted as daily averages per female for each irradiated group provided an intertwined cluster of curves with a characteristic radiation-induced nadir. Also, most of the hatchability data were similar when the comparison was organized on the basis of gametogenic cell types exposed. Although significantly lower than the control values, an exception was the hatchability of eggs from wasps not rotated during irradiation (Group 9). These eggs had hatchability consistently higher than those of other treated groups. A full tabular summary is presented in Grosch (1972).

Data on the stages of death provide details in the patterns of response not met in other types of experiments. Reciprocation between the proportion of stage 1 deaths and those in later embryonic development was observed for the ground controls from the unrecovered first Biosatellite (Grosch, 1968), but a persistent predominance of stage 1 deaths among the progeny of aged wasps seems peculiar to the present type of multistress experiment. In Biosatellite II the highest *proportion* of stage 1 deaths was found for eggs from wasps behind the radiation shield (Grosch, 1970b). Thus flight factors more than radiations were implicated. The predominance of stage 3 rather than stage 4 deaths in data from orbited females differs from results reported in Fig. 5.

Stage 1 deaths are death occurring in embryos during karyokinesis and before blastoderm formation (von Borstel and Rekemeyer, 1959). In stage 3 a variety of morphological changes in the yolk mass occur, while midgut, tracheation, and moderate accumulations of urates are normal aspects of late embryonic development seen in stage 4.

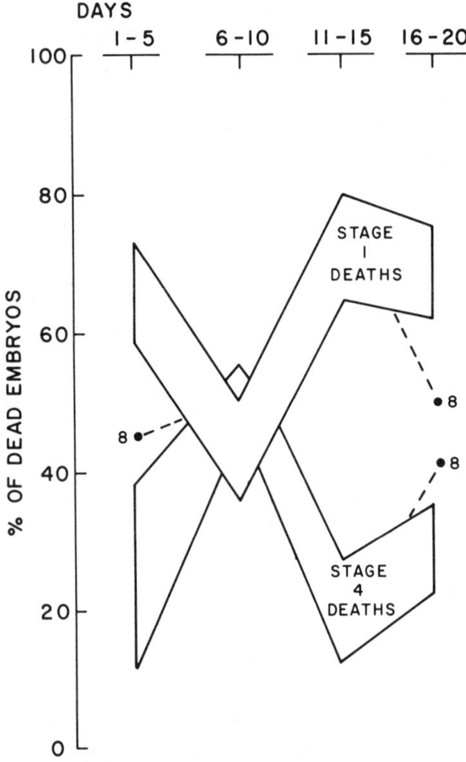

FIG. 5. Outlines of the tight bundles from the composite plot of the mean values from the experimental groups of the simulation experiment performed in the North Carolina State University laboratories. The two major stages considered reveal a reciprocal pattern of embryonic deaths induced by simulated flight stress. Data are organized by successive 5-day periods of oviposition corresponding to youth, adulthood, middle age, and senility of the females. These periods also reflect whether oocytes, transitional cells, or oogonia were the cell type treated.

E. Ingested Radioisotopes

The series of papers on the consequences of feeding ^{32}P to wasps, and a comparison with tests using some other β emitters, ^{89}Sr, ^{35}S, and ^{45}Ca, were covered in the Whiting (1961a) review. Decreased egg production and hatchability were the criteria of damage. Of the three, ^{89}Sr was the most effective and ^{45}Ca the least effective (Grosch *et al.*, 1956).

Subsequently, the distribution and effects on reproduction were studied for three fourth-period cations: ^{58}Co, ^{63}Ni, and ^{65}Zn. In separate single-meal experiments with each isotope the effective half-life was 3 days.

Most of the ^{58}Co burden was abdominal. Even after 20 days 89% of the cobalt still in the wasp was in the abdomen (Grosch, 1965). About half of this was associated with the gut; the rest was in the fat body and the poison apparatus. Most of the cobalt burden was firmly associated with protein tissue components. The only evidence of radiation damage was the low hatchability of the eggs laid on the first day after treatment. Already by the third day hatchability was at control level.

^{63}Ni is another radioisotope with an affinity for the poison apparatus. Autoradiographic studies showed that the unit area burden of radionickel in the poison reservoir exceeded that in all other internal structures. Furthermore, the poison glands pass nickel into the reservoir at a greater rate than excretion by the Malpighian tubules (Grosch *et al.*, 1965). Fat cells had a generally sparse autograph and a low rate of loss. The other major tissues of the abdomen showed similar rates of disposal. Thoracic muscle had an even slower turnover or replacement. Over time the brain showed very slight loss. Only eggs laid the first 2 days were demonstrably radioactive but neither in their hatchability nor in egg deposits did significant alterations in reproductive performance occur, up to the threshold of metal toxicity of somatic tissues exposed to carrier-free ^{63}Ni. However, the use made of wasp toxin suggests a route of cation dispersion through arthropods outside the ordinary pathways of the food web.

In braconids, ^{65}Zn in a single meal decreased daily egg production for the entire 3 weeks of adult life, similar to the toxic effect of nonradioactive zinc. On the other hand, consistently lower hatchability of the eggs occurred due to radiation damage (Grosch, 1962a). An exceptionally low hatchability of the eggs laid on the third day was demonstrated for the most radioactive sample of eggs obtained.

Again as obtained with radiocobalt, nearly all of the ^{65}Zn was in the abdomen after a single feeding. Also, about half of it was associated with the gut and Malpighian tubules. Most of the remainder was carried in the fat body and urate cells. The ovarioles were most radioactive only during the first 3 days, and hatchability was lowest on the third day when egg incorporation of ^{65}Zn peaked.

Wasp tests have been performed with only one ingested transuranic element. The effects of ^{239}Pu on fecundity, fertility, and life-span have been determined after single feedings in citrate-buffered sugar solution (Erdman, 1962). The unpalatability of 0.54 μCi/mg solutions limited intake but these were adequate to cause prompt decline in egg production, permanent infecundity after day 4, and death by day 8. Solutions containing lower concentrations of plutonium were readily eaten and induced a standard pattern of decreased egg production with a nadir point on day 7. A low plateau followed for the rest of the life-span, which was slightly lower than that of the control mean. Decreased hatchability of the eggs also followed a radiation-induced pattern of decrease, despite the short penetration (32 μm) of the α rays from ^{239}Pu. An α particle could reach the egg nucleus from outside the convex side of the egg. Less likely is adequate penetration from outside the ovariole. Isotope inside the egg in proximity to the nucleus would be most effective.

The passage of radioisotopes through food chains is an important concern in radiation ecology. One set of wasp experiments investigated nuclear-bound alkaline earth metals. Host caterpillars were injected orally with several concentrations of ^{45}Ca. Virgin female braconids were placed in the same container and laid eggs on the radioactive hosts. After haploid male larvae hatched they fed for several days on the labeled caterpillars. When adult, the radioactive males were mated to unlabeled females. Fertilized eggs were collected shortly after their deposit and prepared for autoradiography. Subsequent development of the photographic emulsion revealed calcium-labeled sperm nuclei in unlabeled cytoplasm. ^{45}Ca was present in the paternal nucleus both before and during syngamy. Subsequently, the same method was used to demonstrate that the Ca congener ^{90}Sr can be incorporated into and retained by chromatin material (Steffensen and LaChance, 1960). The binding is firm enough to carry through mitotic divisions. The mutational consequences could be due both to the effects of mutation and to transmutation if the latter causes chromosomal breakage. The appreciable amount of strontium in fallout from nuclear explosions lends ecological significance to these results.

F. Ultraviolet Rays

At the molecular site of energy absorption and the lowest level of biological amplification, the results from ultraviolet exposure are expected to differ from those obtained with ionizing radiations. Ultraviolet-sensitive sites are damaged by excitations instead of ionizations. Photoreactivation is a notable example of an effect capable of occurring only after ultraviolet irradiation. Also, merely on the basis of superficial penetration only sites located at the cell periphery are exposed to ultraviolet rays. Those internally are out of range. In a newly laid braconid egg the nucleus is at the convex side. When the nuclear side of the egg is exposed, nuclear inactivation follows an exponential curve reflected in dose–effect plots of hatchability. If only cytoplasm is exposed, hatchability data follow a sigmoidal curve (von Borstel and Wolff, 1955). Chromosome breakage is less extensive after ultraviolet radiation than after ionizing radiation. Survival curves from ultraviolet ray experiments often indicate a higher multiplicity of targets or a higher threshold before the lethal action appears than obtained with ionizing radiation (von Borstel, 1961). In one experiment von Borstel calculated the response to ultraviolet radiation to approximate eight hits.

In addition to the relative radiosensitivity of the egg nucleus a proportion of the nuclear damage is genetically nontransmissible (Löbbecke and von Borstel, 1963). This is caused prezygotically by ultraviolet exposure. If braconid eggs are fertilized before irradiation, the frequency of early deaths is markedly reduced. It cannot be considered a dominant lethal event because the damage is not expressed in the presence of a second nucleus, nor is the damage found by genetic testing of the F_1 females which would reveal deleterious recessive mutations. Unlike cytoplasmic damage, ultraviolet damage responds to photoreactivation. Some type of damage is counteracted by the presence of a second nucleus. Whether this happens before or after syngamy is not known and the defect is either repaired or metabolic support is provided until spontaneous repair can occur.

Amy (1964) extended ultraviolet damage to embryos beyond the early stages. He exposed haploid embryos in 16 developmental stages covering 26 of the 29 hours embryonic period. A wavelength of 2537 Å was used, which is within the maximum range of 2483 to 2753 Å for causing unhatched eggs (Amy and von Borstel, 1957). At each stage tested one group was exposed on the dorsal surface, the other group on the ventral surface. Survival curves were exponential only when the ventral surface was exposed within minutes after the eggs were laid, or within 3 hours during early cleavage if both surfaces were exposed. This indicates a period during which the nature of the damage is considered to be "single hit." At this time a number of newly formed nuclei are produced by mitosis but evidently only one sensitive target needs to be inactivated.

Otherwise, the survival curves were sigmoid for the majority of ages and surfaces exposed, indicating a multihit mechanism. Radiosensitivity decreased as the rate of mitosis declined, but increased again after gastrulation began. During histogenesis and organogenesis the embryo was again radiosensitive, especially so at 23 hours. Despite the low penetration of ultraviolet rays, no demonstrably different sensitive periods were identified for specific vital organs, by dorsal, ventral, or lateral exposures.

V. Experiments with Chemical Agents

A. Altered Egg Production

The braconid provides a diagnostic basis for the cytological analysis of ovarian damage. Typically there are four polytrophic ovarioles, two per ovary, each containing a series of developing units ranging from ova in first meiotic metaphase, through oocytes and accompanying nurse cells, to oogonia. Because of the synchronized egg production by young adults the different effects on specialized, transitional, and primitive stem cells are revealed in the numbers of eggs deposited on particular days when they are collected in sequence.

A seventh-day nadir in the family of curves obtained after exposure to ionizing radiations corresponds to the destruction of mitotic cells producing the oocyte and its clump of nurse cells from a single oogonium (Grosch and Sullivan, 1954). Other patterns of damage to egg production were discovered when adult females were treated with various chemical agents. Most common has been the third-day nadir corresponding with an interference of the vitellogenesis of differentiated oocytes. Initially this type of curve was obtained after a single feeding of the folic acid antagonist methotrexate (Grosch, 1963). Subsequently, we compared the effects of three folic acid antagonists (Mirsalis and Grosch, 1978). Aminopterin was less effective and trimethoprim produced only a slight decline in egg production and hatchability. The expansible crop, which limits the amount ingested at a single feeding, assures quantitative control. Figure 4 shows the different patterns of egg deposit obtained from different agents.

Another agent which interferes with the growth and survival of young oocytes is alanosine, an aspartate analog (Kratsas and Grosch, 1974). The pattern of decreased daily egg production and hatchability indicated that the oocytes actively involved in vitellogenesis are the most vulnerable of the units making up the ovariole sequence. In addition, undersized eggs were laid as the most convincing evidence of disturbed vitellogenesis. Characteristically, the small eggs were nearly normal in width but shortened to about one-third normal length.

Feeding the compound had a more pronounced effect on egg production than injecting a comparable amount of equivalent molarity, but injected alanosine was slightly more toxic as measured by decreased adult life-span. To assure that the moderate starvation preliminary to feeding was not responsible, starved females were also injected. No significant difference was noted for the reproductive performance of injected–starved and injected–unstarved groups of females. Because provisioning the wasp oocytes depends upon accessory cell function and somatic tissue metabolism, the initial action of this inhibitor of adenine synthesis may be somewhat removed from the ultimate site of demonstrable damage. Ovarioles dissected from treated females during the first week revealed deteriorating oocytes and gaps appearing in the oogenetic sequence. Subsequently, ovariole repopulation occurred because the oogonial stem cells proliferated.

Alanosine had no effect on mature sperm as shown by normal numbers of viable biparental offspring in the F_1 and no induced mutations appearing in the F_2 generation.

Alkylating agents have received special attention because they can be used for insect control by chemosterilization. Smith *et al.* (1964) reported that reproductive damage was induced in more than a dozen species from four orders of insects. Our research with *Bracon* added a fifth order, Hymenoptera (Grosch and Valcovic, 1964). Three modes of application to adult females were compared using a hexafunctional aziridine compound, apholate. Contact with a coated surface, topical application, and injection treatments were tested. An eighth-day nadir in egg production resulted from every type of treatment, but injection was the most effective in decreasing the daily egg deposits (Valcovic and Grosch, 1968). This posttreatment pattern in eggs produced is very similar to the effect of ionizing radiation. Mitosis and endomitosis, both nuclear processes leading to the successful differentiation of an oocyte and its associated nurse cells, are vulnerable to DNA alkylation. The differentiated units within follicles and the interphase oogonia are different cell types which are less vulnerable cytologically.

When the monofunctional ethyleneimine (EI) was injected, the overall pattern of oviposition was similar to that obtained with apholate. However, this required 50–100 times the apholate dose. Again, the greatest decline in oviposition was due to damage of the oogenetic cells undergoing mitosis, but in addition the reactivity of EI was evident in damage to differentiated oocytes. Alkyl alkanesulfonates are another class of alkylating agents which decrease braconid egg production to a nadir reached a week after treatment. Ethyl methanesulfonate (EMS) produced the deep valley seen earlier in radiation and apholate experiments. However, with EMS only ingestion and injection were highly effective. Topical application and exposure to an aerosol were inadequate routes of entry (Hoffman and Grosch, 1972). Methyl methanesulfonate was also effective when injected into the braconid abdomen, but myleran (busulfan) could not be used at high enough doses to produce a deep valley. Because of its insolubility in water we had to dissolve it in acetone but evaporation at the tip of the capillary needle at 0.01 *M* stopped up the system.

When fed, the nucleotide analog 5-fluorodeoxyuridine (FUdR) caused early third- or fourth-day lows in eggs laid with or without the presence of thymidine or uridine (Smith, 1962). The most extreme decrease in egg production was obtained from FUdR alone, never approaching control levels after analog ingestion. Added thymidine or uridine in the diet improved egg production after the early nadir. On the other hand, 5-bromodeoxyuridine (BUdR) fed alone lowered daily egg deposits only slightly for the first 10 days after treatment. BUdR fed together with FUdR caused lower daily values than those obtained a meal of FUdR alone.

FUdR inhibits DNA replication in the only mitotically active cells occurring in adult braconids. These give rise to the nests of oocytes and their accompanying nurse cells. Thus at low concentrations FUdR can drastically reduce egg production and alter hatchability without shortening the life-span of mature females. Supplying thymidine caused a significant increase of eggs laid by FUdR-fed females, consistent with the hypothesis that FUdR blocks a step in thymidine synthesis. Uridine may act along the same biosynthetic pathway on the basis that added uridine also alleviated FUdR damage to oogenesis.

An early nadir in braconid egg production can also result from a disturbance in the action of juvenile hormones, although other orders of insects have featured more prominently in investigations into the mode of action of these hormones and their analogs. In Hymenoptera Bracken and Nair (1967) increased the number of eggs matured in female ichneumonids by adding the acyclic terpinoid farnesol methyl ether to chemical diets. Earlier, the quality of the natural diet was shown to influence ichneumonid fecundity. A considerable difference in the number of eggs deposited resulted when an ichneumonid was fed on each of six different species of host caterpillars (Leius, 1962).

On the other hand, two aromatic juvenile hormones applied topically for the control of stable flies did not alter the development or the oviposition of the parasitic wasp *Musoidifurax raptor* (Pteromalidae). Furthermore the wasp eggs were viable and development to F_1 adults occurred (Wright and Spates, 1972). Also, treating a breeding site for stable flies did not interfere with the emergence of *M. raptor* and *Spalangea endrus* adult wasps from fly pupae.

Some molecules related to the juvenile hormones do alter the reproductive performance of wasps. In *Habrobracon* two acyclic terpinoids and two aromatic terpinoid ethers were tested by dissolving each in acetone and applying a droplet topically. Egg production was reduced significantly after treatment of a female with any of the four juvenile hormone analogs (Wissinger and Grosch, 1975). Most effective was a compound carying a 3,4-methylenedioxyphenyl group, prepared originally by W. S. Bowers of Cornell University. This was marketed commercially by Hoffmann-LaRoche Inc. as RO-20-3600. Its complete formula is 6,7-epoxy-3-methyl-7-ethyl-1[3,4-(methylenedioxy)-phenoxy]-2-octene. RO-20-3600 produced a dose-related family of curves in which egg production was initially parallel to control values, but a decline begun on the third day reached a low value on the fifth day. Except at the lethal threshold dose, an upward trend began on the sixth day posttreatment. The majority of the eggs laid on the third and fourth days were undersized. In ovarioles dissected from treatment females a deterioration of oocyte–trophocytes units was correlated with poor egg production (Wissinger and Grosch, 1975).

B. Altered Egg Size

The most convincing evidence of severely disturbed vitellogenesis is the occurrence of small eggs. These diverge from the standard egg length which in turn alters the shape. Most striking are the eggs obtained from feeding or injecting alanosine (L-2-amino-3-hydroxy-nitrosaminepropionic acid). Although egg diameter is normal, their length is only half that of a normal egg (Kratsas and Grosch, 1974). Other examples are not as extreme. Eggs laid by females fed FUdR were 25% shorter than the controls (Smith, 1962). The aromatic terpinoid analogs of juvenile hormones decreased the length of eggs laid by about 10% in a dose-related frequency. The proportion of small eggs laid, not their size, depended on the concentration in identical drops applied topically to the mother's abdomen (Wissinger and Grosch, 1975). Even smaller discrepancies in egg length were observed when females were fed purine analogs (Cassidy and Grosch, 1973). Probably, slight changes in egg length are common when inhibitors of yolk formation are used in insects but these will go unnoticed unless ocular micrometer measurements are obtained. In the braconid cases studied, all short eggs appeared during the first week of oviposition after a single treatment. They were most frequent during the third day, on which day the eggs laid were derived from the smallest oocytes in the vitellarium at the time of treatment.

C. Altered Hatchability

1. *Treated Oocytes*

The eggs deposited by chemically treated braconid females have escaped cytotoxicity in the ovariole but this does not guarantee the completion of embryogenesis. Failure to hatch may be due to either dominant or recessive lethal mutations. A dominant lethal results in the death of a zygote even though contributed by only one of the parents. Either the egg or the sperm may carry the lethal nuclear damage.

A differential response of braconid meiotic oocytes to chemical mutagens was first reported by Löbbecke and von Borstel (1962). Oocytes in the first meiotic metaphase were much more sensitive to EMS and nitrogen mustard than oocytes in prophase when dominant lethality was the criterion. In this respect too, the action of the chemical mutagens resembled that of ionizing radiation. When induced recessive lethality was the criterion only nitrogen mustard caused more damage to metaphase than to prophase oocytes. With EMS no significant difference was obtained between the hatchability of eggs exposed as either metaphase I or prophase oocytes, and the numbers of unhatched eggs were barely above the control level. In a preliminary route-of-entry experiment performed before his M.S. thesis research, Hoffman (1968) found EMS aerosol treatments relatively ineffective. Accordingly, he used feeding or injection of EMS in his study of posttreatment improvement of hatchability.

In braconids, the oviposition process rather than sperm penetration causes the initiation of embryonic development. Thus normal haploid males are produced parthenogenetically. Storing females without hosts postpones oviposition following exposure to mutagen. However, if the cleavage divisions begin promptly after mutagenic treatment, repair processes have not had time to correct genetic damage. LaChance (1955) demonstrated this with X-rayed females, and Hoffman (1972) used the technique in inhibition studies following EMS treatment. Low temperature, low oxygen concentration, 2,4-dinitrophenol, and reduced glutathione inhibited repair during female storage. Although most of the dead embryos were stage 1 in the von Borstel and Rekemeyer classification (1959), the improvement by delayed oviposition also applied to the other stages.

When injected into the female abdomen, methyl methanesulfonate and myleran proved to be additional alkylating agents which caused more dead embryos when metaphase I oocytes were exposed. Prophase oocytes and oogonia were less sensitive (Kratsas, 1972). Furthermore, monofunctional ethyleneimine and hexafunctional apholate, two aziridine types of alkylating agents, caused more damage to metaphase I than to prophase I oogenetic cells when injected into females (Valcovic and Grosch, 1968; Valcovic, 1972). However, a quantitative difference occurred between the persistence of low hatchability of the eggs derived from oocytes. After apholate treatment of the mothers, egg hatchability increased daily until it approached the control level on the fourth day. In contrast, the hatchability in ethyleneimine experiments remained significantly low beyond the fourth day.

In radiation studies and in tests of alkylating agents, the eggs laid on the first 2 days after treatment were characterized by a large number of deaths of the von Borstel and Rekemeyer (1959) type 1. In contrast, the unique carcinostatic drug *cis*-diamminedichloroplatinum (DDP) induced stage 1 lethals in every class of oocytes from which the entire first week's eggs are derived (Grosch and Segreti, 1983).

Most of the dominant lethals induced by fed FUdR occurred in oocytes. The greatest effect on hatchability was observed on days 2 to 5 posttreatment with near-zero values for day 3 eggs. When FUdR was injected most of the dominant lethals (stage 1 death) occurred in the metaphase I eggs laid the first day, but overall, hatchability in eggs derived from the oocytes was decreased because of deaths occurring in later embryonic development. Improvement nearly to control level hatchability after the fifth day resulted when thymidine was supplied to FUdR-treated females. However, the supplement did not interfere with the extreme drop to a third-day nadir. FUdR had only a slight effect on braconid fecundity and fertility (Smith, 1962).

Also, the oocytes progressing through late prophase into first meiotic metaphase expressed the greatest damage from 8-azaguanine and

6-mercaptopurine (Cassidy and Grosch, 1973). The proportion of embryos failing to complete cleavage (stage 1 death) exceeded control values for the first 11 days in eggs laid by 6-MP-fed females.

Experiments with juvenile hormone analogs demonstrated that aromatic turpenoids lowered egg hatchability, especially during the first week. The Bowers compound marketed commercially as RO-20-3600 caused a particularly drastic decrease in which most of the embryos failed to complete early cleavage. The effectiveness of topical applications of analogs dissolved in acetone is especially notable (Wissinger and Grosch, 1975).

In contrast, aflatoxin B_1 was ineffective in topical treatment. Either injection or feeding provided routes of entry which decreased braconid hatchability. Consistent with the necessity for metabolic activation to occur, ingested aflatoxin far exceeds injected equivalent doses in causing low hatchability (Grosch, 1982). A nitrosation experiment identified another example in which the most effective route of entry is the alimentary canal (Grosch, 1980). Ethylurea and $NaNO_2$ fed simultaneously to adult female wasps caused a decrease in hatchability of the eggs derived from oocytes.

There are a variety of toxic agents which do not act as genetic poisons to insects. Egg hatchability was not lowered significantly for eggs deposited by females fed sodium azide, potassium fluoride, or dinitrophenol (Grosch, 1975). Sulfur dioxide in acute exposure did not induce consistent differences from control values (Petters and Mettus, 1982). Although deleterious to egg production, carbamate herbicides and insecticides have little effect on the hatchability of the reduced number of eggs laid (Grosch, 1982). Topical application of the soluble fraction of diesel particulate emissions decreased adult female survival and the hatchability of eggs which were vitellogenic oocytes at the time of treatment. Egg hatchability was decreased for those representing vitellogenic oocytes at the time of treatment but consistency and dose dependence were lacking. An extensive dominant lethal test gave negative results (Petters *et al.*, 1983).

Another kind of negative result was obtained from topical application of the chlorinated hydrocarbon insecticide heptochlor. For the first 2 weeks after treatment the daily mean number of eggs laid was well below that of the control group. However, an examination of individual performances by treated females revealed normal egg production by half the group and greatly decreased daily deposits by the other half. All of the latter died before the fifteenth day of a 20-day experiment (Grosch, 1970a). In other words, in an aspect which partly reflects somatic tissue functions, the moribund wasps lowered the group mean.

Egg hatchability, which depends upon gene and chromosomal normality, was excellent.

2. Treated Oogonia

As summarized above, a great many of the hatchability studies in *Habrobracon* were performed on eggs derived from the ovariole's vitellarium. Originally the interest was in comparing prophase and metaphase chromosomal vulnerability to clastogens. Subsequently, the importance of the accumulation of cytoplasmic materials into these oocytes has emerged in studies on egg production. Also, a comparison between the differentiated oocyte–trophocyte masses and the eggs developing from the oogonia of the germarium has been carried out.

Alanosine, the agent most impressive in producing small eggs as well as oocyte resorption, also decreased the hatchability of eggs derived from the vitellarium. After a single cropful of alanosine, hatchability decreased for all eggs laid for the subsequent 3 weeks of the female's life. The eggs produced during the wasp's last 10 days come from the stem cells and interphase oogonia in the terminal ends of the ovariole (Kratsas and Grosch, 1974). Furthermore, alanosine is one of the agents which cause lower hatchability when it is fed than when it is injected.

In general it is the agents which attack DNA that act upon the oogonia as well as upon oocytes. A good example is the platinum compound DDP (Grosch and Segreti, 1983). Oogonia are not structured to respond to proteinaceous yolk elements and the sites for synthesizing them are outside the ovariole. Accordingly, not all agents which act on oocytes also decrease hatchability of eggs from oogonia. After ingestion of a cropful of methotrexate, hatchability of every day's deposit of eggs was decreased for the life of the female although lowest values occurred for eggs treated as oocytes. On the other hand, egg hatchability after single meals of aminopterin and trimethoprim, two other folic acid analogs, was not lower than control values for those derived from oogonia. No decrease in hatchability occurred after the eighth day after wasps were fed DNP, NaN_2, KF (Grosch, 1975), and ethylurea plus $NaNO_2$ (Grosch, 1980).

The number of recessive lethals induced by either FUdR and BUdR increased significantly for eggs derived from oogonia at the time of ingestion. In these two treated groups the percentage of eggs with at least one recessive lethal was obviously higher in FUdR-fed females (Smith, 1962). When *Bracon* females were treated topically with juvenile hormone analogs the acyclic terpenoids had no effect on the hatchability of the eggs developing from undifferentiated oogonia, but hatchability was decreased by the aromatic compounds, especially RO-20-3526 (Wissinger and Grosch, 1975).

When egg hatchability is the criterion considered, alkylating agents appear to do most of their damage to differentiated or differentiating oocytes. However, at any given time most of the population of oogonial cells are in interphase and they have time for DNA repair before mitotic division. Even if cell damage is not repaired the possibility of cell selection and the elimination of those which could give rise to imperfect eggs would give a false impression.

D. Male Sterility

Male sterility has been summarized above in Section IV,B. Here are presented results on induced dominant lethals and sperm inactivation with emphasis on the latter. In contrast to radiation effects where sperm inactivation is infrequent at doses below that which induces a dominant lethal in every sperm, sperm inactivation is common in males receiving modest doses of alkylating agents.

It was generally assumed that chemosterilants acted primarily on the sperm nucleus until Grosch and Valcovic (1964) reported sperm inactivation in *Habrobracon* treated with the hexafunctional aziridine compound apholate. After hearing our presentation at the 1964 annual meeting of the Entomological Society of America LaChance tested eight chemosterilants: tretamine, tepa, Ent-50990, Ent-50991, hempa, Ent-51254, metepa, hemel. Although they vary considerably in effectiveness as sterilizers, all produced dominant lethal mutations in sperm (LaChance and Leverich, 1968). Tretamine and metepa produced complete sterility without sperm inactivation. In contrast, tepa and its numerically designated analogs produced significant amounts of sperm inactivation. Tepa was the most effective compound but Ent-51254 was nearly as good. Hemel, a non-alkylating analog of tretamine, did not inactivate sperm. Analog 50991 and analog 51254 differ only in a methyl group but its presence makes 50991 much more toxic, while 51254 is a more effective sterilant at low doses. Tepa proved 100 times more effective than hempa in producing recessive lethals, which is a different matter than the one we are considering.

The LaChance treatments were by tarsal contact with a residual film deposited on the wall of a vial. The milligrams per square millimeter and length of time of exposure were varied to provide a series of doses. In Valcovic's experiments 1 μl was delivered to the ventral surface of each wasp.

With apholate an increase in the number of male progeny indicated some degree of sperm inactivation even at low doses (Valcovic, 1966). Other experiments with the monofunctional relative ethylenimine (EI) demonstrated the induction of expected dominant lethals plus sperm inactivation, although the dose of EI required to give comparable results was 50–100 times that of apholate. The high pH of aqueous solutions of EI made it necessary to use the buffer Tris–HCl to bring the mixture to a pH closer to physiological levels. Nevertheless, the monofunctional EI was much more toxic to the adults than was apholate (Valcovic, 1972).

In another set of experiments, the study on *Bracon* sperm was extended by LeChance to 3 other aziridines and 11 additional compounds. The 14 agents were administered to males by feeding, tarsal contact, and topical application. Only 5 of the compounds were effective sterilants at tolerated doses. Two bifunctional aziridines, Ent-50838 and Ent-51909, produced complete sterility without evidence of sperm inactivation. Two sulfonates, ethyl methanesulfonate and 1,3-propanediol methanesulfonate, and a nitrogen mustard, 2,2′-dichloro-*N*-methyldiethylamine hydrochloride, produced sperm inactivation in addition to dominant lethals in the sperm (LaChance and Leverich, 1969).

E. Postembryonic Mutations

The Whiting review (1961a) has summarized the early mutagenesis research on braconids exposed to nitrogen mustard. In the subsequent decade Smith at the Oak Ridge National Laboratory examined other agents. Now in the 1980s Wissinger at St. Bonaventure University, in collaboration with Petters, is testing a variety of genotoxic chemical agents in a validation of *Bracon* as an assay organism.

Mitomycin C has been injected into males in order to study the pattern of induced dominant and recessive mutations in sperm. By mating on three successive days three broods were obtained from pretested females free of recessive lethal mutations. The dose-related percentage of dominant lethals increased with subsequent matings, and most of the embryonic deaths were in early development. Recessive lethals were also obtained but the numbers observed were not sufficient for a valid comparison between broods. Three temperature-sensitive lethal broods were found in the day 1 brood and two in the day 2 brood (Smith, 1969).

The monoalkylating compounds such as ethyl methanesulfonate (EMS) are the most effective chemical mutagens for inducing point mutations, especially those of the temperature-sensitive type. Ten to 20% of all recessive lethal mutations induced by EMS in sperm of *H. serinopae* were temperature sensitive (Smith, 1971). Many of them were rendered homozygous and maintained in the Oak Ridge Biology Division for several years. As in microorganisms, the nature of these conditional mutations permits the study of gene action during development. Typically, death at a particular stage occurs at the restrictive temperature of 35°C, but survival to adulthood occurs at the permissive temperature of 28°C. One mutant showed dosage compensation in that the haploid was lethal at the restrictive temperature but a low frequency of diploids survived to adulthood. Also, some diploids died at later stages of development than the haploid individuals. There are even mutants which can be maintained as heterozygotes but are lethal at both the restrictive and permissive temperatures.

The time and stage at which the gene product becomes necessary for survival can be determined by shifting developing wasps from one temperature to another. Likewise, the stage at which the gene product is no longer necessary can be determined. Most of the mutants were recessive and showed a wide spectrum of variation in times of death and gene action. These aspects were unique for each mutant examined (Smith, 1971).

Mutations in visible qualitative traits like body colors and appendage shape have occurred in the progeny of wasps exposed to chemical mutagens but not at a rate that provided good data for plotting dose–effect lines. Changes in antennae, legs, and wings have been more common than coloration mutations, but the ease in obtaining temperature-sensitive mutations suggests it would be more profitable to seek biochemical and physiological mutants than the obviously visible but relatively infrequent changes.

VI. Concluding Remarks

As stated above, the genus name has been changed back and forth repeatedly during the last two decades. In genetics research *Habrobracon* and *Bracon* should be considered synonymous, and as such the wasp has featured in most of the major research areas. Investigation of gametogenic radiosensitivity and radiation-induced mutations carried over from earlier years into the Biosatellite program. Subsequently, emphasis has changed to the study of actual and suspected mutagens in which the haploidy of normal male offspring can be exploited. Furthermore, in females, the seriation of oocytes of known meiotic stages in the four ovarioles plus the visible evidence of vitellogenesis made it possible to distinguish between chemosterilants active in both sexes and antivitellogens effective only in females.

Developmental matters omitted from the 1961 review have been covered here. Particularly expeditious was the discovery of a mutation which could routinely provide gynandromorphs. This enabled investigation of the clonal development of individual structures and the determination of a morphogenetic fate map. The study of the pattern of gene action needs to include senescence as well as embryogenesis. In braconids, haploid male and diploid biparental offspring provide experimental evidence of a difference between normal aging and the life-span of irradiated organisms.

Other topics in which research became active in recent years include wasp venom and other biochemical phenotypes. In contrast to other organisms, wasps show a very low allozyme variation. Negative results in other areas have not been included in this review because of space limitation. This includes the decreased egg production accompanied by high hatchability, which demonstrated that DDT and other chlorinated hydrocarbon insecticides are not potent mutagens. Heptachlor results were summarized as an example that induced somatic debility of a fraction of a treated group can alter reproductive performance and life-span.

Only some of the ultrastructural results obtained by electron microscopy are included in the text. The molecular biology of braconid nucleic acids has not yet been explored.

References

Abbott, B. D., and Grosch, D. S., (1984). Developmental anomalies in *Habrobracon hebetor* exposed to volatilized agents. *Ann. Entomol. Soc. Am.* **77,** 597–603.

Amy, R. L. (1961). The embryology of *Habrobracon juglandis* (Ashmead). *J. Morphol.* **109,** 199–217.

Amy, R. L. (1964). Ultraviolet sensitivity in the *Habrobracon* embryo. *J. Exp. Zool.* **155,** 43–56.

Amy, R. L. (1975). Mitochondrial and other ultrastructural changes in the developing *Habrobracon* embryo. *J. Embryol. Exp. Morphol.* **34,** 179–190.

Amy, R. L., and von Borstel, R. C. (1957). The effects of different wavelengths of ultraviolet light on the Habrobracon egg. *Proc. Int. Congr. Photobiol., 2nd, Turin,* 419–422.

Anglas, M. J. (1901). Observations sur les metamorphoses internes de la guepe et l'abeille. *Bull. Sci. Fr. Belg.* **34,** 362–463.

Bacci, G. (1965). "Sex Determination." Pergamon, Oxford.

Bacq, Z. M., and Alexander, P. (1961). Chemical protection against X- and gamma rays. "Fundamentals of Radiobiology," 2nd Ed., pp. 457–483. Pergamon, Oxford.

Baird, M. B., and Clark, A. M. (1971). X-ray induced life-shortening mutations in *Habrobracon:* A genetic approach to senescence and duration of life. *Exp. Gerontol.* **6,** 1–8.

Beard, R. L. (1972). Effectiveness of paralyzing venom and its relation to host discrimination by braconid wasps. *Ann. Entomol. Soc. Am.* **65,** 90–93.

Benson, J. F. (1973). Intraspecific competition in the population dynamics of *Bracon hebetor* Say. *J. Anim. Ecol.* **42,** 105–124.

Bodenstein, D. (1950). The postembryonic development of Drosophila. The nervous system. *In* "Biology of Drosophila" (M. Demerec, ed.), pp. 319-325. Wiley, New York.

Bracken, G. K., and Nair, K. K. (1967). Stimulation of yolk deposition in an ichneumonid parasitoid by feeding synthetic juvenile hormone. *Nature (London)* **216,** 483–484.

Buchanan, P. D. (1970). Genetic response of Habrobracon to radiation in combination with clinostat rotation, vibration and centrifugation. Ph.D. thesis, North Carolina State University, Raleigh, North Carolina.

Cassidy, J. D., and Grosch, D. S. (1973). Quantitative effects of purine analogue ingestion on reproduction of *Habrobracon juglandis. J. Econ. Entomol.* **66,** 319–324.

Cassidy, J. D., and King, R. C. (1972). Ovarian development in *Habrobracon juglandis* (Ashmead). I. The origin and differentiation of the oocyte–nurse cell complex. *Biol. Bull.* **143,** 483–505.

Clark, A. M. (1961). Some effects of X-irradiation on longevity in *Habrobracon* females. *Radiat. Res.* **15,** 515–519.

Clark, A. M. (1963). The influence of diet upon the adult life span of two species of *Bracon. Ann. Entomol. Soc. Am.* **56,** 616–619.

Clark, A. M. (1964). Genetic factors associated with aging. *Adv. Gerontol. Res.* **1,** 207–255.

Clark, A. M., and Cole, K. W. (1967). The effects of ionizing radiation on the longevity of ploidy types in the wasp *Mormoniella vitripennis. Ext. Gerontol.* **2,** 89–95.

Clark, A. M., and Egen, R. C. (1975). Behavior of gynandromorphs of the wasp *Habrobracon juglandis. Dev. Biol.* **45,** 251–259.

Clark, A. M., and Gould, A. B. (1972). Evidence for post-clevage fertilization among mosaics in *Habrobracon juglandis. Genetics* **72,** 63–68.

Clark, A. M., and Herr, E. B., Jr. (1955). The effect of certain gases on the radiosensitivity of *Habrobracon* during development. *Radiat. Res.* **2,** 538–543.

Clark, A. M., and Kidwell, R. N. (1967). Effects of developmental temperature on the adult life span of *Mormoniella vitripennis* males. *Exp. Gerontol.* **2,** 79–84.

Clark, A. M., and Osmun, D. E. (1967). Adult life span in *Habrobracon serinopae* after X-irradiation during early development. *Nature (London)* **214,** 717–718.

Clark, A. M., and Rubin, M. A. (1961). The modification by X-irradiation of the lifespan of haploids and diploids of the wasp, *Habrobracon* sp. *Radiat. Res.* **15,** 244–253.

Clark, A. M., and Smith, R. E. (1967a). Egg production and adult life span in two species of *Bracon. Ann. Entomol. Soc. Am.* **60,** 903–905.

Clark, A. M., and Smith, R. E. (1967b). Urate accumulation and adult life span in two species of *Habrobracon. Exp. Gerontol.* **2,** 217–226.

Clark, A. M., and Smith, R. E. (1968). The modification of adult life span in *Bracon hebetor* by irradiation in nitrogen. *Ann. Entomol. Soc. Am.* **61,** 541–542.

Clark, A. M., Bertrand, H. A., and Smith, R. E. (1963). Life span differences between haploid and diploid males of *Habrobracon serinopae* after exposure as adults to X rays. *Am. Nat.* **97,** 203–208.

Clark, A. M., Gould, A. B., and Potts, M. F. (1968). Mosaicism in *Habrobracon juglandis* associated with ebony locus. *Genetics* **58,** 415–422.

Clark, A. M., Gould, A. B., and Graham, S. F. (1971). Patterns of development among mosaics in *Habrobracon juglandis*. *Dev. Biol.* **25,** 133–148.

Clark, A. M., Petters, R. M., and Bryant, P. J. (1973). Patterns of genetic mosaicism in the antennae and legs of *Habrobracon juglandis*. *Dev. Biol.* **32,** 432–445.

Cork, J. M. (1957). Gamma-radiation and longevity of the flour beetle. *Radiat. Res.* **7,** 551–557.

Crozier, R. H. (1977). Evolutionary genetics of the Hymenoptera. *Annu. Rev. Entomol.* **22,** 263–288.

Dalebroux, M. A., and Kojima, K. (1967). An analysis of radiation induced variation on body weight of *Habrobracon juglandis*. *Genetics* **55,** 315–328.

Davey, K. G. (1965). "Reproduction in the Insects." Oliver and Boyd, London.

Davey, W. P. (1919). Prolongation of life of Tribolium confusum apparently due to small doses of X-rays. *J. Exp. Zool.* **28,** 447–458.

Davidson, E. H. (1968). Panoistic and meroistic insect oogenesis. "Gene Activity in Early Development," pp. 180–184. Academic Press, New York.

Drenth, D. (1974). Susceptibility of different species of insects to an extract of the venom gland of the wasp *Microbracon hebetor* (Say). *Toxicon* **12,** 189–192.

Dyson, J. G. (1965). Natural selection of two mutant genes, honey and orange, in laboratory populations of *Habrobracon juglandis*. Ph.D. thesis, North Carolina State University, Raleigh, North Carolina.

Edson, K. M., and Vinson, S. B. (1979). A comparative morphology of the venom apparatus of female braconids. *Can. Entomol.* **111,** 1013–1034.

Edson, K. M., Barlin, M. R., and Vinson, S. B. (1982). Venom apparatus of braconid wasps: Comparative ultrastructure of reservoir and gland filaments. *Toxicon* **20,** 553–562.

Egen, R. C. (1974). A study of mosaicism in the abdomen of Habrobracon juglandis. Ph.D. thesis, University of Delaware.

El-Sawaf, B. M., and Zohdy, N. Z. M. (1976). Host–parasite relationship. 2. Cholinesterase activity of the larvae of the rice moth *Corcyra cephalonica* parasitized by *Bracon hebetor*. *Entomophaga* **21,** 99–101.

Erdman, H. E. (1961). Analyses of the differential radiosensitivity of developing reproductive tissues in *Habrobracon juglandis* (Ashmead) to ionizing radiation. *J. Radiat. Biol.* **3,** 183–204.

Erdman, H. E. (1962). Effects of ingested Pu^{239} on fecundity, fertility and life span of Habrobracon (Hymenoptera: Braconidae). *Health Phys.* **8,** 635–638.

Farish, D. J. (1972). The evolutionary implications of qualitative variation in the grooming behaviour of the Hymenoptera. *Anim. Behav.* **20,** 662–676.

Fischer, I. (1936). Über den Wachstumsrhythmus des Follikelepithels der Läuse und Federlinge und seine Beziehungen zum Arbeitsrhythmus der Zelle und zur Amitose. *Z. Zellforsch.* **23,** 219–243.

Fisher, R. C. (1961). A study in insect multiparasitism. II. The mechanism and control of competition for possession of the host. *J. Exp. Biol.* **38,** 605–628.

Frost, S. W. (1959). Death feigning. "Insect Life and Natural History," pp. 468–470. Dover, New York.

Geitler, L. (1953). Endomitose und endomitotische Polyploidisierung. *Protoplasmatol. Handb. Protoplasmaforsch.* **6,** 1–86.

Genieys, P. (1925). *Habrobracon brevicornis,* Wesm. *Ann. Entomol. Soc. Am.* **18,** 143–202.

Glover, P. M. (1934). The developmental stages of *Bracon tachardiae*. *Bull. Entomol. Res.* **25,** 521–539.

Green, R. F., Gordh, G., and Hawkins, B. A. (1982). Precise sex ratios in highly inbred parasitic wasps. *Am. Nat.* **120,** 653–665.

Griggs, R. C. (1959). A study of succinic dehydrogenase activity in the wasp Habrobracon juglandis (Ashmead). *Bios* 202–207.

Grosch, D. S. (1947). The importance of antennae in mating reaction of male Habrobracon. *J. Comp. Physiol. Psychol.* **40,** 23–29.

Grosch, D. S. (1948). Experimental studies on the mating reaction of male Habrobracon. *J. Comp. Physiol. Psychol.* **41,** 188–195.

Grosch, D. S. (1949). Histological observations on the metamorphosis of male Habrobracon. *J. Elisha Mitchell Sci. Soc.* **65,** 205–206.

Grosch, D. S. (1950a). Cytological aspects of growth in impaternate (male) larvae of Habrobracon. *J. Morphol.* **86,** 153–176.

Grosch, D. S. (1950b). Starvation studies with the parasitic wasp Habrobracon. *Biol. Bull.* **99,** 65–73.

Grosch, D. S. (1951). Octonucleate and uninucleate structural units: Cytological contrasts in the larval and adult midguts of the parasitic wasp Habrobracon. *J. Elisha Mitchell Sci. Soc.* **67,** 184–185.

Grosch, D. S. (1952). The spinning glands of impaternate (male) Habrobracon larvae: Morphology and cytology. *J. Morphol.* **91,** 221–236.

Grosch, D. S. (1956a). Induced lethargy and the radiation control of insects. *J. Econ. Entomol.* **49,** 629–631.

Grosch, D. S. (1956b). The restoration of egg production after a radiation-induced lapse in Habrobracon. *J. Elisha Mitchell Sci. Soc.* **72,** 198.

Grosch, D. S. (1960). Protective effects on fecundity and fertility from feeding cysteine and glutathione to *Habrobracon* females before X-irradiation. *Radiat. Res.* **12,** 146–154.

Grosch, D. S. (1962a). Distribution of zinc-65 in the wasp, Habrobracon, and its effects on reproduction. *Nature (London)* **195,** 356–358.

Grosch, D. S. (1962b). Entomological aspects of radiation as related to genetics and physiology. *Annu. Rev. Entomol.* **7,** 81–106.

Grosch, D. S. (1963). Insect fecundity and fertility: Chemically induced decrease. *Science* **141,** 732–733.

Grosch, D. S. (1965). Distribution and effective half-life of cobalt-58 in Habrobracon. *Nature (London)* **208,** 906–907.

Grosch, D. S. (1968). Reproductive performance of female braconids compared after (A) brief and (B) protracted exposures to ionizing radiations. *Proc. Symp. Isotopes Radiat. Entomol. IAEA, Vienna,* pp. 377–389.

Grosch, D. S. (1970a). Reproductive performance of a braconid after heptachlor poisoning. *J. Econ. Entomol.* **63,** 1348–1349.

Grosch, D. S. (1970b). Egg production and embryo lethality for Habrobracon from Biosatellite II and associated postflight vibration experiments. *Mutat. Res.* **9,** 91–108.

Grosch, D. S. (1971). The response of the female arthropod's reproductive system to radiation and chemical agents. *Proc. Symp. Steril. Principle Insect Control Eradicat., IAEA, Vienna,* pp. 217–228.

Grosch, D. S. (1972). Final report: The utilization of Habrobracon and Artemia as experimental materials in bioastronautic studies. Contract No. NAS2-6684. Ames Research Center, NASA.

Grosch, D. S. (1974). Environmental aspects: Radiation. *Physiol. Insecta, 2nd Ed.* **2,** 85–126.

Grosch, D. S. (1975). Combined effects of radiation and chemical agents in altering the fecundity and fertility of a braconid wasp. *Proc. Symp. Steril. Principle Insect Control., IAEA, Vienna,* pp. 243–259.

Grosch, D. S. (1981). Alterations in the reproductive performance of Habrobracon females following combined treatments with ethylurea and sodium nitrate or nitrite. *Mutat. Res.* **88,** 179–189.

Grosch, D. S. (1982). Complex approaches to insect sterilization including metabolic activation and radiation adjuvants. *Proc. Symp. Sterile Insect Technique Radiat. Insect Control., IAEA, Vienna,* pp. 309–320.

Grosch, D. S., and Clark, A. M. (1949). Cytological investigations of the gut epithelium in haploids and diploids of Habrobracon. *Biol. Bull.* **97,** 237–238.

Grosch, D. S., and Clark, A. M. (1961). Nitrogen protection of fecundity and fertility in female *Habrobracon* treated with X-rays. *Nature* (*London*) **190,** 546–547.

Grosch, D. S., and Segreti, W. O. (1983). The pattern of damage to the oogenetic series of cells after a single feeding of *cis*-diamminedichloroplatinum to Habrobracon females. *Mutat. Res.* **117,** 153–162.

Grosch, D. S., and Sullivan, R. L. (1954). The quantitative aspects of permanent and temporary sterility induced in female *Habrobracon* by X-rays and β-radiation. *Radiat. Res.* **1,** 294–320.

Grosch, D. S., and Valcovic, L. R. (1964). Topical application of apholate to male braconids. *Bull. Entomol. Soc. Am.* **10,** 163.

Grosch, D. S., Sullivan, R. L., and LaChance, L. E. (1956). The comparative effectiveness of four beta-emitting isotopes fed to Habrobracon females on production and hatchability of eggs. *Radiat. Res.* **5,** 281–289.

Grosch, D. S., Lin, J. C-H., and Smith, R. H. (1965). The distribution of nickel 63 in Habrobracon. *Radiat. Res.* **25,** 194.

Grosch, D. S., Kratsas, R. G., and Petters, R. M. (1977). Variation in *Habrobracon juglandis* ovariole number. I. Ovariole number increase induced by extended cold shock of fourth-instar larvae. *J. Embryol. Exp. Morphol.* **40,** 245–251.

Hagstrum, D. W., and Smittle, B. J. (1977). Host-finding ability of *Bracon hebetor* and its influence upon adult parasite survival and fecundity. *Environ. Entomol.* **6,** 437–439.

Hamilton, W. D. (1967). Extraordinary sex ratios. *Science* **156,** 477–488.

Hase, A. (1923). Beiträge zur Kenntnis des Geschlechtsleben männlicher Schlupfwespen. *Arb. Biol. Reichsanst. Land. Forstwirtsch.* **12,** 339–346.

Heidenthal, G., Nelson, W., and Clark, L. (1972). Fecundity and longevity of F_1 females of Habrobracon from sperm X-rayed with 3000r. *Genetics* **71,** 349–365.

Henschen, W. (1928). Über die Entwicklung der Geschlechtsdrüsen von Habrobracon juglandis Ash. *Z. Morphol. Oekol. Tiere* **13,** 144–178.

Herr, E. G. (1953). Studies on the phosphomonoesterases of haploid and diploid Habrobracon. *Anat. Rec.* **117,** 546–547.

Hochberg, V. B. (1976). Genetic mosaicism of the eye—a study of the small eye mutation in *Habrobracon juglandis.* Ph.D. thesis, University of Delaware.

Hoffman, A. C. (1968). The effects of ethyl methane sulfonate on the fertility, fecundity, and life span of the female parasitic wasp, *Habrobracon juglandis.* M.S. thesis, North Carolina State University, Raleigh, North Carolina.

Hoffman, A. C. (1972). A characterization of repair mechanisms operating subsequent to genetic damage induced by ethyl methanesulfonate and gamma radiation in *Bracon hebetor. Mutat. Res.* **16,** 175–188.

Hoffman, A. C., and Grosch, D. S. (1972). The effects of ethyl methane sulfonate on the fecundity and fertility of *Bracon* (Habrobracon) females. *Pestic. Biochem. Physiol.* **1,** 319–326.

Hotta, Y., and Benzer, S. (1973). Mapping of behavior in *Drosophila* mosaics. *In* "Genetic Mechanisms of Development" (F. H. Ruddle, ed.), pp. 129–167. Academic Press, New York.

Keller, E. C., Jr. (1970). Xanthine dehydrogenase activity in parental and F_1 Drosophila and Habrobracon under conditions of hypogravity. *BioScience* **20,** 1045–1049.

King, P. E. (1962). The structure and action of the spermatheca in *Nasonia vitripennis* (Walker). *R. Entomol. Soc. London Ser. A* **37,** 73–75.

King, R. C., and Cassidy, J. D. (1973). Ovarian development in *Habrobracon juglandis* Ashmead. II. Observations on growth and differentiation of component cells of egg chamber and their bearing upon interpretation of radiosensitivity data from *Habrobracon* and *Drosophila. Int. J. Insect Morphol. Embryol.* **2,** 117–136.

Kratsas, R. G. (1972). Comparison of the effects of alanosine, methyl methane sulfonate, and myleran, three different types of carcinostatic agents, on the reproductive performance of the parasitic wasp *Habrobracon juglandis.* M.S. thesis, North Carolina State University, Raleigh, North Carolina.

Kratsas, R. G., and Grosch, D. S. (1974). Contrasts in cell type sensitivity to alanosine demonstrated by altered patterns of *Bracon hebetor* oviposition, hatchability, and egg morphology. *J. Econ. Entomol.* **67,** 577–583.

LaChance, L. E. (1955). Effects of delayed oviposition on X-ray-induced sterility. *Nucleonics* **13,** 49–50.

LaChance, L. E. (1958). Ingestion of ethylenediaminetetraacetic acid and its effect on life span of irradiated and control *Habrobracon* females. *Nature (London)* **182,** 870–871.

LaChance, L. E., and Leverich, A. P. (1968). Chemosterilant studies on *Bracon* sperm. I. Sperm inactivation and dominant lethal mutations. *Ann. Entomol. Soc. Am.* **61,** 164–173.

LaChance, L. E., and Leverich, A. P. (1969). Chemosterilant studies on *Bracon* sperm. II. Studies of selected compounds for induction of dominant lethal mutations or sperm inactivation. *Ann. Entomol. Soc. Am.* **62,** 790–796.

Leibenguth, F. (1967). Regulation of tryptophan metabolism in the parasitic wasp *Habrobracon juglandis. Experientia* **23,** 1069–1074.

Leibenguth, F. (1970). Concerning non-darkening of mutant *Habrobracon (Bracon hebetor)* eyes as a consequence of a new chromagen-reducing mechanism in insect larvae. *Experientia* **26,** 659–660.

Leibenguth, F. (1971). Zur Pleiotropie des *wh*- und *el*-Locus bei *Habrobracon juglandis. Z. Naturforsch. Sect. B* **26,** 53–60.

Leius, K. (1962). Effects of the body fluids of various host larvae on fecundity of female *Scambus buolianae* (Htg.). *Can. Entomol.* **94,** 1078–1082.

Lester, L. J., and Selander, R. K. (1979). Population genetics of haplodiploid insects. *Genetics* **92,** 1329–1345.

Lin, J. C.-H. (1965). The genetic and physiological diversification of two related species of parasitic wasps, *Habrobracon juglandis* Ashmead and *Habrobracon serinopae* Ramkr. Ph.D. thesis, North Carolina State University, Raleigh, North Carolina.

Lin, J. C.-H., and Grosch, D. S. (1982). The genetic and physiological diversification of two species of braconids, *Habrobracon juglandis* Ashmead and *H. serinopae* Ramkr. *Biol. Bull. Natl. Taiwan Normal Univ.* No. 17, June, 1–5.

Löbbecke, E. A., and von Borstel, R. C. (1962). Mutational response of Habrobracon oocytes in metaphase and prophase to ethyl methanesulfonate and nitrogen mustard. *Genetics* **47,** 853–864.

Löbbecke, E. A., and von Borstel, R. C. (1963). Genetically nontransmissible nuclear damage induced by ultraviolet radiation in the wasp Habrobracon. *Genetics* **48,** 1313–1322.

Manning, A. (1967). Genes and the evolution of insect behavior. *In* "Behavior–Genetic Analysis" (J. Hirsch, ed.), pp. 44–60. McGraw-Hill, New York.
Matthews, R. W. (1974). Biology of Braconidae. *Annu. Rev. Entomol.* **19,** 15–32.
Matthews, R. W. (1975). Courtship and parasitic wasps. *In* "Evolutionary Strategies of Parasitic Insects and Mites" (P. W. Price, ed.), pp. 66–86. Plenum, New York.
Metcalf, R. A., Marlin, J. C., and Whitt, G. S. (1975). Low levels of genetic heterozygosity in Hymenoptera. *Nature (London)* **257,** 792–794.
Mettus, R. V., and Petters, R. M. (1981). Genetic investigation of a behavior in response to mechanical shock in the parasitic wasp, *Bracon hebetor. Genetics* **97** (Suppl.), s72–s73.
Milne, C. P., Jr. (1976). Morphogenetic fate map of prospective adult structures of the honey bee. *Dev. Biol.* **48,** 473–476.
Mirsalis, J. C., and Grosch, D. S. (1978). The effects of three folic acid antagonists on reproduction of *Habrobracon juglandis. Ann. Entomol. Soc. Am.* **71,** 559–563.
Mizianty, T. J. (1967). Succinoxidase activity in two species of *Bracon. Ann. Entomol. Sci. Am.* **60,** 1092–1094.
Murr, L. (1930). Über den Geruchsinn der Mehlmottenschlupfwespe Habrobracon juglandis Ashmead. *Z. Vergleich. Physiol.* **11,** 210–270.
Noirot, C., and Quennedey, A. (1974). Fine structure of insect epidermal glands. *Annu. Rev. Entomol.* **19,** 61–80.
Pamilo, P., Varvio-Aho, S.-L., and Pekkarinen, A. (1978). Low enzyme gene variability in Hymenoptera as a consequence of haplodiploidy. *Hereditas* **88,** 93–99.
Parks, H. B. (1936). Cleavage patterns in Drosophila and mosaic formation. *Ann. Entomol. Soc. Am.* **29,** 350–392.
Petters, R. M. (1977). A morphogenetic fate map constructed from *Habrobracon juglandis* gynandromorphs. *Genetics* **85,** 279–287.
Petters, R. M., and Grosch, D. S. (1976). Increased production of genetic mosaics in *Habrobracon juglandis* by cold shock of newly oviposited eggs. *J. Embryol. Exp. Morphol.* **36,** 127–131.
Petters, R. M., and Grosch, D. S. (1977). Reproductive performance of *Bracon hebetor* females with more or fewer than the normal number of ovarioles. *Ann. Entomol. Soc. Am.* **70,** 577–582.
Petters, R. M., and Mettus, R. V. (1980). Decreased diploid male viability in the parasitic wasp, *Bracon hebetor. J. Hered.* **71,** 353–356.
Petters, R. M., and Mettus, R. V. (1982). Reproductive performance of *Bracon hebetor* females following acute exposure to sulphur dioxide in air. *Environ. Pollut. Ser. A* **27,** 155–163.
Petters, R. M., Grosch, D. S., and Olson, C. S. (1978). A flightless mutation in the wasp *Habrobracon juglandis. J. Hered.* **69,** 113–116.
Petters, R. M., Kendall, M., and Gilchrist, K. (1982). Genetic relatedness and mating time in a laboratory stock of the parasitic wasp, *Bracon hebetor. Genetics* **100** (Suppl.), s54.
Petters, R. M., Mettus, R. V., and Casey, J. N. (1983). Toxic and reproductive effects of the soluble fraction from diesel particulate emissions on the parasitoid wasp, *Bracon hebetor. Environ. Res.* **32,** 37–46.
Petters, R. M., and Stefanelli, J. (1983). Developmental arrest of endoparasitoid wasp larvae (*Nemeritis canescens* Grav.) caused by an ectoparasitoid wasp (*Bracon hebetor* Say) *J. Exp. Zool.* **225,** 459–465.
Piek, T., and Engels, E. (1969). Action of the venom of *Microbracon hebetor* Say on larvae and adults of *Philosamia cynthia* Hübn. *Comp. Biochem. Physiol.* **28,** 603–618.
Postlethwait, J. H., and Schneiderman, H. (1971). A clonal analysis of development in

Drosophila melanogaster. Morphogenesis, determination and growth of wild type antenna. *Dev. Biol.* **24,** 477–519.

Prohammer, L. A., and Wade, M. J. (1981). Geographic and genetic variation in death feigning behavior in the flour beetle, *Tribolium castaneum*. *Behav. Genet.* **11,** 395–401.

Rasch, E. M., Cassidy, J. D., and King, R. C. (1977). Evidence for dosage compensation in parthenogenetic Hymenoptera. *Chromosoma* **59,** 323–340.

Reinert, J. A., and King, E. W. (1971). Action of *Bracon hebetor* Say as a parasite of *Plodia interpunctella* at controlled densities. *Ann. Entomol. Soc. Am.* **64,** 1335–1340.

Rotary, N., and Gerling, D. (1973). The influence of some external factors upon the sex ratio of *Bracon hebetor* Say. *Environ. Entomol.* **2,** 134–138.

Schlüter, J. (1933). Die Entwicklung der Flügel bei der Schlupfwespe Habrobracon juglandis Ash. *Z. Morphol. Oekol. Tiere* **27,** 488–516.

Schmieder, R. G. (1928). Observations on the fat body in Hymenoptera. *J. Morphol. Physiol.* **45,** 121–185.

Scossiroli, R. E., and von Borstel, R. C. (1962). Esperimento di selezione in *Habrobracon* dopo inincrocio. *Atti Assoc. Ital.* **7,** 191–197.

Scossiroli, R. E., and von Borstel, R. C. (1963). Selection experiments after inbreeding in Habrobracon. *Proc. Int. Cong. Genet., 11th, The Hague* **1,** 152.

Seurat, L. G. (1899). Contributions à l'étude des hymenoptères entomophages. *Ann. Sci. Nat. 8e Ser.* **10,** 1–159.

Shenefelt, R. D. (1965). A contribution towards knowledge of the world literature regarding Braconidae. *Beitr. Entomol.* **15,** 243–500.

Shenefelt, R. D. (1969). Storage and retrieval of entomological information as applied to Braconidae. *Bull. Entomol. Soc. Am.* **15,** 246–250.

Shoup, J. R. (1966). The development of pigment granules in the eyes of wild type and mutant *Drosophila melanogaster*. *J. Cell Biol.* **29,** 223–249.

Smith, C. N., LaBreque, G. C., and Borkovec, A. B. (1964). Insect chemosterilants. *Annu. Rev. Entomol.* **9,** 269–284.

Smith, R. H. (1962). The effect of thymidine analogues alone and in combination with gamma radiation upon adult *Habrobracon* females. M.S. thesis, North Carolina State University, Raleigh, North Carolina.

Smith, R. H. (1969). Induction of mutations in Habrobracon sperm with mitomycin C. *Mutat. Res.* **7,** 231–234.

Smith, R. H. (1971). Induced conditional lethal mutations for the control of insect populations. *Proc. Symp. Steril. Principle Insect Control Eradicat. IAEA, Vienna,* pp. 453–465.

Smith, R. H., and Whiting, A. R. (1966). X-radiation sensitivity of Habrobracon oocytes at diakinesis. *Genetics* **54,** 364.

Snyder, T. P. (1974). Lack of allozymic variability in three bee species. *Evolution* **28,** 687–689.

Spanjer, W., Grosu, L., and Piek, T. (1977). Two different paralyzing preparations obtained from a homogenate of the wasp *Microbracon hebetor* (Say). *Toxicon* **15,** 413–421.

Speicher, B. R. (1936). Oögenesis, fertilization and early cleavage in Habrobracon. *J. Morphol.* **59,** 401–421.

Sridhara, S. (1981). Macromolecular changes during insect metamorphosis. *In* "Metamorphosis, a Problem in Developmental Biology" (G. and E. Freiden, eds.), Chap. 6, pp. 177–216. Plenum, New York.

Steffensen, D., and LaChance, L. E. (1960). Radioisotopes and the genetic mechanism: Cytology and genetics of divalent metals in nuclei and chromosomes. *Symp. Radioisot. Biosphere* pp. 132–145.

Stern, C. (1968). "Genetic Mosaics and Other Essays." Harvard Univ. Press, Cambridge, Massachusetts.

Sturtevant, A. H. (1929). The claret mutant type of *Drosophila simulans*. A study of chromosome elimination and of cell lineage. *Z. Wiss. Zool.* **135,** 323–356.

Sullivan, R. L., and Grosch, D. S. (1953). The radiation tolerance of an adult wasp. *Nucleonics* **11,** 21–23.

Thelen, E., and Farish, D. J. (1977). An analysis of the grooming behavior of wild and mutant strains of *Bracon hebetor*. *Behaviour* **62,** 70–102.

Thompson, S. N. (1977). Lipid nutrition during larval development of the parasitic wasp, *Exeristes*. *J. Insect Physiol.* **23,** 579–583.

Thompson, S. N., and Barlow, J. S. (1974). The fatty acid composition of parasitic Hymenoptera and its possible biological significance. *Ann. Entomol. Soc. Am.* **67,** 627–632.

Togashi, I. (1963). A comparative morphology of the poison glands in the adults of Ichneumon-flies. *Kontyu* **31,** 297–304.

Valcovic, L. R. (1966). The effects of apholate, an alkylating agent, on the fecundity, fertility, and life span of the adult wasp, *Habrobracon juglandis*. M.S. thesis, North Carolina State University, Raleigh, North Carolina.

Valcovic, L. R. (1972). Genetic response of adult *Bracon hebetor* Say to alkylation by ethyleneimine. *Mutat. Res.* **15,** 67–75.

Valcovic, L. R., and Grosch, D. S. (1968). Apholate-induced sterility in *Bracon hebetor*. *J. Econ. Entomol.* **61,** 1514–1517.

Van Pelt, G. S. (1969). A study of eye-color mutants of *Habrobracon*. Ph.D. thesis, University of Tennessee, Knoxville, Tennessee.

Visser, B. J., Spanjer, W., de Klonia, H., Piek, T., van der Meer, C., and van der Drift, A. C. M. (1976). Isolation and some biochemical properties of a paralyzing toxin from the venom of the wasp *Microbracon hebetor* Say. *Toxicon* **14,** 357–370.

von Borstel, R. C. (1961). Induction of nuclear damage by ionizing and ultra-violet radiation. *Proc. Int. Congr. Photobiol., 3rd,* pp. 243–250.

von Borstel, R. C. (1963). Effects of radiation on germ cells of insects: Dominant lethals, gamete inactivation, and gonial-cell killing. *Proc. Symp. Radiat. Radioisotopes Appl. Insects Agric. Importance, IAEA, Vienna,* pp. 367–385.

von Borstel, R. C., and Rekemeyer, M. L. (1959). Radiation-induced and genetically contrived dominant lethality in *Habrobracon* and *Drosophila*. *Genetics* **14,** 1053–1074.

von Borstel, R. C., and St. Amand, W. (1963). Stage sensitivity to X-radiation during meiosis and mitosis in the egg of the wasp *Habrobracon*. *In* "Repair from Genetic Radiation Damage and Differential Radiosensitivity in Germ Cells" (F. H. Sobels, ed.), pp. 87–100. Macmillan, New York.

von Borstel, R. C., and Wolff, S. (1955). Photoreactivation experiments on the nucleus and cytoplasm of Habrobracon eggs. *Proc. Natl. Acad. Sci. U.S.A.* **41,** 1004–1009.

von Borstel, R. C., Smith, R. H., Whiting, A. R., and Grosch, D. S. (1971). Mutational and physiologic responses of *Habrobracon* in Biosatellite II. *In* "NASA SP-204 The Experiments of Biosatellite II" (J. F. Saunders, ed.), pp. 17–39. Scientific and Technical Office NASA, Washington, D.C.

Walker, W. F. (1980). Sperm utilization strategies in nonsocial insects. *Am. Nat.* **115,** 780–799.

Waller, J. B. (1964). Bracon venom—a naturally occurring selective insecticide. *Proc. Int. Congr. Entomol. 12th,* pp. 509–511.

Walther, C., and Rathmayer, W. (1974). The effect of *Habrobracon* venom on excitatory neuromuscular transmission in insects. *J. Comp. Physiol.* **89,** 23–38.

Walther, C., and Reinecke, M. (1983). Block of synaptic vesicle exocytosis without block of Ca^{2+}-influx. An ultrastructural analysis of the paralyzing action of *Habrobracon* venom on locust motor nerve terminals. *Neuroscience* **9,** 213–224.

Whiting, A. R. (1961a). Genetics of *Habrobracon. Adv. Genet.* **10,** 295–348.

Whiting, A. R. (1961b). Temperature effects on lethal mutation rates of Habrobracon oöcytes X-irradiated in first meiotic metaphase. *Genetics* **46,** 811–816.

Whiting, A. R. (1963). X-ray induced visible mutations in Habrobracon oocytes. *Genetics* **48,** 491–495.

Whiting, P. W. (1932). Reproductive reactions of sex mosaics of a parasitic wasp Habrobracon juglandis. *J. Comp. Physiol. Psychol.* **14,** 345–363.

Whiting, P. W., Greb, R. J., and Speicher, B. R. (1934). A new type of sex-intergrade. *Biol. Bull.* **66,** 152–165.

Wissinger, W. L., and Grosch, D. S. (1975). Influence of juvenile hormone analogues on reproductive performance in the wasp, *Habrobracon juglandis. J. Insect Physiol.* **21,** 1559–1564.

Wright, J. E., and Spates, G. E. (1972). A new approach in integrated control: Insect juvenile hormone plus a hymenopteran parasite against the stable fly. *Science* **178,** 1292–1293.

Ziegler, I. (1961). Genetic aspects of ommochrome and pterin pigments. *Adv. Genet.* **10,** 349–403.

Zohdy, N. Z. M., Abdu, R. M., and El-Sawaf, B. M. (1976). Host–parasite relationship—alkaline and acid phosphatase distribution in the silk glands of parasitized larvae of the rice moth *Corcyra cephalonica. Entomophaga* **21,** 93–97.

Note Added in Proof

Shortly after submitting this manuscript, Dr. Grosch suffered a stroke and was forced to retire. A former student and colleague, Robert M. Petters, revised and proofread the manuscript. Correspondence may be directed to Dr. Petters, Department of Animal Science, School of Agriculture and Life Sciences, North Carolina State University, Box 7621, Raleigh, North Carolina 27695-7621.

THE VARIABLE MITOCHONDRIAL GENOME OF ASCOMYCETES: ORGANIZATION, MUTATIONAL ALTERATIONS, AND EXPRESSION

Klaus Wolf* and Luigi Del Giudice†

*Institut für Genetik und Mikrobiologie der Universität München, D-8000 Munich 19, Federal Republic of Germany
†Istituto Internazionale di Genetica e Biofisica, Consiglio Delle Ricerche, 80125 Naples, Italy

ADVANCES IN GENETICS, Vol. 25

I. Introduction

Our present knowledge of the mitochondrial (mt) genome is mainly based on two efforts: the complete sequencing of the human mt genome by Anderson *et al.* (1981) and the establishment of a genetic mapping system due to the various mt mutants in *Saccharomyces cerevisiae.* With the availability of molecular techniques, it was possible to analyze and compare a variety of fungal mt genomes. Many new insights, generalizations, and developments were achieved concerning their structure and their role in biogenesis. The mt genome is an ideal experimental system because of its small size, and the increasing number of original articles, review articles, symposia proceedings, and books suggests an inverse proportionality between genome size and amount of published information. Many colloquia and symposia have been devoted to this subject, but only a few recent ones will be mentioned here (Slonimski *et al.,* 1982; Schweyen *et al.,* 1983; Quagliarello *et al.,* 1986). A careful compilation of relevant literature is provided by various contributions over the years in *Progress in Botany.* Numerous reviews have appeared dealing with the genetics and biogenesis of the mt genome, but again only a few recent ones will be listed here (Dujon, 1981; Grivell, 1982; Nagley and Novitski, 1982; Wallace, 1982; Grivell, 1983; Tabak *et al.,* 1983a; Borst *et al.,* 1984; Lukins *et al.,* 1984; Sederoff, 1984; Clark-Walker, 1985; Grossman and Hudspeth, 1986; de Zamaroczy and Bernardi, 1986; Tzagoloff and Myers, 1986; Wolf, 1987a,b; Munz *et al.,* 1988).

The theme of this review is an overview on the mt genomes of ascomycetes, since in contrast to the mt genomes of mammals, there is a tremendous variability among genomes of this fungal group. This article deals with the different principles of genome organization, the structure of genes—especially the mosaic genes—and the replication and transcription of these genomes. Most of our knowledge is derived from studies on *S. cerevisiae,* but due to molecular techniques, progress in characterizing other funal genomes has been rapid.

Despite the small number of genes encoded in the mt genome, no consensus on a genetic nomenclature could be obtained. Thus, the diversity in genome organization among fungi is also reflected by the diversity of nomenclature. The nomenclature, proposed by Kohli (1987) for *Schizosaccharomyces pombe* (*S. pombe*), which is used in this article is simple: genes are designated according to their gene products. Following a three letter code, we shall call the gene encoding subunit 1 of cytochrome *c* oxidase the *cox1* gene and that encoding subunit 6 of ATPase the *atp6* gene. Mt tRNA genes are abbreviated *tmx,* where *t* stands for *t*ransfer, *m* for *m*itochondrial, and *x* for the respective amino acid in the one letter code.

II. Mitochondrial Genome Organization

A. The Economy and Invariable Gene Order in Mammalian Mitochondrial Genomes

The mt genomes of animals (excluding protozoa) have similar sizes in a range between 16 and 19 kilobase pairs (kbp) and also the same gene order. In order to illustrate the extreme variability of mt genomes in ascomycetes (Clark-Walker *et al.,* 1985a), we shall briefly discuss the mt genome of mammals, and as example, the human mt genome. Human, bovine, and mouse mt genomes have been completely sequenced (Anderson *et al.,* 1981, 1982a,b; Bibb *et al.,* 1981). The human mt genome is small (16,569 bp) and characterized by its extreme economy (Attardi, 1981a,b; Borst and Grivell, 1981b). Protein coding sequences, rRNA genes, and tRNA genes are virtually contiguous, and one end of a functional gene is usually overlapping with the beginning of the following one (Montoya *et al.,* 1981). Mammalian mt genomes contain the genes for the large and small rRNA (*rnl* and *rns*) 22 tRNAs, 3 subunits of cytochrome *c* oxidase (*cox1, cox2,* and *cox3*), 2 subunits of ATPase (*atp6* and *atp8*), 7 subunits of the NADH dehydrogenase (formerly *urf1, urf2, urf3, urf4, urf4l, urf5,* and *urf6,* now *nd1*, etc.; Chomyn *et al.,* 1985a,b, 1986). (In order to maintain the three letter code, we propose to abbreviate these genes *ndh1–ndh6*.) All urfs are expressed in mouse mitochondria (Michael *et al.,* 1984). Most genes are encoded by the H strand; some tRNA genes and one urf are located on the L strand (Dawid and Rastl, 1979; Attardi *et al.,* 1980; Bibb *et al.,* 1981). All animal mtDNAs are circular (protists again excluded), and intact molecules are covalently closed (Wallace, 1982; Clayton, 1982). The only variable region in mammalian mtDNA is that of the origin of H strand replication (D loop) (Montoya *et al.,* 1982). From the tightly packed organization of the genome and the overlapping reading frames the animal mt genomes are highly constrained with respect to rearrangements of structural variations. Another reason for the absence of rearrangements of genes in the mammalian genomes is the lack of recombination, whereas extensive recombination takes place between mt genomes of lower eukaryotes (see Sena *et al.,* 1986).

B. The Basis of Diversity in Genome Organization and Gene Order in Ascomycetes

Mt genomes in fungal genera and even among yeasts are highly variable both in size and organization. The following catalog of features may account for the structural divergence among fungal genomes; some of them will be discussed in detail below: linearity or circularity of the genome, intergenic "spacers," different size of the coding parts of a gene, variable number of repetitive sequences, duplication of genome segments, location of homologous genes in the nuclear chromosomes, additional intergenic open reading frames (orfs), optionality of introns, variable gene order.

In ascomycetes, circular mt genomes are the rule. In the genus *Hansenula,* however, *H. mrakii* possesses a linear mt genome (Wesolowski and Fukuhara, 1981), wheras *H. wingei* has a circular mt genome (O'Connor *et al.,* 1975). Similarly, in the genus *Candida, C. utilis* possesses a circular genome (Wills *et al.,* 1985), whereas the genome of *C. rhaqii* is linear (Kovac *et al.,* 1984).

MtDNA could be isolated as circles in reasonable yield from a series of ascomycetes, which are listed in Table 1. Most of the methods of mtDNA preparation, however, have yielded exclusively linear molecules in *S. cerevisiae,* and only a few circular molecules with a circumference of about 25 μm have been observed (Hollenberg *et al.,* 1970; Christiansen and Christiansen, 1976). The mt genome of *S. cerevisiae* is exceptional because of its low GC content (18%) and its clustering of GC- and AT-rich regions. The physicochemical properties of this unusual DNA have been described in various papers, which are reviewed in Dujon (1981).

Based on digestion of the wild-type genome with spleen acid DNase and micrococcal nuclease, Bernardi and co-workers have derived a structural model of the mt genome of *S. cerevisiae* (Bernardi *et al.,* 1972; Prunell and Bernardi, 1974, 1977; Bernardi, 1976; Prunell *et al.,* 1977; Fonty *et al.,* 1978; de Zamaroczy and Bernardi, 1985a). Half of the genome is made up of AT-rich sequences containing less than 5% CG: these sequences are named "spacers." Two to three percent of the genome consist of short sequences with a GC content higher than 50%. Half of these GC-rich sequences contain a *Hae*III restriction site and are called "site clusters," the other half harbors a *Hpa*II site and is named "GC clusters." The rest of the genome exhibits a GC content of approximately 26%: these sequences are supposed to be the genes. Thus, the genome consists of the regular arrangement of these four elements: <GC cluster><site cluster><gene><spacer> and so forth.

By DNA sequencing it turned out that Bernardi's proposed genes were indeed coding sequences. The GC-rich sequences, however (Prunell and Bernardi, 1977; de Zamaroczy and Bernardi, 1986), were not at the beginning of the genes, but rather within AT-rich intergenic or intragenic regions (Cosson and Tzagoloff, 1979; Sor and Fukuhara, 1982a; Farelly *et al.*, 1982). High AT content is common to other mt genomes, but not to such an extent as in *S. cerevisiae*. A possible explanation for the evolution toward AT-rich genomes might be the lack of a mechanism for uracil excision.

The comparison of *S. cerevisiae* and *Torulopsis glabrata* is of interest (Clark-Walker *et al.*, 1980). These two yeasts differ in their genome size by roughly 60 kbp and their genic regions are more than 80% homologous. Thus, most of the size difference is due to the absence of AT-rich spacers (Clark-Walker, 1985; de Zamaroczy and Bernardi, 1987). The genomes may even be variable among *S. cerevisiae* laboratory strains. Morimoto and Rabinowitz (1979a,b) have reported restriction site differences between strains, located in the region of the *atp6* gene. Nagley *et al.* (1981) and Cobon *et al.* (1982) have classified yeast strains according to the presence or absence of a DNA sequence of 1800 bp in an AT-rich spacer downstream of *atp6*.

The variable number of GC-rich clusters and the *ori*/*rep* sequences (see Section VI,A) are also responsible for the differences in genome size in *S. cerevisiae* strains. Many of these regions have been sequenced (Sor and Fukuhara, 1982a): each of these GC-rich regions is flanked by an AT-rich segment. Nearly perfect repeats of GC clusters have been found (Cosson and Tzagoloff, 1979; Sor and Fukuhara, 1982a; Farelly *et al.*, 1982; Bernardi, 1982b; Butow *et al.*, 1982). Since an optional GC cluster is also present in the *rns* gene, Sor and Fukuhara (1982a) have proposed that the constant 3′ end of the GC-rich sequences might play a role in the insertion of this element. It starts with TAGT and ends with AAGGAG.

The mt genome of *Neurospora crassa* (a detailed restriction site map has been established by Taylor and Smolich, 1985) contains 50–100 highly conserved GC-rich palindromic sequences that include two *Pst*I sites (Yin *et al.*, 1981). These palindromes flank most of the mt genes in this organism (Browning and RajBhandary, 1982; Burger *et al.*, 1982; Burger and Werner, 1983; Macino and Morelli, 1983). Based on the analysis of transcripts of the region encoding cytochrome *b* (*cob*), subunit 1 of cytochrome *c* oxidase (*cox1*), and the unassigned reading frame 1 (*urf1* = *ndh1*), Burger *et al.* (1985), Breitenberger *et al.* (1985a,b), and de Vries *et al.* (1985b) have suggested that they do not serve as signals for RNA processing as proposed previously (Yin *et al.*,

TABLE 1
Sizes of Mitochondrial Genomes in Different Ascomycetes

Organisms	Size		References
	kbp	μm	
Aspergillus nidulans	32–33	10.6	Lopez Perez and Turner (1975); Küntzel *et al.* (1982); Turner *et al.* (1982)
A. nidulans var. *echinulatus*	38		Turner *et al.* (1982)
A. nidulans var. *quadrilineatus*	31		Turner *et al.* (1982)
Brettanomyces anomalus CBS 577	57.7	18.15	Clark-Walker and Sriprakash (1982)
Brettanomyces custersii	108		Clark-Walker and Sriprakash (1982)
Candida parapsilosis		11.1	O'Connor *et al.* (1975)
Cochliobolus heterostrophus	115		Garber and Yoder (1984)
Dekkera bruxellensis CBS 74	75		Clark-Walker and Sriprakash (1982)
Dekkera intermedia CBS 4914	63.5		Clark-Walker and Sriprakash (1982)
Hanseniospora vineae CBS 2171	26.7		Clark-Walker and Sriprakash (1982)
Hansenula mrakii	55	17.5	Wesolowski and Fukuhara (1981)
Hansenula petersonii	42		Falcone (1984)
Hansenula wingei	25.5	8.2	O'Connor *et al.* (1975)
Kloeckera africana CBS 277	26.5	8.33	Clark-Walker and Sriprakash (1982)
Kluyveromyces lactis	37	11.4	O'Connor *et al.* (1975)
Neurospora crassa	62	19–20	Küntzel *et al.* (1976); Terpstra *et al.* (1976)

Podospora anserina	95	31	Kück and Esser (1982); Cummings *et al.* (1979a,b)
Pachytichospora transvaalensis CBS 2186	41.4		Clark-Walker and Sriprakash (1982)
Saccharomyces cerevisiae KL14-4A	77.8		Sanders *et al.* (1977)
S. cerevisiae NCYC 74	68	24.7	Sanders *et al.* (1977) Christiansen and Christiansen (1976)
S. cerevisiae (Danish maltese cross)		26.6	Christiansen and Christiansen (1976)
S. carlsbergensis		25.6	Hollenberg *et al.* (1970); Sanders *et al.* (1975a, 1976)
S. exiguus	23.7	7.47	Clark-Walker and Sriprakash (1982)
Saccharomycopsis (*Candida*) *lipolytica*	44		Wesolowski *et al.* (1981a)
S. telluris CBS 2685	34.8		Clark-Walker and Sriprakash (1982)
S. unisporus CBS 398	27.4		Clark-Walker and Sriprakash (1982)
Schizosaccharomyces pombe EF1	17.3		Zimmer *et al.* (1984)
Schizosaccharomyces pombe 50	19.43	6.00	O'Connor *et al.* (1975); Manna *et al.* (1981)
Schizosaccharomyces pombe EF4	22.3		Wolf (1983)
Schizosaccharomyces pombe EF2	23		K. Wolf *et al.* (unpublished)
Torulopsis colliculosa		14.7	Kojo (1976)
Torulopsis glabrata CBS 138	18.9	5.95	Clark-Walker and Sriprakash (1982)
Torulopsis glabrata Phaff 71-91	20.3		Clark-Walker and Sriprakash (1982)

1981; Macino and Morelli, 1983). Such repetitive sequence elements have not been found in other yeasts or filamentous fungi, not even in the closely related *Aspergillus nidulans.*

Another source of size variation is the length polymorphism of genes, which has been reported for the rRNA genes by Grant and Lambowitz (1981) and Küntzel and Köchel (1981).

The longest genomes found among ascomycetes so far are those of *Brettanomyces custersii* with 108 kbp (Clark-Walker, 1985; McArthur and Clark-Walker, 1983) and *Cochliobolus heterostrophus* with 115 kbp (Garber and Yoder, 1984), and the shortest that of *Schizosaccharomyces pombe* strain EF1 with about 17.6 kbp (Zimmer *et al.,* 1984). There is a considerable size variation within the same genus: the mt genome of *Saccharomyces exiguus* is 23.4 kbp and that of *S. cerevisiae* up to 85 kbp, depending on the strain (Borst, 1981; Clark-Walker, 1985; de Zamaroczy and Bernardi, 1985a).

Yeasts are classified according to their ability to produce *rho*$^-$ petites in a petite-positive and a petite-negative group (Bulder, 1964a,b). Based on an examination of various petite-positive and petite-negative yeasts, no correlation could be established between genome size or duplications of sequences and the ability to produce *rho*$^-$ petites. Both *S. cerevisiae* and *T. glabrata* are petite positive, despite their extreme difference in genome length (Clark-Walker *et al.,* 1981a; Clark-Walker, 1985). The mt genome of the yeast *Kloeckera africana* carries a long inverted duplication that contains a part of the *rnl* gene and some tRNA genes (Clark-Walker *et al.,* 1981b). Since at least 1000 bp of the *rnl* gene is missing, the duplication is not functional. Despite this duplication, the genome is stable and the organism produces very low amounts of petites. Two inverted duplications have also been described for *Candida albicans* (Wills *et al.,* 1985), but no correspondence to specific genes has been established so far.

Other cases of duplication have been observed in *A. nidulans.* The first 36 amino acids of the *urfx* (*atp8*) gene product (Grisi *et al.,* 1982) are repeated at the start of a second reading frame, *urfa3,* with only 6 mismatches. Examination of the segment at the 5′ end of the two genes showed that the conserved peptides were part of a 300-bp duplication, which also includes the *tmn* gene and a conserved 116-bp intergenic region (Brown *et al.,* 1983b). The genes *tmc* and *tmn* are present in duplicate, separated by about 8 kbp. These duplications are part of a larger duplications (153 bp in the case of *tmc* and 300 bp for *tmn,* as mentioned before; Brown *et al.,* 1985). Other shorter duplications have been found elsewhere on the *A. nidulans* genome, e.g., a 52-bp segment duplicated in the region between *cox1* and *atp9*, but these segments do not carry genetic information. (For the gene order see Table 3.)

The *tmm* gene and the *tmm*$_f$ gene (formyl methionine) are present in duplicate in *N. crassa* (Yin *et al.*, 1982). The *tmm*$_f$ genes are identical, but only one gene is expressed. The *tmm* genes differ by 1 bp and are both expressed. The two *tmm* segments show no homology at the 5′ flanking segments, but the 3′ adjacent segments are highly homologous, suggesting a recent duplication event. One of the tRNA gene duplications is a part of a 400-bp repeat which flanks the bulk of tRNA genes and both rRNA genes. This duplication includes the N-terminal coding region of *urf2* (*ndh2*). It is interesting, but has not been analyzed in detail, that *S. pombe* possesses three tRNAs with the anticodon CAU (Lang *et al.*, 1983; Merlos-Lange and Wolf, 1986).

Another source of variability among mt genomes is the differential presence of genes (see Table 2). The gene coding for subunit 9 or dicyclohexylcarbodiimide (DCCD) binding protein of the ATPase complex (*atp9*) is present and also translated in *S. cerevisiae* (Tzagoloff *et al.*, 1979) and *S. pombe* (Goffeau *et al.*, 1977), but the homologous gene is located in the nucleus of mammals (Anderson *et al.*, 1981). In *N. crassa, atp9* gene sequences are found in both nuclear and mitochondrial DNA (Macino, 1980), but only the nuclear sequence is transcribed and translated (Van den Boogaart *et al.*, 1982a,b). The mt sequence has been denoted the *mal* (*m*t *A*TPase–*p*roteolipid-like) gene. It is not unique to the commonly used *N. crassa* strain, since four other strains of *N. crassa* and three other *Neurospora* species also contain this sequence (De Vries *et al.*, 1983). A similar situation was found in *A. nidulans* (Brown *et al.*, 1984, 1985); the authors suggested that both versions of this gene might be expressed at different stages of the life cycle.

A further case of differential presence of genes is found for the *vari* gene. In *S. cerevisiae,* the *vari* sequence is transcribed and translated (Hudspeth *et al.*, 1982). The gene is lacking in *A. nidulans* (Brown *et al.*, 1985) and *S. pombe* (Lang and Wolf, 1984), but present and transcribed in *T. glabrata* (Clark-Walker and Sriprakash, 1983a,b; Ainley *et al.*, 1985; Clark-Walker *et al.*, 1985b).

Comparing the physical maps of *Saccharomyces* strains, Sanders *et al.* (1977) found that four insertions between 0.9 and 0.6 kbp could be responsible for the size variations among laboratory strains. Besides these large insertions/deletions, smaller insertions of 25–50 bp were found. As in *S. cerevisiae,* variation in genome size in other ascomycetes implies insertions and deletions. Close relatives of *A. nidulans* (Croft *et al.*, 1979; Earl *et al.*, 1981; Kozlowski *et al.*, 1982; Turner *et al.*, 1982) and *N. crassa* (Collins *et al.*, 1981b) differ considerably in size of their mt genomes. Two variant types of *N. crassa* differ by a 2.1-kbp fragment which is tandemly repeated (Manella *et al.*, 1979a).

TABLE 2
Genes and Unassigned Reading Frames in Man and Various Ascomycetes[a]

Gene	*HeLa*	*Neurospora crassa*	*Saccharomyces cerevisiae*	*Aspergillus nidulans*	*Schizosaccharomyces pombe*	*Torulopsis glabrata*
Protein synthesis						
rnl	+	+ (1)	+ (0–1)	+ (1)	+	+
rns	+	+	+	+	+	+
tmx	+	+	+	+	+	+
var1	–	?	+	–	–	+
tsl	–	?	+	–	–	+ ?
Respiratory chain						
cox1	+	+	+ (5–9)	+ (3)	+ (2–4)	+ (+)
cox2	+	+	+	+	+	+
cox3	+	+	+	+	+	+
cob	+	+ (2)	+ (2–5)	+ (1)	+ (0–1)	+
urf1 (*ndh1*)	+	+ (1)	–	+	–	–
urf2 (*ndh2*)	+	+	–	+	–	–
urf3 (*ndh3*)	+	+	–	+	–	–
urf4 (*ndh4*)	+	+	–	+	–	–
urf4l (*ndh4l*)	+	+ (1)	–	–	–	–

urf5 (*ndh5*)	+	+ (2)	–	+	–	–
urf6 (*urfa1*, *ndh6*)	+	+	–	+	–	–
Oxidative phosphorylation						
atp6	+	+ (2)	+	+	+	+
atp8 (*urfx*, *urfa6l*)	+	+	+	+	+	+
atp9 (*mal*)	–	+	+	+	+	+
Unknown function						
urfa2	–	?	–	+	–	–
urfa3	–	?	–	+	–	–
urfa4	–	?	–	+	–	–
urfn	–	+	–	–	–	–
urfu	–	+	–	–	–	–
urfa	–	?	–	–	+	–
rf1 (*orf1*)	–	?	+	?	–	–
rf2 (*orf2*)	–	?	+	?	–	–
rf3 (*orf3*)	–	?	+	?	–	–
rf4 (*orf4*)	–	?	+	?	–	–
rf5 (*orf5*)	–	?	+	?	–	–

[a] The nomenclature of genes is according to Kohli (1987): *rnl* and *rns*, large and small rRNA; *tmx*, transfer RNA, where *x* represents the respective amino acid in the one letter code; *var1*, protein of the small subunit of the mt ribosome; *tsl*, *t*RNA *s*ynthesis *l*ocus; *cox1*, *cox2*, *cox3*, subunits of cytochrome *c* oxidase; *cob*, cytochrome *b*; *urf*, *u*nassigned *r*eading *f*rame; *ndh*, *N*ADH *deh*ydrogenase; *atp6*, *atp8*, *atp9*, subunits of ATPase; *rf*, *r*eading *f*rame; *orf*, *o*pen *r*eading *f*rame. The assignment of urfs in ascomycetes to *ndh* genes is tentative at present. *urf6* (*ndh6*) is very likely homologous to *urfa1* in *N. crassa* and *A. nidulans*.

Mt genomes of such variant *N. crassa* strains may have insertions up to 8 kbp (Collins *et al.*, 1981b; Collins and Lambowitz, 1983).

Most size variations may be attributed to the presence of intergenic open reading frames (intergenic orfs) and intragenic open or closed reading frames (introns). Any mt genome studied so far contains a varying number of yet unassigned intergenic orfs (urfs). In *S. cerevisiae,* several additional orfs have been identified (see Table 3). Reading frame 1 (*rf1* = *orf1*) is located downstream of *cox2* (Coruzzi *et al.*, 1981). Sequence data by Thalenfeld and Tzagoloff (1980) and Macino and Tzagoloff (1980) suggested additional orfs downstream of *cox3* and *atp6*. The sequence of the segment downstream of *cox3* (*rf2* = *orf2*) has been corrected by Michel (1984). *orf2* is rather closely related to *orf1* and furthermore to the major family of intronic orfs (see below). Due to interruptions by GC clusters, creating a −1 frameshift, the reading frame is not continuous, but nevertheless this gene may be expressed by ribosome slippage during translation. *orf3* is located downstream of *rns* and has been demonstrated to be a functional gene in *Saccharomyces uvarum* (Seraphin *et al.*, 1987). *orf4* (between *atp6* and *ori7*) (see Fig. 1) has been sequenced (Seraphin *et al.*, 1985). It contains four fairly large orfs which overlap within GC-rich sequences. A shift of +1 base was found between each pair of consecutive reading frames. If these shifts were overcome, a 500-amino acid polypeptide could be produced, which also resembles maturases. In the region between *ori7* and *ori2* is located *orf5*, which is transcribed into a 900-base RNA (Colin *et al.*, 1985). The putative gene product has an amino acid composition similar to *var1* and other intergenic orfs.

The mt genome of *A. nidulans* contains at least seven urfs (Brown *et al.*, 1985). Four of these are homologous to the mammalian mt *urf1, urf3, urf4,* and *urf5,* (*ndh1, ndh3, ndh4,* and *ndh5*), which encode subunits of the NADH dehydrogenase in human mitochondria. Each *urf* shows an amino acid sequence homology with the human genes greater than 25%, and in *urf1* and *urf4* extensive similarities in predicted secondary structure of the gene product could be found (Brown *et al.*, 1983a). *urf2* of *A. nidulans* possesses some homology with its human counterpart *nhd2*. *urfa1* could correspond to *ndh6* = *urf6* of HeLa (Netzker *et al.*, 1982). This means that this fungus has equivalents to all *urfs* encoded by the H strand (and possibly the L strand) in human mitochondria. In addition, *A. nidulans* contains another orf, called *urfa3,* which is not homologous with any mt orf. The origin of the *urfa3* reading frame lies in one of the duplicated regions mentioned before. Therefore, the first 36 codons are highly homologous with the start of *atp8* gene. It may be assumed that it encodes another

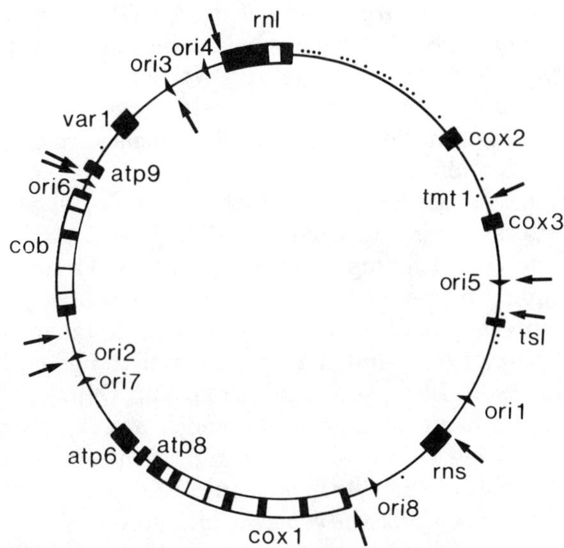

FIG. 1. Map of the mt genome of *S. cerevisiae,* showing the known genes, the origins of replication (*ori*), and their orientation. Black bars indicate exons, and white bars indicate introns. tRNA genes are symbolized by dots. Sites of transcription–initiation are indicated by arrows; the arrow near *ori3* indicates a transcription–initiation site at the complementary strand. *tmt1* is the only tRNA which maps on the complementary strand. For nomenclature of the genes, see footnote *a* in Table 2.

ATPase subunit (Brown *et al.,* 1983b). This is supported by the labeling in cycloheximide-treated cells of a polypeptide with a size expected for this *urf* (Turner *et al.,* 1979). Other possible *urfs* have been deduced from DNA sequencing: the short *urfa2* (Netzker *et al.,* 1982) and *urfa4,* which contains some unusual codons near the 3′ end. Maturase-like sequences like *rf1* and *rf2* in *S. cerevisiae* are not present in *A. nidulans.*

In *N. crassa,* six subunits of NADH-dehydrogenase were shown to be synthesized in the mitochondria (Ise *et al.,* 1985a,b). The corresponding *urfs* (*1, 2, 3, 4l, 5,* and *a1*) are very likely homologous to *ndh1, ndh2, ndh3, ndh4l, ndh5,* and *ndh6* (U. RajBhandary and S. Werner, personal communication). The gene product of *urf1* (*ndh1*) has been identified as a hydrophobic polypeptide of 30 kDa (Zauner *et al.,* 1985). *urf4l* and *urf5* (*ndh4l* and *ndh5*) are located downstream of *cob,* overlap each other, and form a single transcriptional unit. *urf4l* is interrupted by one, *urf5* by two group I introns (Nelson and Macino, 1985, 1987a,b). *urfa1* (*ndh6*?, Breitenberger *et al.,* 1985) is located in the middle of the tRNA cluster between *cox3* and *rnl*. *urfu* and *urfn* are flanking the *cox1* gene (Burger and Werner, 1983). The gene product of *urfn* (Burger *et al.*, 1985) has been shown to be a long polypeptide with highly repetitive structure (Burger and Werner, 1986).

Schizosaccharomyces pombe contains only one *urf* (*urfa*) (Lang *et al.*, 1983), located between the *atp6* and *atp8*, which shows no homology to any known mt sequence.

All ascomycetous genomes in which major parts have been sequenced contain introns (Bernardi, 1978; Borst, 1980) which are lacking completely in mammalian mt genomes. These introns may have open reading frames or not. In *S. cerevisiae*, the *cob* gene of strain D273-10B contains two introns (Nobrega and Tzagoloff, 1980b,c; Dieckmann *et al.*, 1982), and the long form of the gene, five introns (Grivell *et al.*, 1980; Lazowska *et al.*, 1980; Van Ommen *et al.*, 1980). *cox1* is composed of up to nine exons which are separated by eight introns, so the gene covers about 10 kbp (Hensgens *et al.*, 1983a). *Omega*$^+$ strains contain an intron of about 1100 bp in the *rnl* gene (Bos *et al.*, 1978). In *A. nidulans*, the *cob* gene has one and the *cox1* gene two introns (Waring *et al.*, 1981, 1982, 1984; Brown *et al.*, 1985). The *rnl* gene is interrupted by a 1678-bp intron at a position similar to that in *S. cerevisiae* (Brown *et al.*, 1985). An intron in the *rnl* gene has also been found in *N. crassa* (Yin *et al.*, 1982). *cob* and *atp6* are interrupted by two introns each (Helmer-Citterich *et al.*, 1983a,b). The introns in *urf1* (Burger, 1985), *urf4l* and *urf5* (Nelson and Macino, 1985), have already been mentioned. In *Podospora anserina*, the size of the *cox1* gene is 23 kbp (Jamet-Vierny *et al.*, 1984; Kück *et al.*, 1985a,b), indicating the presence of several introns. The first intron has been identified as plasmid-like (pl) DNA, and is the causative agent of senescence in this fungus (Kück *et al.*, 1985b). The *rnl* gene possesses two intervening sequences (Wright and Cummings, 1983c). The gene which is equivalent to the mammalian *nhd1* gene is interrupted by at least three introns (Michel and Cummings, 1985).

In *Hansenula petersonii* there is evidence for an intron in the *rnl* gene and also for introns in *cob* and *cox1* (Falcone, 1984). Despite its small size of only 19.4 kbp, *S. pombe* strain *50* contains two introns in *cox1* and one in *cob* (Lang *et al.*, 1983; Lang, 1984). In *S. pombe* strain EF1, (17.6 kbp), *cox1* contains two additional short introns (Trinkl and Wolf, 1986), but *cob* is continuous (Trinkl *et al.*, 1985). In *T. glabrata* only one transcript each maps to the *rnl, cox2, cox3, cob, atp6,* and *atp9* genes, which shows that these genes are continuous. In contrast, several transcripts map to the *cox1* gene, indicating that this gene is mosaic. This is supported by the finding that intron 4 of *cox1* from

S. cerevisiae hybridizes with mtDNA from *T. glabrata* and also from *K. africana* (Clark-Walker and Sriprakash, 1983b).

The mtDNA size diversity in *Dekkera/Brettanomyces* yeasts may also be due to the presence or absence of introns. There is a graded range from the 28-kbp molecule in *Brettanomyces custersianus* to the 100-kbp genome of *B. custersii. Dekkera intermedia* (72 kbp) and *Dekkera bruxellensis* (82 kbp) differ in size by 10 kbp, but their restriction patterns are very similar (McArthur and Clark-Walker, 1983; Hoeben and Clark-Walker, 1986). Like in the mammalian mt genomes, genes coding for subunits of the same enzyme complex are usually not adjacent. Distances bewteen individual genes are highly variable among ascomycetes. *rnl* and *rns* are adjacent in mammalia and also in *S. pombe,* and are separated by 1–2 kbp in *Kluyveromyces lactis* (Groot and Van Harten-Loosbroek, 1980; Wesolowski *et al.,* 1981a), *K. africana, T. glabrata,* and *H. mrakii,* by 10 kbp in *Saccharomycopsis lipolytica* (Kück *et al.,* 1980; Wesolowski *et al.,* 1981a), and 27 kbp in *S. cerevisiae* (Sanders *et al.,* 1975b). In mammalian mt genomes, one tRNA gene and in *S. pombe* three tRNA genes separate two rRNA genes; in *K. africana* the *cob* gene maps in between those genes.

The gene order is variable among ascomycetes (Clark-Walker and McArthur, 1978; Clark-Walker and Sriprakash, 1982): Some rearrangements of blocks of genes can be seen in Table 3.

Some similarities have been found in the organization of tRNA genes. In all ascomycetes there is some clustering of tRNAs. The location of specific tRNA genes has been compared in *A. nidulans* and *N. crassa* (Köchel *et al.,* 1981). In both fungi, the main tRNA gene clusters are located around the *rnl* gene; in *S. cerevisiae* the main cluster of 19 tRNAs is situated downstream from the *rnl* (Borst and Grivell, 1978). In *N. crassa,* the upstream cluster contains 9 tRNA genes, which are grouped around *urfa1* (*urf6, ndh6*), and 11 tRNA genes are located in the downstream cluster. In *A. nidulans,* the upstream and downstream clusters contain the same number of tRNA genes as in *N. crassa* (Brown *et al.,* 1985). In *A. nidulans,* the order of tRNA genes is *tmt, tme, tmv,* tmm_f (*tmm3*), *tml1, tma, tmf, tml2, tmq, tmm;* in *N. crassa,* the homologous cluster reads *tmt, tme, tmi,* tmm_f *tml1, tma, tmf, tml2, tme, tmh, tmm2.* Thus, there is conservation in most tRNA genes.

In *S. pombe,* there are two main tRNA clusters between *cob* and *atp6* and *atp8* and *atp9;* the other tRNA genes are localized between rRNA- and protein-coding genes.

TABLE 3

Gene Order in Mitochondrial Genomes of Selected Ascomycetes[a]

Ascomycete	Gene order[b,c]																
HeLa	rnl	urf1	urf2	cox1	cox2	atp8	atp6	cox3	urf3	urf4l	urf4	urf5	urf6	cob	rns		
Aspergillus nidulans	rnl	urfa2	cox1	atp9	urf3	cox2	urf5	urf2	urfa4	urfa3	cob	urf1	urf4	atp8	atp6	rns	urfa1 cox3
Neurospora crassa	rnl	urf2	urf3	urf4l	urf5	cob	urfu	cox1	urfn	urf1	atp8	atp6	mal	cox2	rns	cox3	urfa1
Saccharomyces cerevisiae	rnl	cox2	orf1	cox3	orf2	tsl	rns	orf3	cox1	atp8	atp6	orf4	orf5	cob	atp9	var1	
Saccharomyces exiguus	rnl	rns	cox3	cob	cox1		atp6	atp9	cox2								
Torulopsis glabrata	rnl			cob	cox1	atp8	atp6	atp9	cox2	cox3	rns	var1					
Schizosaccharomyces pombe	rnl	rns	cox1	cox3	cob	atp6	urfa	atp8	atp9	cox2							

[a] Derived from references mentioned in the text.

[b] Segments of conserved gene order are indicated by blocks and different lines.

[c] For nomenclature of genes, see footnote *a* in Table 2.

C. Genetic and Physical Mapping of the Mitochondrial Genome in *Saccharomyces cerevisiae*

The discovery that antibacterial antibiotics selectively inhibit mt functions (Clark-Walker and Linnane, 1966; Lamb *et al.,* 1968) allowed the isolation of mutants resistant to chloramphenicol, spiramycin, and erythromycin (Linnane *et al.,* 1968a,b; Thomas and Wilkie, 1968; Bunn *et al.,* 1970; Molloy *et al.,* 1973; Trembath *et al.,* 1973). In the following years mutants conferring resistance to various antibiotics or drugs have been isolated and analyzed. A compilation of mutations, genes, and genetic and physical maps covering the status of mt research up to mid-1977 has been published by Dujon *et al.* (1977). (For a review see also Michaelis and Somlo, 1976; and Borst *et al.,* 1977.) The breakthrough in the genetic analysis of mitochondria came by the discovery of mt point mutations (and deletions of limited size) that do not affect protein synthesis, but are deficient for one (or only few) specific functions (Handwerker *et al.,* 1973; Flury *et al.,* 1974; Storm and Marmur, 1975; Foury and Tzagoloff, 1976b; Coruzzi *et al.,* 1979a,b).

Another group of mutants are the *syn*$^-$ mutants, in which point mutations lead to a deficiency in mt protein synthesis. These mutants map at the *var1* locus, which specifies a protein of the small ribosomal subunit (see Section III,D), within rRNA genes (Kaldma, 1975; Bolotin-Fukuhara *et al.,* 1976, 1977, 1978, 1983; Mason *et al.,* 1976, 1979; Devenish *et al.,* 1978, 1979; Singh *et al.,* 1978; Spithill *et al.,* 1978; Bolotin-Fukuhara, 1979; Sor and Faye, 1979; Contamine and Bolotin-Fukuhara, 1984; Joulou and Bolotin-Fukuhara, 1982; Joulou *et al.,* 1984) or tRNA genes (Faye *et al.,* 1976a; Trembath *et al.,* 1977; Wesolowski and Fukuhara, 1979; Berlani *et al.,* 1980b).

The last class of mutants, *mim,* are characterized by their ability to suppress *mit*$^-$ mutations and restore a wild-type or pseudo-wild-type phenotype (Dujardin *et al.,* 1980b; Kruczewska, 1982; Kruczewska and Slonimski, 1984a,b).

The most extensively used and first discovered type of mutation, the *rho*$^-$ mutation, will be discussed in detail later.

Mainly, the antibiotic resistance mutations have allowed the study of mt transmission genetics (Birky, 1978). Evaluating numerous crosses (reviewed by Dujon, 1981), Dujon and co-workers (Dujon, 1974, 1976; Dujon *et al.,* 1977; Dujon and Slonimski, 1976) have developed a model for mt crosses which is analogous to that proposed for bacteriophage genetics by Visconti and Delbrück (1953). The limitation of recombinational analyses was due to the fact that markers in a distance of more than 1000 bp appear genetically unlinked e.g., show the maximum of 25% recombinants (see also Zinn *et al.*, 1987). For linked markers, values between 0.06 and 0.14% wild-type recombinants for distances around 30 bp (Sebald *et al.*, 1979b; Lazowska *et al.*, 1980) have been reported.

Drug resistance markers have also been applied to the genetic analysis of individual zygotic clones, which has provided insight into segregation, recombination, and uniparental inheritance of mt genomes (reviewed by Birky, 1983). By analysis of retention or loss of mutational sites in *rho*$^-$ clones, a determination of the map order and relative distances of a few mt loci has been achieved (Deutsch *et al.*, 1974; Molloy *et al.*, 1975, 1976; Schweyen *et al.*, 1976a,b, 1978). Physical mapping of the genomes from several wild-type strains and defined *rho*$^-$ mutants by DNA–DNA hybridization (Sriprakash *et al.*, 1976a,b; Linnane *et al.*, 1976; Linnane and Nagley, 1978; for a review see Borst *et al.*, 1979), DNA–RNA hybridization (Morimoto *et al.*, 1978, 1979b), and restriction site mapping (Sanders *et al.*, 1975a, 1976; Choo *et al.*, 1977; Morimoto *et al.*, 1977; Lewin *et al.*, 1978; Morimoto and Rabinowitz, 1979b) led to the construction of a detailed physical and genetic map of the mt genome. The entire genome, unfortunately from different strains, has been sequenced (see de Zamaroczy and Bernardi, 1985, 1986).

D. Mitochondrial Mutants and Maps in the Other Ascomycetes

The first instances of cytoplasmic inheritance in *A. nidulans* were reported at the same time as the beginning of mt genetics in *S. cerevisiae* (Arlett, 1957). *Aspergillus nidulans* and *Aspergillus amstelodami* have been screened for sensitivity to a large number of inhibitors of mt functions, but most of them had little or no influence on growth (for a review see Turner and Rowlands, 1977). Finally, extrachromosomal mutants resistant to oligomycin and chloramphenicol have been isolated (Rowlands and Turner, 1973, 1974a–c, 1975, 1976, 1977; Gunatilleke *et al.*, 1975; Turner and Watson, 1976; Lazarus and Turner, 1977). Waldron and Roberts (1973) isolated a cold-sensitive mutant, in which the mutation can be suppressed by mt suppressors (Waring and Scazzocchio, 1980, 1983). Cold-sensitive mutants have also been characterized as affected in ribosome production (Waldron and Roberts, 1974a,b). A direct screening for respiratory-deficient extranuclear mutants has not been successful, but some cold-sensitive and oligomycin-resistant mutants showed alterations in their cytochrome spectra and cyanide-resistant respiration (Turner and Rowlands, 1976). Three- and four-point crosses have been performed (Rowlands and Turner, 1974a–c, 1975, 1976; Mason and Turner, 1976; Waring and Scazzocchio, 1983).

Progress in characterizing the genome of *A. nidulans* has been made by hybridization of *S. cerevisiae* gene probes with *A. nidulans* mtDNA fragments (Macino *et al.,* 1980). After the mt genome was cloned (Bartik *et al.,* 1979, 1981; Stepien, 1982), most of the genome was sequenced (Küntzel *et al.,* 1976, 1980, 1982; Scazzocchio *et al.,* 1983; Brown *et al.,* 1985).

Restriction maps were also established for some other members of the *Aspergillus* group and of mt hybrids between species (Croft *et al.,* 1979; Earl *et al.,* 1981; Turner *et al.,* 1982).

Mt mutants of *N. crassa* date back to 1952 (Mitchell and Mitchell, 1952), 3 years after the discovery of the *rho*$^-$ mutation in *S. cerevisiae,* and numerous papers have been published since on the *poky* mutant and related mutants. Since most of these mutations are due to deletions or rearrangements of the mt genome, they will be dealt with separately (Section VI,B). Some other extranuclear (presumably mt) mutants will be mentioned here. Chang *et al.* (1969) reported a non-Mendelian pattern of inheritance of UV sensitivity, and Pittenger and West (1979) described an extranuclear ATPase mutant. A mutant with a temperature-sensitive defect of mt protein synthesis and ATPase was described by Collins *et al.* (1981a).

MtDNA from *N. crassa* has been characterized by denaturation mapping and by restriction enzyme analysis (Bernard *et al.,* 1975a–c, 1976; Bernard and Küntzel, 1976, 1978; Terpstra *et al.,* 1976, 1977a–c). Sequence homologies between the mt genomes of *N. crassa* and *S. cerevisiae* (Agsteribbe *et al.,* 1980) permitted a gene map to be constructed by hybridization of *S. cerevisiae* gene probes with restriction fragments of the mt genome of *N. crassa* (Macino, 1980).

The first extranuclear traits described in *P. anserina* were the barrage phenomenon (Rizet, 1952; reviewed by Esser and Raper, 1965; and Esser, 1966) and senescence. Whereas very little is known about the barrage phenomenon, considerable progress has been made in elucidating the molecular mechanism of senescence, which will be dealt with in a separate section (Section VI,D). Cytoplasmic mutants have been isolated (Belcour, 1975): one resistant to chloramphenicol and four developmental mutants. Crosses involving these five mutations have shown that they map at three loci, two of which are closely linked (Belcour and Begel, 1977). Using gene probes from *S. cerevisiae,*

Kück and Esser (1982) and Wright *et al.* (1982b) have constructed a gene map.

In *S. pombe,* various drug-resistant mutants have been isolated and characterized (Lang *et al.,* 1975; Wolf *et al.,* 1976a–d; Burger *et al.,* 1976a,b, 1977; Lang *et al.,* 1976a,b; Seitz *et al.,* 1977a,b; Del Giudice *et al.,* 1977; Burger and Wolf, 1981; Massardo *et al.,* 1982; for a review see Egel *et al.,* 1980 and Wolf, 1987b). Crosses involving these mutants revealed differences in transmission genetics between *S. pombe* and *S. cerevisiae:* segregation of mt genomes is much slower, due to slow cytoplasmic mixing (Seitz-Mayr *et al.,* 1978; Lückemann *et al.,* 1979, 1988). The phenomenon of uniparental inheritance, described for *S. cerevisiae,* has also been observed and analyzed in *S. pombe* (Wolf and Kaudewitz, 1977; Wolf *et al.,* 1979; Thrailkill *et al.,* 1980). A cytoplasmic gene has been described which suppresses antimycin sensitivity, but not respiratory deficiency in a nuclear mutant (Colson *et al.,* 1976; Labaille *et al.,* 1977). A detailed restriction map was established by Anziano *et al.* (1983). Using gene probes from *S. cerevisiae,* the ubiquitous mt genes could be localized in two different *S. pombe* strains (Wolf, 1983; Wolf *et al.,* 1982; Lang *et al.,* 1983; Lang and Wolf, 1984; Zimmer *et al.,* 1984). Due to a cytoplasmic mutator which very likely influences the stability of the mt genome (Seitz-Mayr and Wolf, 1982), mt respiratory-deficient mutants could be isolated (Wolf *et al.,* 1976b) and characterized (Seitz *et al.,* 1977a,b) which carry point mutation and deletions up to 1500 bp (Ahne *et al.,* 1984). These deletions fall in the gene loci *cox1, cox2, cox3,* and *cob.* The entire mt genome has been cloned as one piece (Del Giudice, 1981), as restriction fragments (Lang and Wolf, 1984), and sequenced entirely (Lang *et al.,* 1983, 1985; Lang, 1984; B. F. Lang, unpublished results).

In the petite-negative yeast *K. lactis,* extrachromosomal drug-resistant and respiratory-deficient mutants have been isolated and characterized (Celis *et al.,* 1975; Algeri *et al.,* 1977; Brunner *et al.,* 1977; Del Giudice and Brunner, 1977; Guerrini *et al.,* 1977; Morgan and Whittaker, 1978; Marmiroli and Puglisi, 1980).

In *T. glabrata,* drug-resistant mutants have been used in a study on segregation and transmission of mt genomes (Sriprakash and Batum, 1981).

In several yeasts the approach by heterologous hybridization with DNA probes from *S. cerevisiae* or *N. crassa* has led to a more or less precise gene map: *T. glabrata* (Clark-Walker and Sriprakash, 1981); *S. exiquus* (Clark-Walker *et al.,* 1983); *K. lactis* (Kück *et al.,* 1980; Wesolowski *et al.,* 1981); *K. africana* (Clark-Walker and Sriprakash, 1981); *H. petersonii* (Falcone, 1984); *C. heterostrophus* (Garber and Yoder, 1984).

III. Mitochondrial Gene Organization

A. Ribosomal RNA Genes and Their Introns

Except for the partial duplication of *rnl* in *K. africana,* all ascomycetes contain one copy each of the *rnl* and the *rns* gene (for a review see Borst, 1978). All ascomycete *rns* genes are continuous, whereas some organisms harbor mosaic *rnl* genes. Most studies at the beginning of mt genetics in *S. cerevisiae* have focused on the *rnl* gene locus (illustrated in Fig. 2), since the first antibiotic-resistant mutants map there (for a review see Dujon, 1981). Three linked loci have been described: *rib1, rib2,* and *rib3* (Bolotin *et al.,* 1971; Netter *et al.,* 1974). Chloramphenicol-resistant mutants (*cap*^{r} or C^R) map at *rib1;* mutants resistant to erythromycin (*ery*^{r} or E^R) or spiramycin are located in *rib2* and *rib3*. *In vitro* experiments with isolated ribosomes from mutants have shown that the *rib* loci are part of *rnl* (Grivell *et al.,* 1973). *Ery*^{r} and *cap*^{r} mutants not allelic to those used in the first genetic analyses have been described (Kleese *et al.,* 1972; Knight, 1980; Knight *et al.,* 1982). Physical mapping has allowed the precise location of the *rib* loci (Faye *et al.,* 1974; Dujon and Michel, 1976; Heyting and Sanders, 1976; Jacq *et al.,* 1977; Morimoto *et al.,* 1978; Atchison *et al.,* 1979; Heyting and Meijlink, 1979; Heyting and Menke, 1979; Heyting *et al.,* 1979a,b; Michel *et al.,* 1979). DNA sequencing of *cap*^{r} mutants has revealed that single base changes in the distal region of *rnl* determine resistance to chloramphenicol, which inhibits peptidyl transfer in bacterial and organelle ribosomes (Dujon, 1980). The nucleotide changes occur in two short sequences of 10 bp each, which are highly conserved among mt and bacterial *rnl* genes. The same blocks are also responsible for *cap* resistance in mammals (Kearsey and Craig, 1981; Blanc *et al.,* 1981). (A third site for *cap* resistance has been found in rat by Koike *et al.,* 1983.) RNA secondary structure models have been proposed for bacteria, chloroplasts, and mitochondria (Glotz *et al.,* 1981; Maly and Brimacombe, 1983). The large rRNA of *S. cerevisiae* can also be folded into a secondary structure similar to that of *Escherichia coli* (Sor and Fukuhara, 1982b, 1983a). There is evidence that the RNA is directly involved in peptidyl transferase action, since *cap*^{r} mutations occur within unpaired regions and therfore do not destabilize the secondary structure (Dujon and Jacquier, 1983). Mutation conferring *ery* resistance have also been identified by DNA sequencing. These mutations are A→G transitions and map about 500 bases downstream from the *cap*^{r} mutations (Sor and Fukuhara, 1982b). The *rnl* gene shows a strain-dependent polymorphism: in *omega*$^+$ strains the gene is interrupted by a 1143-bp intron, which is absent in *omega*$^-$ strains (Fig. 2)

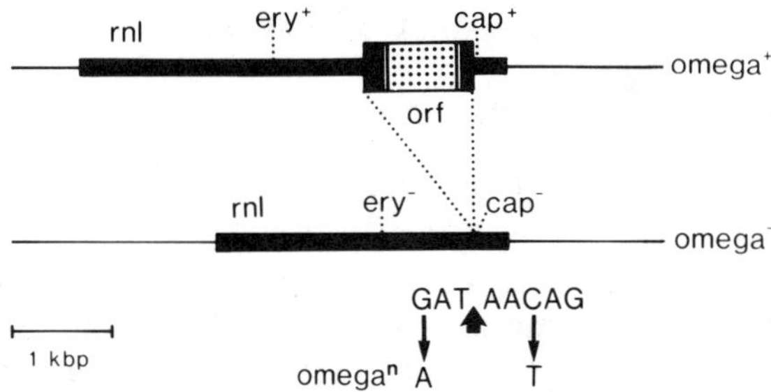

FIG. 2. Molecular structure of the *rnl* gene in *omega*$^+$ and *omega*$^-$ strains. *cap*$^+$/*cap*$^-$ and *ery*$^+$/*ery*$^-$ are alleles of mt loci conferring resistance–sensitivity to chloramphenicol and erythromycin. In the recognition site for the double-strand break (heavy arrow), the two base changes in *omega*^{n} mutants are indicated.

(Bos *et al.,* 1978, 1980b; Faye *et al.,* 1979). A second insert of 66 bp is located upstream of the intron insertion site (Faye *et al.,* 1979). In a screening of numerous yeast species (including the genera *Kluyveromyces, Saccharomyces, Brettanomyces,* and *Rhodosporidium),* Jaquier and Dujon (1983) found that the large intron is present in all *Kluyveromyces* species, some of the *Saccharomyces* species, and none of the other yeasts tested. In comparison with *S. cerevisiae,* the intron in *Kluyveromyces thermotolerans* is inserted at the same place, its reading frame also starts with AUG, and it can be aligned over the entire length with that of the *S. cerevisiae* intron. In *Kluyveromyces fragilis,* the *rnl* intron has been transposed to the *atp9* gene, and it can be argued that the new intron homing site is relevant to this transposition event (Dujon *et al.,* 1986; see also Wolf and Del Giudice, 1987). Presence or absence of this intron in *S. cerevisiae* is correlated with the phenomenon of polarity. The phenomenon of polarity in the drug-resistant mutants *cap*^{r}, *ery*^{r}, and *spi*^{r} has been analyzed. In the crosses *omega*$^+$ × *omega*$^+$ or *omega*$^-$ × *omega*$^-$ no polarity is observed, i.e., the two parental types and the two recombinant types are equally frequent. When an *omega*$^+$ strain is crossed with an *omega*$^-$ strain, the two parental types and the two recombinant types occur with extremely unequal frequencies (Coen *et al.,* 1970; Bolotin *et al.,* 1971; Rank and Bech-Hansen, 1972a; Howell *et al.,* 1973; Dujon *et al.,* 1976). This nonreciprocality also influences the presence of the *omega* alleles: the large majority of recombinants are *omega*$^+$ and the *omega*$^-$ allele is nearly eliminated from the progeny. This polar effect decreases for *rib1* over *rib2* to *rib3*. Thus, the *omega*$^+$ character is associated with the presence, and the *omega*$^-$ character with the absence of the intron. Point mutations (Dujon *et al.,* 1976, 1985; Jacquier and Dujon, 1985;

Macreadie *et al.*, 1985a,b) and deletions in the region between *cap* and *ery* (Devenish *et al.*, 1978) were shown to influence polarity in crosses. Except for the *var1* gene which shows some polar effect, no other case of polarity has been observed so far. Certainly, polarity is not due to insertions or deletions, since optional introns do not confer polarity in crosses (Williamson, 1978).

Before there was knowledge of the DNA sequence in this region, Perlman and Birky (1974) proposed a model for polarity based on the assumption that *omega*$^-$ strains possess an insert in the *rnl* gene relative to *omega*$^+$ strains. This could lead to the formation of a single-stranded loop after heteroduplex formation. A single-strand-specific endonuclease would break the *omega*$^-$ strand at the *omega* locus, and an exonuclease would degrade the *omega*$^-$ strand. DNA synthesis would repair the gap, using the *omega*$^+$ strand as template.

Recently, the nature of the *omega* locus has been elucidated by use of mutations and DNA sequencing. The intron in the *rnl* gene of *omega*$^+$ strains is inserted at a sequence motif GAT/AACAG (Fig. 2). The orf inside this intron can encode a basic protein of 235 amino acids. In crosses between *omega*$^+$ (intron-plus) and *omega*$^-$ (intron-minus) variants, this intron determines a gene conversion phenomenon, which results in the integration of the intron sequence in nearly all intron-minus copies of the gene. The absence of this phenomenon of polarity in *rho*$^-$ strains (devoid of mt protein synthesis; see paragraph below) and mutations in this orf (Jacquier and Dujon, 1985; Macreadie *et al.*, 1985a,b; Zinn *et al.*, 1987) demonstrate that it encodes a protein active in gene conversion that promotes spreading of this intron (therefore this gene has been termed *fit1*: *f*actor for *i*ntron *t*ransposition). Expression of the intronic orf (i.e., mt protein synthesis) is not needed for splicing of the intron (Tabak *et al.*, 1981). The involvement of a mt function in polarity is in contradiction with the finding of Strausberg and Birky (1979) that mt recombination does not require mt protein synthesis. The gene product of the intron orf has been shown to produce a double-strand break in *omega*$^+$ DNA at or close to the intron insertion site. The nature of the *omega*^{n} (neutral) mutations that abolish polarity can now be explained: they destroy the target sequence which is recognized by the intron orf gene product. In the wild type, repair of the double-strand break using *omega*$^+$ DNA as template would result in a conversion of the *omega*$^-$ allele into the *omega*$^+$ allele (Zinn and Butow, 1984, 1985; for a review: Butow, 1985). This view of the conversion has been inspired by the double-strand break–gap repair model proposed by Szostak *et al.* (1983) to explain gene conversion. It is worth mentioning that the *HO* endonuclease involved in mating type interconversion introduces a 4-bp staggered cut within the very similar sequence GCAACAG near the Y–Z junction of *MAT* (Kostriken *et al.*, 1983). A universal code equivalent of the intron orf can be expressed in *E. coli* as a specific double strand endonuclease (Colleaux *et al.*, 1986).

The large rRNA in *A. nidulans* has been characterized by Edelmann *et al.* (1970, 1971) and Verma *et al.* (1971, 1974). Physical mapping of the *rnl* gene (Netzker *et al.*, 1982) has shown that the gene is interrupted by a 1678-bp intron and by a 91-bp miniinsert, both at positions homologous to the similar-sized intervening sequences in the *rnl* gene of *S. cerevisiae omega*$^+$ strains. The RNA can be folded according to the general secondary structure model of prokaryotic and eukaryotic rRNAs (Köchel and Küntzel, 1982).

The *rnl* gene in *N. crassa* has been physically characterized (Kroon *et al.*, 1976; De Vries *et al.*, 1979; Green *et al.*, 1981; Yin *et al.*, 1982) and cloned (RajBhandary *et al.*, 1979). The gene is interrupted by an 2295-bp intron (Heckman and RajBhandary, 1979), which encodes the ribosomal protein S5 (LaPolla and Lambowitz, 1979, 1981; Breitenberger and RajBhandary, 1985b; see also Section III,D).

In *P. anserina* the *rnl* gene carries two introns of 1.65 and 2.73 kbp (Wright *et al.*, 1982b). The larger one is in the position of the conserved introns in the mt genomes of other fungi (Wright and Cummings, 1983c).

The *rnl* gene in *S. pombe* is continuous and is much shorter than that of *S. cerevisiae* and *Saccharomyces carlsbergensis* (Reijnders *et al.*, 1973) and can be folded into the general secondary structure model (Lang *et al.*, 1987). Comparison of the *S. cerevisiae* sequence with that of *S. pombe* revealed a perfect alignment including the 5.8 S-like region in the 5′ part, but no 4.5 S-like sequence at the 3′ end.

The *rns* gene in *S. cerevisiae* was localized by use of *rho*$^-$ petites (Faye *et al.*, 1975, 1976b), found to carry the paromomycin locus (Kutzleb *et al.*, 1973; Wolf *et al.*, 1973; Spithill *et al.*, 1979), characterized by restriction enzyme mapping (Tabak *et al.*, 1979), and sequenced (Sor and Fukuhara, 1980, 1982a,c). Secondary structures can be built corresponding to the universal structure proposed by Stiegler *et al.* (1980, 1981), Woese *et al.* (1980), Gray *et al.* (1984), and Noller (1984). Sor and Fukuhara (1982c) confirmed data from Morimoto *et al.* (1979b) and Osinga *et al.* (1981) that the gene is transcribed in a precursor form that is 95 bp longer than the mature rRNA. There is a strain-dependent presence or absence of a short GC-rich segment (Sor and Fukuhara, 1982c) which is transcribed. Similar types of insertions have been described by Dujon (1980) for the *rnl* gene in *S. cerevisiae*. Tabak *et al.* (1982) reported that paromomycin resistance is transmitted by petites that do not contain the *rns* gene region, where this locus maps (Li *et al.,* 1982). The authors speculate that illegitimate recombination between the *rho*⁻ petite and the genome of the wild-type parent could occur.

A suppressor of ochre mutations in the mt genome of *S. cerevisiae* has been reported that maps in or near the *rns* gene (Fox and Staempfli, 1982), and a mt frameshift suppressor has been mapped in *rns* (Weiss-Brummer *et al.,* 1987).

The *rns* genes of *A. nidulans* and *S. pombe* have been completely sequenced, and their secondary structures show a high degree of correspondence with the universal secondary structure model (Köchel and Küntzel, 1981; H. Trinkl *et al.,* unpublished results).

B. Transfer RNA Genes and the Genetic Code

Mitochondria encode a set of tRNAs which are different from those in the cytoplasm (Accoceberry and Stahl, 1981; Halbreich and Rabinowitz, 1971; Schneller *et al.,* 1975d; R. P. Martin *et al.,* 1976a,b, 1977) and like in bacteria, *fmet* is used as start (Smith and Marcker, 1968; Epler *et al.,* 1970; Heckman *et al.,* 1978; for a review see R. P. Martin *et al.,* 1983). There is only one tRNA, a *lys* tRNA in *S. cerevisiae* (R. P. Martin *et al.,* 1979) that cannot be charged; it is very likely imported from the cytoplasm and may be used for purposes other than protein synthesis. In many cases, two or more major and/or minor isoacceptor species have been found (Schneller *et al.,* 1975b; Baldacci *et al.,* 1976a,b, 1977, 1978, 1979; Fukuhara *et al.,* 1976a,b; N. C. Martin *et al.,* 1976, 1979a,b; R. P. Martin *et al.,* 1976b, 1977; N. C. Martin and Rabinowitz, 1978; Coletti *et al.,* 1979; Macino and Tzagoloff, 1979a; Wesolowski *et al.,* 1980).

In *S. cerevisiae,* tRNA genes have been mapped by hybridization with labeled tRNAs, using defined petites or genetically by deletion mapping (Fukuhara, 1967; Halbreich and Rabinowitz, 1971; Casey *et al.,* 1972, 1974a–c; Cohen and Rabinowitz, 1972; Jakovcic *et al.,* 1975; Fukuhara *et al.,* 1976a,b; N. C. Martin *et al.,* 1977; Trembath *et al.,* 1977; Van Ommen *et al.,* 1977; Wesolowski and Fukuhara, 1979; Berlani *et al.,* 1980a,b; Miller and N. C. Martin, 1981; Wesolowski *et al.,* 1981a,b) and sequenced (Bos *et al.,* 1979; Li and Tzagoloff, 1979; N. C. Martin *et al.,* 1979a,b, 1980; Bonitz and Tzagoloff, 1980; Miller *et al.,* 1979, 1980; Newman *et al.,* 1980; Nobrega and Tzagoloff, 1980a). Some tRNAs have also been sequenced at the RNA level (Schneller *et al.,* 1975a; R. P. Martin *et al.,* 1976a, 1978, 1983; Sibler *et al.,* 1979, 1983, 1985; Canady *et al.,* 1980; R. P. Martin and Dirheimer, 1983). In *S. cerevisiae* all tRNA genes except *tmt1* (Li and Tzagoloff, 1979) are located on the same DNA strand, none of them contain intervening sequences, and the terminal sequence CCA is always added posttranscriptionally.

Several mutations in tRNA genes have been identified (Faye *et al.,* 1976a; Miller *et al.,* 1981b, 1983; Najarian *et al.,* 1986; Norbrega and Nobrega, 1986). The effect of a mutation in the extra arm of *tmy* is discussed in Section VI,A (Bordonne *et al.,* 1987).

In *N. crassa* tRNAs have been characterized biochemically (Epler *et al.,* 1969; De Vries *et al.,* 1978), and at least 27 tRNA genes (specifying 25 different tRNA species, since both *tmm* and tmm_f are duplicated) have been mapped and sequenced (Terpstra *et al.,* 1977; De Vries *et al.,* 1979; Heckmann and RajBhandary, 1979; Heckman *et al.,* 1979a,b; RajBhandary *et al.,* 1979; Yin *et al.,* 1980, 1982, 1983; Breitenberger and RajBhandary, 1985a,b).

In *A. nidulans* there is uncertainty about the precise number of tRNA genes, since one tRNA predicted by decoding principles, the *arg* tRNA (CGN), has not been located so far. In *S. pombe,* 25 tRNA genes have been identified by DNA sequencing and in some cases by expression of a 4 S RNA species (B. F. Lang, unpublished results).

Most components necessary for the biosynthesis of tRNA are encoded by nuclear genes, but one function necessary for the correct processing of tRNA precursors is encoded by the mt genome in *S. cerevisiae* (Morimoto *et al.,* 1979b). By deletion mapping and restriction enzyme analysis this locus (*tsl* for *t*RNA *s*ynthesis *l*ocus) has been confined to within 780 bp between tmm_f and *tmp* (N. C. Martin and Underbrink-Lyon, 1981; N. C. Martin *et al.,* 1982; Miller *et al.,* 1983; Underbrink-Lyon *et al.,* 1983). This region is very AT-rich with one GC-rich element. It encodes a transcript of about 450 bases since the activity of this enzyme is both protease and micrococcal nuclease sensitive, and since petite mutants never make mt proteins, but some do have RNase P activity, the protein must be coded by a nuclear gene (Miller and N. C. Martin, 1983; N. C. Martin *et al.,* 1985a,b; Hollingsworth and N. C. Martin, 1986). The function of *tsl* has been demonstrated by use of a cloned *tms* (UCN) gene as probe. In RNA from wild-type cells, the *ser* tRNA is the most prominent transcript of the gene rho^- petites lacking *tsl* accumulate high molecular weight transcripts containing *ser* tRNA, but no mature *ser* tRNA is made. These results have been confirmed and extended using a primary transcript of tmm_f. There is now convincing evidence that *tsl* encodes a

5′ tRNA processing function (N. C. Martin *et al.*, 1985a). A tRNA synthesis locus might also be encoded by *T. glabrata* mt DNA (Clark-Walker *et al.*, 1985b).

It had been suspected earlier (Reijnders and Borst, 1972; Schneller *et al.*, 1975c) that the mitochondrion encodes fewer tRNA genes as required by the wobble rules, and it is now evident that mitochondria code for a full set able to charge all 20 amino acids. The lower number of tRNAs (about 24) is compensated by deviations from the universal code. Sequence analysis from a variety of mt genomes has revealed genetic codes different from the universal and also different among species. Mammalian mt genomes code for 22 tRNAs, *S. cerevisiae* for 24. The reduced number of tRNAs is compensated by having a single tRNA recognize all codons in each of the four codon boxes: *leu* (CUN), *val* (GUN), *ser* (UCN), *pro* (CCN), *thr* (ACN), *ala* (GCN), *arg* (CGN), and *gly* (GGN). In *S. cerevisiae, N. crassa, A. nidulans,* and *S. pombe,* the reduction is achieved by the presence of U in the wobble position of the anticodon (Barrell *et al.*, 1980; Bonitz *et al.*, 1980a; Bibb *et al.*, 1981; Köchel *et al.*, 1981; Lang *et al.*, 1983; Sibler *et al.*, 1986; B. F. Lang, unpublished results). This U is unmodified in *N. crassa* tRNAs (Heckman *et al.*, 1980). In the universal code the modification of the U in the wobble position only allows A or G to be read.

Mammalian and fungal mitochondria read the *trp* codon (UGG) and the *opal* triplet (UGA) as tryptophan (Fox, 1979b; Heckman *et al.*, 1979; Barrell *et al.*, 1980; Bonitz *et al.*, 1980a; Young and Anderson, 1980; Bibb *et al.*, 1981; Köchel *et al.*, 1981; Breitenberger and RajBhandary, 1985b). This is achieved by changing the *trp* tRNA anticodon from ACC to ACU (Macino *et al.*, 1979; De Ronde *et al.*, 1980; N. C. Martin *et al.*, 1980b; Anderson *et al.*, 1981; R. P. Martin *et al.*, 1981; Tuite and McLaughlin, 1982). In *S. pombe,* the anticodon of the *trp* tRNA is ACC (Lang *et al.*, 1983; Lang, 1984). UGA condons, however, are only found in two of the three introns and in *urfa.* It is not yet clear, how efficiently, if at all, this tRNA is able to read UGA codons.

In addition to these general deviations of the mt from the universal genetic code, there are differences among mt genomes from various species. Unlike mammalia, *A. nidulans, N. crassa,* and *S. pombe, S. cerevisiae* read the CUN family as *thr* instead of *leu* (Sebald and Wachter, 1979; Bonitz *et al.*, 1980a; Köchel and Küntzel, 1981; Burger and Werner, 1983; Lang *et al.*, 1983). Li and Tzagoloff (1979) have identified a tRNA gene (the *tmt1* gene) where the corresponding tRNA has eight nucleotides in the anticodon loop instead of seven. They concluded that this tRNA is charged the *thr,* but possesses the anticodon GAU. *Saccharomyces cerevisiae* and mammals differ in the use of *arg* codons. Mammalian mitochondria lack a tRNA recognizing AGA and AGG, which therefore are termination codons. Instead, they possess a tRNA with a GCU anticodon that is able to read the four members of the *arg* family.

Saccharomyces cerevisiae has two *arg* tRNAs (anticodons UCU and GCA); however, the anticodon GCA can only recognize CGU and CGC, but not CGA and CGG. Since the two latter codons are never used in *S. cerevisiae,* they are probably also termination codons.

An AUA codon for *met* has been reported for *S. cerevisiae* (Hudspeth *et al.,* 1982): mammals and all other fungi read AUA as *ile*. In general, fungal mt genomes select codons that terminate in A or U, which explains the high AT content.

The most interesting fact is the different codon usage in exons and introns of mosaic genes (see below). In the introns of *S. cerevisiae* and in the second *cox1* intron in *S. pombe* strain 50, there is an excess of UUU versus UUC, which is not seen in exons. This may be due to the peculiarities of the *phe* tRNA (R. P. Martin *et al.,* 1978), which produces frameshifts in a run of Us (Fox and Weiss-Brummer, 1980). This situation, however, can be avoided in the exons by the choice of UUC codons.

C. Continuous Protein Coding Genes

In *S. cerevisiae,* mt mutants (*mit*$^-$) deficient in cytochrome *c* oxidase map in three loci (Tzagoloff *et al.,* 1975a,b, 1980; Foury and Tzagoloff, 1976b; Cabral *et al.,* 1977). *cox2* and *cox3* are continuous in all ascomycetes studied so far, whereas *cox1* is mosaic in *S. cerevisiae, A. nidulans, P. anserina, S. pombe,* and other yeasts. In *N. crassa* strain 74A, *cox1* is also continuous (Burger *et al.,* 1982; Burger and Werner, 1983; De Jonge and De Vries, 1983; De Vries *et al.,* 1983), while it is mosaic in natural wild-type strains (Collins and Lambowitz, 1983).

Cabral *et al.* (1978) showed by use of chain termination mutants that the *cox2* encodes subunit 2 of cytochrome *c* oxidase. This was apparent by the colinearity between the map position of mutants and the size of the polypeptides (Fox, 1979a,b; Weiss-Brummer *et al.,* 1979; Kruszewska *et al.,* 1980). The gene has been sequenced (Coruzzi and Tzagoloff, 1979; Fox, 1979b; Coruzzi *et al.,* 1981), and its transcripts identified (Fox and Boerner, 1980; Coruzzi *et al.,* 1981). The protein is synthesized as a precursor (Sevarino and Poyton, 1980).

In *N. crassa, cox2* has been localized (van den Boogaart *et al.,* 1982a) and shown to be synthesized as a precursor protein (van den Boogaart,

1982c). DNA sequencing (Macino and Morelli, 1983) and transcript analysis have revealed that the gene contains a leader sequence. The *cox2* gene has also been sequenced in *Hansenula saturnus* (Lawson and Deters, 1985).

The *cox3* gene in *S. cerevisiae* has been identified genetically (Cabral *et al.,* 1978) and sequenced by Thalenfeld and Tzagoloff (1980). Mapping and transcript studies have been carried out by Stephenson *et al.* (1981b), Baranowska *et al.* (1983), and Thalenfeld *et al.* (1983b). In *N. crassa, cox3* was localized and sequenced by Browning and RajBhandary (1982).

In *S. cerevisiae* and *S. pombe,* three ATPase subunits (6, 8, and 9) are encoded in the mt genome. In *A. nidulans* and *N. crassa* the *atp9* sequence is present, but very likely not translated. Mt mutants (for reviews see Kovac, 1974; Michelis and Somlo, 1976) resistant to oligomycin (oli^r), venturicidin (ven^r), ossamycin (oss^r), and miconazole (mic^r; linked to *oli1*) have been isolated and mapped genetically and physically in *S. cerevisiae.* Mutants belonging to the loci *oli1, oli3, ven1, oss2,* and probably *oli5* are located in the *atp9* gene, and mutants at the loci *oli2, oli4, oss1* map in the *atp6* gene (Parker *et al.,* 1968; Stuart, 1970; Wakabayashi and Gunge, 1970; Wakabayashi, 1972, 1976, 1978; Avner *et al.,* 1973; Shannon *et al.,* 1973; Wakabayashi and Kamei, 1973, 1974; Griffiths and Houghton, 1974; Lancashire *et al.,* 1974; Somlo *et al.,* 1974; Griffith *et al.,* 1975; Lancashire and Griffith, 1975; Trembath *et al.,* 1975, 1976; Tzagoloff *et al.,* 1975b, 1976a; Clavilier, 1976; Somlo and Cosson, 1976; Murphy *et al.,* 1978, 1980; Lancashire and Mattoon, 1979; Darlison and Lancashire, 1980; Novitski *et al.,* 1984; Connerton *et al.,* 1984; Ooi *et al.,* 1985; Ooi and Nagley, 1986; John and Nagley, 1986; Jean-Francois *et al.,* 1986; Nagley *et al.,* 1986; Willson *et al.,* 1986; Willson and Nagley, 1987; Hefta *et al.,* 1987).

Mutants deficient in mt oligomycin-sensitive ATPase activity (pho^-) map in three genes: *pho1* is closely linked to *oli2* (*atp6*) and *pho2* is closely linked to *oli1* (*atp9*) (Foury and Tzagoloff, 1976a,b; Coruzzi *et al.,* 1978, 1979; Darlison and Lancashire, 1980; Roberts *et al.,* 1979; Stephenson *et al.,* 1981a; Somlo *et al.,* 1985). A group of mit^- mutants maps in the *aap1* gene (the equivalent to *urfa6l* in mammalia = *atp8*) (Macreadie *et al.,* 1982, 1983; Novitski *et al.,* 1983; Velours *et al.,* 1984).

atp6, 8, and *9* have been sequenced (Hensgens *et al.,* 1979; Macino and Tzagoloff, 1979b,c, 1980; Tzagoloff *et al.,* 1980b; Novitski *et al.,* 1984).

For a few *oli1* mutants the amino acid sequence had been determined (Wachter *et al.*, 1977; Sebald *et al.*, 1979b; see also Sebald and Hoppe, 1981; and Sebald *et al.*, 1979) and the exchanges have been localized near the DCCD binding site (Partis *et al.*, 1979). A mutation in the 5′-untranslated region of *atp9* has been shown to result in an impairment of translation (Ooi *et al.*, 1987).

In *A. nidulans, atp6* has been sequenced and found to be continuous (Grisi *et al.*, 1982; Netzker *et al.*, 1982). *atp8* (previously *urfx,* Brown *et al.*, 1985, or *urfb,* Netzker *et al.*, 1982) has been sequenced (Netzker *et al.*, 1982) and identified as *atp8* gene through its homology to the homologous gene in *S. cerevisiae* (Brown *et al.*, 1985).

In *N. crassa* a sequence homologous to *atp8* was detected 700 bp upstream of *atp6* (Morelli and Macino, 1984). The sequence of the *atp9* equivalent genes in *N. crassa* (*mal,* van den Boogaart *et al.*, 1982b) and *A. nidulans* (Brown *et al.*, 1984) have been determined.

In *S. pombe,* all three mt ATPase genes gave been sequenced (B. F. Lang, unpublished results).

D. Genes for Ribosomal Proteins

In *S. cerevisiae* (Groot *et al.*, 1979; Terpstra *et al.*, 1979) and *N. crassa* (Lambowitz *et al.*, 1979) one protein of the small subunit of the mt ribosomes is encoded by the mt DNA, whereas all other ribosomal proteins are chromosomally encoded (Lambowitz *et al.*, 1976).

In *S. pombe* there was no hybridization with a *var1* DNA probe from *S. cerevisiae* (Lang and Wolf, 1984).

The *T. glabrata* mt genome hybridizes with a *var1* probe from *S. cerevisiae,* and the gene is transcribed (Clark-Walker and Sriprakash, 1983a,b; Clark-Walker, 1985; Clark-Walker *et al.*, 1985b).

The *var1* gene product exists in a series of different strain-dependent electrophoretic mobility variants ranging in apparent size between 40 and 44 kDa (Douglas and Butow, 1976; Douglas *et al.*, 1976a,b; Butow *et al.*, 1977, 1979, 1980, 1982; Butow, 1985). In various combinations, these *var1* forms differ in 26 amino acid residues, but these variations do not affect mt function to a measurable extent. It is, however, an essential gene, since upon inhibition of mt protein synthesis, small ribosomal subunits are not assembled (Terpstra and Butow, 1979). The assembly of the large subunit of mt ribosomes is not affected (Maheshwari *et al.*, 1982). Mutants mapping to or in this regions are impaired in respiration and mt protein synthesis (Zassenhaus and Perlman, 1982; Butow *et al.*, 1985). Genetic analyses of the *var1* polymorphism has led to the assignment of the position between *rnl* and the *atp9* gene (Perlman *et al.*, 1977). By crosses (Douglas *et al.*, 1976a,b; Perlman *et al.*, 1977; Strausberg *et al.*, 1978; Vincent *et al.*, 1980; Strausberg and Butow, 1981) and complementation analysis (Strausberg and Butow, 1977; Lopez *et al.*, 1981) two types of genetic elements could be identified that account for all observed forms of the *var1* gene product. The different alleles originate from combinations of these elements, designated a and b and variant forms of b (b_p). The genotypes of strains forming the different variants are listed in Table 4. In particular, the recombination of the a element has the characteristics of an asymmetric gene conversion, whereby the "short" alleles are in most cases converted to "long" forms (Strausberg *et al.*, 1978; Williamson, 1978), a situation which is reminiscent of the *omega* alleles (see Section III,D). The *var1* 40.0 allele has an uninterrupted reading frame of 1186 bp, which consists of 89.6% AT, and encodes a protein of 46,786 Da (Hudspeth *et al.*, 1982). Forty percent of the G and C residues are clustered in a GC palindrome which is found in all *var1* alleles. The elements a, b, and b_p are indeed in-frame insertions, providing additional genetic information and expanding the gene (Hudspeth *et al.*, 1984). The a^+ insert is an additional GC cluster, 146 bp downstream from the common GC cluster. The GC clusters are also transcribed and present in the presumptive mature *var1* mRNA (Zassenhaus and Butow, 1984). Both GC clusters encode amino acids in the *var1* protein. The flanking sequences of the two GC clusters are identical. The variability in the location of GC clusters in different mt genomes of *S. cerevisiae* strains may be analogous to insertion elements.

Secondary structure predictions show that the *var1* gene product

TABLE 4
Genotypes of Forms of *var1* Protein

var1 allele[a] ($MW \times 10^{-3}$)	Genotype
40.0	a^-b^-
41.8	a^+b^-
42.0	$a^-b_p^+$
42.2	a^-b^+
43.8	$a^+b_p^+$
44.0	a^+b^+

[a] The *var1* alleles are designated by the apparent molecular weights ($\times 10^{-3}$) of the *var* protein as estimated by SDS–polyacrylamide gel electrophoresis. (Taken from Butow *et al.*, 1985.)

does not resemble the structural predictions for known ribosomal proteins (Ainley *et al.,* 1984). Farrelly and Butow (1983) found a contiguous sequence in the nuclear genome with extensive homology to noncontiguous mt sequences including *var1*. The authors proposed that these sequences may have originated from *rho*⁻ mtDNA which integrated into the nuclear genome.

E. Mosaic Genes

In contrast to *S. cerevisiae, S. pombe,* and *A. nidulans,* the *atp6* gene of *N. crassa* is split by two introns (Helmer-Citterich *et al.,* 1983; Morelli and Macino, 1984; Lambowitz *et al.,* 1985). The first intron is 93 bp long and two-thirds of it consist of palindromic sequences. The second intron (1370 bp) contains an orf in phase with the upstream exon. This intron could be assigned to group I (according to the nomenclature of Michel and Dujon, 1983). In *S. cerevisiae, cox1* has been identified by the absence of the polypeptide in *mit*⁻ mutants and by *in vitro* translation of transcripts. It is located between *rns* and *atp6* and has been characterized by petite deletion, transcript, and restriction site mapping and by DNA sequencing (Tzagoloff *et al.,* 1975a,b; Cobon *et al.,* 1976; Slonimski and Tzagoloff, 1976; Grivell and Moorman, 1977; Eccleshall *et al.,* 1978; Moorman *et al.,* 1978; Carignani *et al.,* 1979, 1983; Keyhani, 1979; Morimoto *et al.,* 1979a; Bonitz *et al.,* 1980a–c; Keyhani and Keyhani, 1982; Netter *et al.,* 1982a; Hensgens *et al.,* 1983a,b, 1984a,b). It contains a 400-bp 5′ leader and shows a remarkable strain-dependent variation in structure: 10 exons are separated by 9 introns in strain KL14-4A. At least 4 are optional, since 2 of them (splitting exon 5 into three parts in strain KL14-4A) are absent in strain D273-10B (Bonitz *et al.,* 1980b) and another 2 (the intron between exons 1 and 2, and the intron between exons 4 and 5 in strain D273-10B) are lacking in *S. carlsbergensis* (Fig. 3). More than 60 different RNA species map to this locus in its longest form, 20 and 11 (respectively) in the shorter versions (Hensgens *et al.,* 1983a, 1984a,b; Carignani *et al.,* 1986). In the long form, three intron transcripts exist as lariats. The first five introns in the long form of the gene contain long orfs in phase with the upstream exons. It has been demonstrated by use of *mit*⁻ mutants that the first intron encodes a maturase (Groudinsky *et al.,* 1983; Carignani *et al.,* 1983). Mutations in intron 1 are phenotypically suppressed by paromomycin (which is known to enhance ribosomal misreading) (Dujardin *et al.,* 1984) and also by the mt suppressor *mim3-1* (Kruczewska, 1982; Groudinsky, 1983; Zagorski *et al.,* 1987; Miezczak and Zagorski, 1987). The sup-

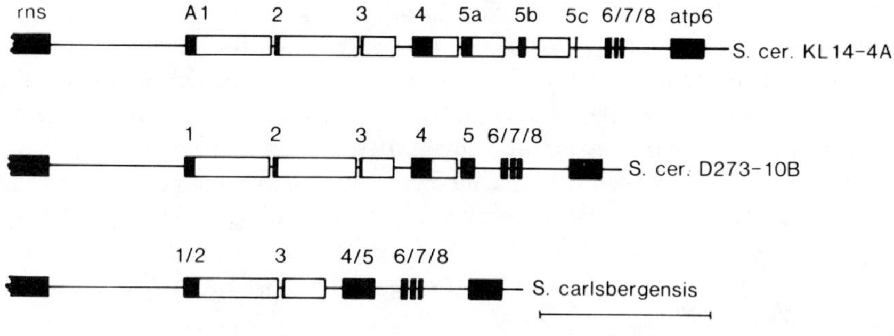

FIG. 3. Three forms of the *cox1* gene of the *S. cerevisiae* strains KL14-4A and D273-10B and of *S. carlsbergensis*. Black bars represent exons, white bars represent intronic open reading frames, and lines represent closed reading frames.

pressor *mim2*, which maps in intron 4 of *cox1* (Dujardin *et al.,* 1982), suppresses intron mutations in the *box7* region of *cob* (see below). Mutations in a short DNA sequence in intron 4 also affect splicing due to a disturbed interaction with a highly homologous intron in the *cob* gene (Netter *et al.,* 1982a; see below). In strain KL14-4A, with the longest form of the *cox1,* exon 5 is split into three parts. A large number of sequence differences has been found in exons 5α and 5β (in strains KL14-4A) related to exon 5 in the shorter version of the gene in strain D273-10B. In the latter strain the nucleotide changes lead to the formation of codons rarely, if ever, used in the yeast mt genome. The origin of this sequence variation may be due to recombinational events between strain D273-10B and a distantly related mtDNA. This could have happened quite recently, before sufficient time has elapsed to permit backmutation to more commonly used codons (Hensgens *et al.,* 1983). Analysis of transcript patterns of *mit*$^-$ mutants in *cox1* has led to the identification of some steps in the RNA processing of this gene (Grivell *et al.,* 1979, 1982; Hensgens *et al.,* 1984).

A similar case of sequence variation at the exon–intron boundary has been found for the optional group II intron in the *cob* gene of the *S. pombe* strain 50 (Lang *et al.,* 1985) and UCD FstI (Zimmer *et al.,* 1987) and its intronless counterpart in strain EF1 (Zimmer *et al.,* 1984; Trinkl *et al.,* 1985).

In *A. nidulans, cox1* contains three introns. Since an AUG codon is not present at the start, possibly AUGA is used for the start, or there is additional intron between the true start and the beginning of the coding region. Introns 2 and 3 possess long orfs (Waring *et al.,* 1984). The third intron is highly homologous to the second *cox1* intron of

S. pombe (Waring *et al.,* 1984; Lang, 1984) and both introns are inserted at the same position of the otherwise less conserved *cox1* genes of both fungi. Intron 3 in *A. nidulans* has an *in-frame* insert of a 37-bp GC-rich sequence, flanked by a 5-bp repeat, a structure which is reminiscent of transposons. All three introns can be assigned to group I. The finding that both introns are inserted at the same place, and that their putative gene products share a 70% homology over a stretch of 253 amino acids, makes it likely that these introns are the result of a recent horizontal gene transfer (Lang, 1984; Wolf and Del Guidice, 1987; Wolf *et al.,* 1987).

The *cox1* gene in *S. pombe* is polymorphic: in strain 50 it contains two group I introns with orfs (*aI1* and *aI2b*). In contrast to the second intron, where two TGA codons are found, only TGG is used in the first intron. Strain EF1 carries two additional group I introns (*aI2a* and *aI3*) without orfs. Introns *aI2a* and *aI3* are inserted at the same sites as the orf containing introns *aI4* and *aI5β* in the *cox1* gene of *S. cerevisiae.* Moreover, the *S. pombe* introns share characteristic features and a high degree of sequence homology with the corresponding *S. cerevisiae* introns (Trinkl and Wolf, 1986). Intron *aI2a,* which is only 258 pb long, shows a remarkable sequence homology with intron bI4 of the *cob* gene in *S. cerevisiae.* The occurrence of similar introns in two different genes could be explained by the fact that the surrounding exon sequences are very similar. The mosaic *cox1* gene of *P. anserina* contains several group I introns (Karsch *et al.,* 1987); the first intron, which is of group II, will be discussed in context with the phenomenon of senescence (Section VI,D).

The *cob* region is *S. cerevisiae* has been located by use of *mit*$^-$ mutants between *atp6* and *atp9* (Tzagoloff *et al.,* 1975a, 1976b, 1978; Slonimski and Tzagoloff, 1976; Schweyen *et al.,* 1977; Dobres *et al.,* 1985). The mosaic nature of this gene has been demonstrated genetically (Slonimski *et al.,* 1978a,b; Grivell *et al.,* 1979; Haid *et al.,* 1979; Alexander *et al.,* 1980; Nobrega and Tzagoloff, 1980c; for reviews see Borst, 1980; Perlman *et al.,* 1980; Jacq *et al.,* 1982; Dieckmann *et al.,* 1982). In its long form, the *cob* gene consists of six exons, separated by five introns, spanning a segment of roughly 7500 bp; in its short form, exons 1 to 4 are fused, and only the last two introns are present (Fig. 4). Both versions of the gene have been physically mapped (Grosch *et al.,* 1981); the short version has been completely sequenced, as well as large parts of the long version (Nobrega and Tzagoloff, 1980c; Tzagoloff and Nobrega, 1980). According to the historical (and undoubtedly confusing) order, the region has been divided into the loci *box1* to *box10* (Fig. 4) (Jacq *et al.,* 1980a,b; Lazowska *et al.,* 1980).

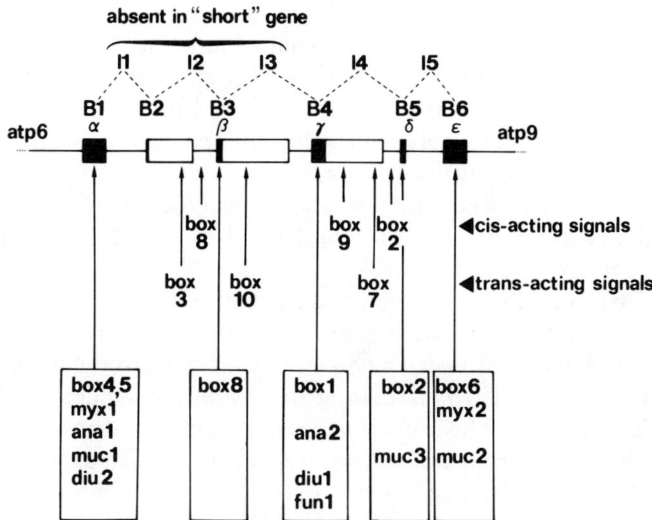

FIG. 4. Presentation of the two forms of the *cob* gene in *S. cerevisiae*. B1–B6 represent exons (α to ε in the old nomenclature) introns I1–I5. The gene is divided into *box* loci 1 to 10 in historical order, in which cis- and trans-acting mutations conferring respiratory deficiency (*mit*$^-$) are localized. The mutational sites conferring resistance to myxothiazol (*myx1, myx2*), antimycin (*ana1, ana2*), mucidin (*muc1, muc2, muc3*), diuron (*diu1, diu2*), and funiculosin (*fun1*) are indicated. (Adapted from Jacq *et al.*, 1982.)

Different exon mutants and their revertants have been used to study the function of the spectral forms of cytochrome *b* (Meunier-Lemesle *et al.*, 1980; Chevillotte-Brivet and Meunier-Lemesle, 1980; Burger, 1984; Burger and Hofner, 1984; Chevillotte-Brivet *et al.*, 1987).

Several drug resistance mutations map in *cob* exons, conferring resistance to mucidin, antimycin, diuron, funiculosin, and myxothiazol (Fig. 4) (Howell *et al.*, 1973; Subik *et al.*, 1974, 1977, 1981; Convent and Briquet, 1975; Subik, 1975, 1976; Burger *et al.*, 1976a,b; Lang *et al.*, 1976a,b; Michaelis, 1976; Pratje and Michaelis, 1976, 1977; Colson *et al.*, 1977, 1979; Colson and Slonimski, 1977, 1979; Groot Obbink *et al.*, 1977; Michaelis and Pratje, 1977; Convent *et al.*, 1978; Subik and Takacsova, 1978; Colson and Wonters, 1980; Roberts *et al.*, 1980; Subik and Goffeau, 1980, 1981; Takacsova *et al.*, 1980; Thierbach and Reichenbach, 1981; Thierbach and Michaelis, 1982; Clejan *et al.*, 1983; Chevillotte-Brivet *et al.*, 1983; di Rago *et al.*, 1986).

Strains with varying numbers of introns could be constructed by genetic crosses (Perlman *et al.*, 1980; Labouesse and Slonimski, 1983) or by intron DNA subtraction (Gargouri *et al.*, 1983). A strain completely devoid of introns has been reported recently (Seraphin *et al.*, 1987). In *S. pombe,* a strain devoid of functional introns has also been constructed by genetic crosses between different *S. pombe* strains and intron subtraction (Merlos-Lange *et al.*, 1987).

The apocytochrome *b* polypeptide has an apparent molecular weight of about 30,000. Mutations in exons may lead to premature chain termination and consequently to shorter polypeptides (Claisse *et al.*, 1977, 1978; Spyridakis and Claisse, 1978; Haid *et al.*, 1979; Hanson *et al.*, 1979; Solioz and Schatz, 1979; Kreike *et al.*, 1979). Other mutants synthesize a complete but inactive apocytochrome *b* polypeptide due to a missense mutation (Claisse *et al.*, 1977, 1978). A third group of *mit*$^-$ mutants does not synthesize a detectable apocytochrome *b* polypeptide at all (Haid *et al.*, 1979; Kreike *et al.*, 1979; Alexander *et al.*, 1980; Claisse *et al.*, 1980). The most interesting class of mutants synthesizes polypeptides that are translational hybrids from both exon and intron sequences (Alexander *et al.*, 1980; Perlman *et al.*, 1980).

The mature *cob* transcript is an 18 S RNA with a long untranslated leader (about 1000 bases) and a short (about 50 bases) untranslated trailer (Church *et al.*, 1979; Grivell *et al.*, 1979, 1980; Haid *et al.*, 1980; Halbreich *et al.*, 1980, 1981; van Ommen *et al.*, 1980). The mature mRNA arises by a series of processing steps (Church *et al.*, 1979; Haid *et al.*, 1980; Halbreich *et al.*, 1980; van Ommen *et al.*, 1980; Bonitz *et al.*, 1982), and a pathway of splicing could be worked out by use of splice-deficient mutants. The only intron that can be spliced out in *rho*$^-$ mutants (i.e., in the absence of mt protein synthesis) is the first one, which belongs to group II. After excision it forms a stable covalently closed circle (a lariat) of 10 S (Fig. 5). Circular RNAs (lariats) have also been found for some introns of the *cox1* gene (Grivell *et al.*, 1979; Arnberg *et al.*, 1980; see the following paragraph).

Complementation studies and sequence analysis have shown that intron mutations belong to two distinct classes: trans-recessive and cis-dominant mutations (Kochko *et al.*, 1979; Lamouroux *et al.*, 1980). The first class corresponds to sequences that encode trans-acting proteins involved in splicing (maturases) and the second class corresponds to critical sequences in the splicing substrate (the pre-mRNA). There is ample evidence that some introns encode functions required for the splicing of the precursor mRNA of *cob* (and also for the precursor mRNA of *cox1*; see below).

First, it has been hypothesized that introns code for a "guide RNA" (Church *et al.*, 1979; Halbreich, 1980), which serves to align the exon–intron boundaries for correct processing. The absence of splicing of several introns in *rho*$^-$ petites and the formation of hybrid polypeptides resulting from readthrough with introns provided evidence for proteins involved in splicing. DNA sequencing supported this finding, since long orfs were found in three introns (Lazowska *et al.,* 1980; Nobrega and Tzagoloff, 1980b). Church and Gilbert (1980) postulated that the orfs in introns 2, 3, and 4 encode splicing enzymes (maturases, spligases) that splice out their own intron RNA (for a review see Grivell and Borst, 1982). After Jacq *et al.* (1980a,b) and Lazowska *et al.* (1980) sequenced the first two introns in the long form of the gene, a model for the action of maturases was proposed (Lazowska *et al.,* 1980), which has been confirmed and extended by other groups (Fig. 5). Schmelzer and Schweyen (1982) provided evidence that maturation of mRNAs is coupled with translation. In a first splicing step, which is independent of mt functions, intron 1 is spliced out of the precursor and circularized (there is evidence that a complementary RNA strand is synthesized at this circular RNA; Halbreich *et al.,* 1984). After the first splice, exons 1 and 2 are fused with the orf from intron 3. From this template, a 42-kDa protein is produced (maturase). Its activity is required for the second step, which ligates the third exon in phase with the two preceding exons. After this event, no new maturase molecules can be synthesized (Fig. 5) (Mahler *et al.,* 1977, 1982; Bechmann *et al.,* 1980, 1981; Jacq *et al.,* 1980a,b; Lazowska *et al.,* 1980; Schmelzer *et al.,*

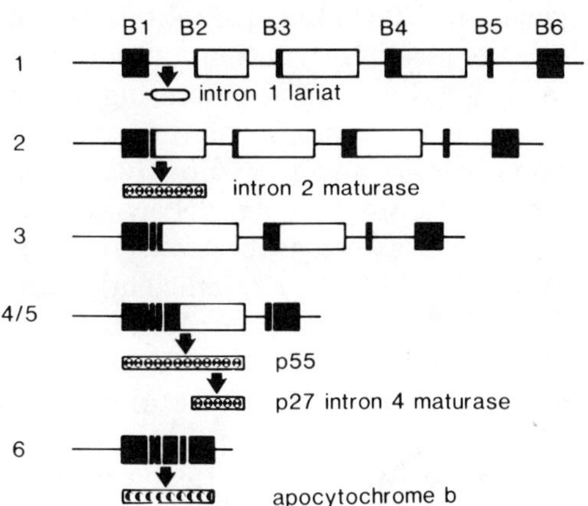

FIG. 5. Possible pathway of transcript processing in the *cob* gene of *S. cerevisiae.* Exons are represented by black bars, open intron reading frames are represented by white bars, and closed intron reading frames are represented by lines. The intron 1 lariat is indicated. The two polypeptides p55 and p27 are so designated by the apparent molecular weight ($\times 10^{-3}$) as estimated by SDS–polyacrylamide gel electrophoresis.

1981; Hanson *et al.,* 1982; Schweyen *et al.,* 1982; for reviews see Dujon, 1979; Lewin, 1980; Borst and Grivell, 1981a).

By use of antibodies against synthetic oligopeptides constructed according to the DNA sequence of the second intron, the putative intron 2 maturase could be identified (Guiso *et al.,* 1984). The involvement of a maturase in splicing is supported by the evidence that mutants in intron 2 can be suppressed by the *mim1* mutation in the *rnl* gene (Dujardin *et al.,* 1980a) and by paromomycin (Jacq *et al.,* 1982; Kruczewska, 1983). A long orf in intron 4 was detected by Nobrega and Tzagoloff (1980b,c), which was found to encode a 27-kDa polypeptide (Claisse *et al.,* 1980; Anziano *et al.,* 1982; Netter *et al.,* 1982a,b; Weiss-Brummer *et al.,* 1982; Lamb *et al.,* 1983). Weiss-Brummer *et al.* (1982) provided evidence that this polypeptide is a cleavage product of a precursor protein (55 kDa) encoded by exon 1 to 4 and the intron 4 orf (Fig. 5). By use of antibodies a 27-kDa protein could be identified in mutants (Jacq *et al.,* 1984).

Mutants in intron 4 of the *cob* gene show the so-called "box effect," i.e., they are unable to synthesize cytochrome *b* and cytochrome oxidase (Cobon *et al.,* 1976; Kotylak and Slonimski, 1976, 1977a,b; Pajot *et al.,* 1976, 1977; Schweyen *et al.,* 1977; Slonimski *et al.,* 1978a,b). It could be demonstrated that the 27-kDa product (the intron 4 maturase) is needed to splice out intron 4 of the *cox1* gene (Alexander *et al.,* 1979; Dhawale *et al.,* 1981; De la Salle *et al.,* 1982; Labouesse *et al.,* 1984). Hensgens *et al.* (1984a,b) found that processing of intron 4 in *cox1* is disturbed in most *cob* mutants analyzed so far. Some mutations show additional defects in processing of the pre-mRNA of *cox1,* which could be a consequence of the structural similarities between the various introns and therefore the proteins involved in splicing. Constructs containing a heterologous presequence and the maturase gene have been shown to be translated on cytosolic ribosomes (Banroques *et al.,* 1986) and to efficiently control splicing of introns in *cob* and *cox1* (Banroques *et al.,* 1987).

In *A. nidulans* the *cob* gene is split by an intron of approximately 1.1 kbp (Waring *et al.,* 1981) which is inserted at the same site as intron 3 in the long form of the *cob* gene in *S. cerevisiae* (Lazowska *et al.,* 1981). Both introns display an orf in phase with the preceding exon and a blocked region at the 3′ end.

The *cob* gene in *N. crassa* is interrupted by two introns of about 1260 bp. The exon parts are highly homologous with *A. nidulans* (79%). The introns, however, are located at sites different from those of introns in *A. nidulans* and *S. cerevisiae* (Helmer-Citterich *et al.,* 1983; Burke *et al.,* 1984).

The 2526-bp *cob* intron in *S. pombe* strain 50 and other *S. pombe* strains is inserted at a position different from those found in the *cob* genes of *S. cerevisiae, A. nidulans,* or *N. crassa* (Lang *et al.,* 1985; Zimmer *et al.,* 1987).

The intron in *urf1* of *N. crassa* is like most introns, inserted in a highly conserved region of the gene. It belongs to group I and contains an orf of 915 bp, not continuous with the upstream exon. The predicted polypeptide polarity pattern of the intronic orf resembles that of the *cox1* intron *aI5β* of *S. cerevisiae* (Burger and Werner, 1985).

The DNA sequence of three mt plasmids associated with senescence of *P. anserina* (Cummings *et al.,* 1985) revealed that the plasmid termed ε senDNA contains three group I introns within a sequence equivalent to the mammalian *urf1* gene (*ndh1*). The first intron is closely related to the self-splicing intron in the *rnl* gene of *Tetrahymena* (see next section). The second intron is the longest member of group I introns and contains remnants of two protein coding sequences, one of which is split by the other (Michel and Cummings, 1985).

F. Secondary Structure, Function, and Evolution of Introns

Fungal mt intron sequences can be attributed to two groups (Michel *et al.,* 1982). Group II includes the first *cob* intron of *S. cerevisiae,* introns 1, 2, and 7 of the longest form of the *cox1* gene of *S. cerevisiae,* the *cob* intron in *S. pombe* strain 50 (Lang *et al.,* 1985), the *plDNA* (*cox1* intron 1) of *P. anserina* (Kück *et al.,* 1985b; see below), and the mt plasmid of the Mauriceville-1c strain of *N. crassa* (Collins *et al.,* 1981b; Nargang *et al.,* 1983, 1984; Stohl *et al.,* 1984). Due to some sequence characteristics, this intron was originally assigned to group I.

The larger group I contains the *cob* introns 2 to 5, the *cox1* introns 3 and 4, and the *rnl* intron in *S. cerevisiae,* as well as all the other introns of the ascomycetes analyzed so far. Most group I introns contain orfs. Common to all orfs are two highly conserved decapeptide motifs with a defined 5′ to 3′ order, but with varying distances between them. According to the one-letter code for amino acids, they are called *LAGLI* and *DADG*. Members of the same intron family possess short similar sequences and can be folded into similar secondary structures.

On the basis of nucleotide sequences and genetic data, a model for RNA splicing in fungal mitochondria has been developed (Waring *et al.,* 1982, 1983; Davies *et al.,* 1982, 1983; Waring and Davies, 1984). The widespread group I is characterized by a particular RNA secondary structure, based upon four conserved nucleotide sequences of 10 to

12 bases, which have been named *P, Q, R,* and *S*. Two other sequence motifs, *E* and *E′*, are highly conserved in their location, but not in sequence. The "core" of the intron consists of the base-paired regions *P3* to *P9* with the associated single-stranded loops (*P2* is present in most cases). Introns in which the core is close to the 5′ splice site can form an internal guide sequence (*IGS*) which is built up by the pairings *P1* and *P10*. Studies on mutants and revertants finally made it possible to test the validity of the predicted RNA secondary structures for the *cob* introns 1, 4, and 5 in *S. cerevisiae* (Dujardin *et al.,* 1982; Schmelzer *et al.,* 1982, 1983; Rödel *et al.,* 1983; Weiss-Brummer *et al.,* 1983; Bonjardin and Nobrega, 1984; Holl *et al.,* 1985a,b; Perea and Jacq, 1985). Structural conventions for group I introns have been published by Burke *et al.* (1987).

The basis for a systematic search for critical sequences involved in the process of splicing was provided by the rRNA intron in *Tetrahymena thermophila*. In this organism, a 413-bp intron interrupts the *rnl* gene. This group I intron (Cech *et al.,* 1983) has been shown to perform self-splicing (for references, see Grabowski *et al.,* 1981; Kan and Gall, 1982; Kruger *et al.,* 1982; Zaug and Cech, 1982, 1986; Brehm and Cech, 1983; Cech *et al.,* 1983; Zaug *et al.,* 1983; Bass and Cech, 1984; Waring and Davies, 1984; Been and Cech, 1986, 1987; Burke *et al.,* 1986; Garriga *et al.,* 1986; Inoue *et al.,* 1986; Szostak, 1986; Waring *et al.,* 1986; Kay and Inoue, 1987; Price, 1987; Price *et al.,* 1987; for reviews, see Abelson, 1982; Lewin, 1982; Been and Cech, 1985; Nielsen and Engberg, 1985; Price *et al.,* 1985; Tanner and Cech, 1985a,b; Altman *et al.,* 1986; Cech, 1986a,c; Cech and Bass, 1986) and to act like a RNA restriction endonuclease (Zaug *et al.,* 1986). Waring *et al.* (1985) developed an *in vivo* RNA splicing assay for this intron *E. coli*. Point mutations were made in the conserved sequences *P, Q,* and *S*. All showed reduced splicing, agreeing with the data on mutation and reversion in the mt introns of *S. cerevisiae*. The results support the findings that pairing of *R* with *S* and *P* with *Q* is important for intron structure and function. The *cob* intron 1 in *N. crassa* is also able to perform self-splicing (Grimm *et al.,* 1981; Wollenzien *et al.,* 1983; Garriga and Lambowitz, 1984). Tabak *et al.* (1984) studied splicing of precursors of the large rRNA of *S. cerevisiae* and found intermediates with the same properties as the self-splicing rRNA intron of *Tetrahymena*. In subsequent papers (Tabak *et al.,* 1985, 1987; van der Horst and Tabak, 1985, 1987; Arnberg *et al.,* 1986; reviews, see Tabak and Arnberg, 1986; Tabak and for Grivell, 1986). Self-splicing has also been described for the terminal intron (*bI5*) of *cob* (Gampel and Tzagoloff, 1987); they showed that the intron in the *rnl* gene (and possibly some of the *cox1* introns) are able to perform self-catalyzed splicing via splicing intermediates. Self-splicing has also been demonstrated for the group II intron *I5γ* in the *cox1* gene and the first *cob* intron in *S. cerevisiae* (Van der Veen *et al.*, 1986, 1987; Peebles *et al.*, 1986; Schmelzer and Schweyen, 1986; Jacquier and Rosbash, 1986; Jacquier and Michel, 1987; Schmelzer and Müller, 1987). The generality of the self-splicing reaction and its relationship to nuclear mRNA splicing were reviewed by Cech (1986). The fact that introns contain their own catalytic center for RNA splicing suggests that the maturases may not function as enzymes in the strict sense, but rather as cofactors to keep the intron in a correct position for splicing *in vivo.*

Group II introns show a sequence within their last 100 or so bases at their 3′ end that can be folded into a 14-bp hairpin with a constant terminal loop and a CG bulge on its 3′ side (Michel and Dujon, 1983).

The biological role of introns and their evolution have been the subject of a variety of reviews. Here we will concentrate mainly on intron-encoded proteins, which are special for lower eukaryotes. More than 20 different orfs have been located in introns of mt genes of ascomycetes. Experimental evidence for at least four different functions of intron-encoded proteins is now provided:

1. *Intron encoded proteins are essential for the correct splicing of some intervening sequences in pre-mRNAs.* It is reasonable to conclude that the transesterification reaction in the self-splicing process is the basic mechanism and that maturases induce or stabilize RNA folding. The introns devoid of orfs may be self-splicing *in vivo* or dependent on extraneous proteins. For group I introns in *N. crassa,* it could be shown that the mt tyrosyl-tRNA synthetase or a derivative thereof is necessary for splicing (Garriga and Lambowitz, 1986; Akins and Lambowitz, 1987).

2. *Intron-encoded proteins may be involved in clean intron DNA deletion.* Gargouri *et al.* (1983) discovered that defective introns can be removed from the gene by clean DNA excision. The authors hypothesized that this is the result of a DNA–DNA recombination between a cDNA reverse-transcribed from a mature mRNA with an intron-containing gene. Since multiple intron deletions within a gene are frequent, this is a strong argument that the spliced RNA is the template. Michel and Lang (1985) showed that proteins encoded by group II introns bear a significant homology to reverse transcriptases of retroviruses. One of these proteins, the product of intron *aI1* in *S. cerevisiae,* has a maturase function (Carignani *et al.*, 1983). It is worth mentioning here that Zimmern (1983) already has found a homology

between some proteins of tobacco mosaic virus with four mitochondrially encoded intron proteins. Akins *et al.* (1986) have provided evidence for reverse transcription in *N. crassa* mitochondria. Steinhilber and Cummings (1986) detected a DNA polymerase possessing properties of a reverse transcriptase in *P. anserina*. It will be of interest to see if this activity is encoded by the first *cox1* intron (*plDNA*) of this fungus. Furthermore, it would be interesting to investigate whether in *vivo* or *in vitro* a reverse transcriptase activity of the intron encoded proteins of *S. cerevisiae* or *S. pombe* could be demonstrated (Flavell, 1985).

3. *An intron-encoded protein is involved in the transposition of its own intron.* The transposase-like function of the *rnl* intron of *omega*$^+$ strains in *S. cerevisiae* has already been described (Jacquier and Dujon, 1985; Macreadie *et al.,* 1985a,b). It may serve for stimulating discussions to view the introns as relict transposons (Hickey and Benkel, 1986).

4. *An intron-encoded protein of one gene induces homologous recombination in the exons of another gene.* This activity was described by Kotylak *et al.* (1985). It concerns the induction of genetic recombination in homologous exons and is therefore referred to as "DNA recombinase." In crosses between mt mutants of *S. cerevisiae* and *Saccharomyces douglasii,* it could be demonstrated that a rearrangement of the *cox1* gene is induced by the protein product of *cob* intron 4 (p27 protein). The DNA recombinase function is important, since it does not stimulate the transposition of the intron that encodes it, but acts on another gene. In contrast to the *omega* transposition, the recombinase promotes a general rather than a site-specific recombination event. The finding of Delahodde *et al.* (1985) that the p27 protein has an affinity *in vitro* to single-stranded DNA supports the idea of a DNA recombinase.

Kotylak *et al.* (1985) have termed the intron-encoded proteins "nucleic acid wielding" proteins, since they appear important for both gene expression and genomic evolution. We feel it appropriate to quote some sentences of Kotylak *et al.* (1985) on the role of introns:

> We believe that they are very ancient in evolution and may represent relics of even more primaeval proteins that could mainpulate both types of nucleic acids. Why are the genes coding for these proteins located at present in introns of the basic set of mt genetic information? We can think of two reasons which are not necessarily mutually exclusive:
> (i) Mosaic genes and RNA molecules may have been the primitive forms of heredity and they could have remained shielded in organelles of lower eukaryotes because the selective pressure was not strong enough to eliminate them;

(ii) Nucleic acid wielding proteins are potentially dangerous for the integrity of nucleic acids in a manner similar to the proteins encoded by other transposable elements. Their location in introns provides an automatic feed-back mechanism that regulates their abundancy at a very low level and provides at the same time a possibility to overexpress them under specific circumstances. Somewhere, at the root of metazoa, the introns and the extra set of genes were lost from the organelle (possibly translocated in the nucleus?) and have never been regained since. As a consequence of this loss the genetic recombination of mt genomes was lost and the overall organization of mt DNA remained unchanged or only slightly modified. (pp. 17–18)

An intronless strain of *S. cerevisiae* has been constructed by Seraphin *et al.* (1987) which could be used to test this hypothesis directly.

IV. Organization and Replication of Mitochondrial Chromosomes

MtDNA represents approximately 5–25% of the total DNA of *S. cerevisiae* (Williamson *et al.*, 1970). The copy number is about 50 molecules per haploid cell and about twice that in diploid cells (Grimes *et al.*, 1974). MtDNA content is under the control of nuclear genes and is variable under different physiological conditions and during the cell cycle. MtDNA is associated with a lysine- and arginine-rich protein (HM) (Caron *et al.*, 1979), which resembles the histones. Mt chromosomes are organized in "chondriolites" (Williamson and Fennell, 1975; Williamson, 1976; Williamson *et al.*, 1977; Rickwood and Chambers, 1981; Rickwood *et al.*, 1981), in which the number of mtDNA molecules varies between one and eight. This organization partially explains the rapid segregation of mt genomes (for a review see Birky, 1983). In *S. cerevisiae,* replication of mtDNA takes place continuously throughout the cell cycle, and even during mitosis (Williamson and Moustacchi, 1971; Dawes and Carter, 1974a,b; Küenzi and Roth, 1974; Pinon *et al.*, 1974; Wells, 1974; Sena *et al.*, 1975, 1978). In contrast, Smith *et al.* (1968) reported that mtDNA synthesis is periodic during synchronous growth of *Saccharomyces lactis,* and occurs at a time different from that at which nuclear DNA is synthesized. The same observation has been made for mtDNA replication in *S. pombe* (Del Giudice and Wolf, 1980; Del Giudice *et al.*, 1981a,b). In *S. cerevisiae,* mtDNA replication continues in α-factor arrested cells and also after inhibition of cytoplasmic protein synthesis (Grossman *et al.*, 1969; Petes and Fangman, 1973). It continues in cells with cell division cycle (*cdc*) mutations affecting initiation of nuclear DNA synthesis, but not in those with mutations affecting continuation (Cottrell *et al.*, 1973; Newlon and Fangman, 1975). There is evidence that the whole mtDNA population replicates synchronously in a yeast cell, i.e., any molecule starts replication at a given point at the same time (Cottrell and Lee, 1981). In *S. pombe,* inhibition of mtDNA synthesis by ethidium bromide also inhibits cell growth and mtDNA synthesis immediately ceases on inhibition of cytoplasmic protein synthesis (Del Giudice *et al.,* 1981a). This difference may reflect a closer interaction of nucleus and mitochondrion in petite-negative rather than in petite-positive yeast. In a Meselson- and Stahl-type experiment, Williamson and Fennell (1974) and Sena *et al.* (1975) observed an apparent dispersive mtDNA replication, which they interpreted as the result of multiple recombination events. In contrast, Mattick and Hall (1977) and Leff and Eccleshall (1978) observed discrete classes of density using BUdR as label. A mtDNA polymerase has been reported by Iwashima and Rabinowitz (1969) and Wintersberger and Wintersberger (1970), and characterized further (Wintersberger and Blutsch, 1976) as a γ-polymerase (Bolden *et al.,* 1977; Zimmermann *et al.,* 1980). Since *rho*$^-$ mutants replicate their mtDNA faithfully, the mtDNA polymerase is encoded in the nucleus. MtDNA synthesis, like that of bacteria, seems to be associated with the inner membrane (Hall *et al.,* 1975). As will be discussed below in the section on the *rho*$^-$ mutation in *S. cerevisiae* (Section VI,A), sequences have been identified which promote efficient replication of a *rho*$^-$ genome carrying this segment of DNA. This phenomenon is called suppressiveness (Bernardi, 1979, 1982a; Bernardi *et al.,* 1980; detailed discussion below), and these sequences are called *rep* (for replication) or *ori* (for origin). All together, eight such sequences have been identified on the mt genome, and five of them (*ori*1 to 5) are active. At least one of these active *ori* sequences is present in the mt genome of most spontaneous *rho*$^-$ petites; they are discussed in detail in the section on *rho*$^-$ petites.

V. Transcription: Initiation, Transcript Processing, and Termination

Transcription and transcript processing in *S. cerevisiae, N. crassa, S. pombe,* and *T. glabrata* have been reviewed (Grivell *et al.,* 1979, 1982; Levens *et al.,* 1979; van Ommen *et al.,* 1979; Edwards *et al.,* 1983a; Lang *et al.,* 1983; Tabak *et al.,* 1983a,b; Breitenberger *et al.,* 1985; Burger *et al.,* 1985; Clark-Walker *et al.,* 1985; Breitenberger and RajBhandary, 1985). A DNA-dependent RNA polymerase from yeast mitochondria has been reported by Tsai *et al.* (1971), Wintersberger (1970, 1972), Scragg (1976), Levens (1981a,b), Lustig *et al.* (1982), and

Winkley *et al.* (1985). The early preparations were of very low specific activity and showed different composition and sensitivity to α-amanitin and rifampicin. The mtRNA polymerase of Levens had a high specific activity: a nascent RNA polymerase–DNA complex completes *in vitro* the synthesis of rRNAs and initiates on mtDNA fragments in a manner similar to *in vivo* studies (Edwards *et al.*, 1982). The preparation of Winkley *et al.* (1985) had two distinctive properties on poly[d(AT)] and cloned DNA containing mt promoters. Using promoter-containing DNA as a template, the preparation was dissociated into two nonfunctional components. Selective transcription was restored when the two components were recombined. Schinkel *et al.* (1986) could also resolve the RNA polymerase in two components: one retaining the capacity to synthesize RNA, the other conferring the correct specificity of initiation to the catalytic component. A specific RNA polymerase has also been isolated from *N. crassa* (Küntzel and Schäfer, 1971; Wintersburger, 1972).

In human mitochondria, and possibly also in other mammals, both DNA strands are transcribed completely from a region near the origin of H-strand replication (Montoya, 1982), and the primary transcripts are subsequently processed (Clayton, 1984; Chang *et al.*, 1985). Large primary transcripts have also been observed in *S. cerevisiae* mitochondria, but at least 19 promoters have been identified (Edwards *et al.*, 1982, 1983b): 14 initiation sites have been mapped precisely (Christiansen and Rabinowitz, 1983; Osinga *et al.*, 1984a,b), and another 5 have been demonstrated by *in vitro* capping experiments (Fig. 1) (Levens *et al.*, 1981a–c; Christiansen and Rabinowitz, 1983).

Transcription starts after a consensus sequence, the nonanucleotide ATATAAGTA, which has been demonstrated for the two rRNA genes (Osinga and Tabak, 1982; Christiansen *et al.*, 1982), *cox1* (Osinga *et al.*, 1984a,b), *cox3* (Thalenfeld *et al.*, 1983b), *var1* (Zassenhaus *et al.*, 1983, 1984), *ATPase9* (Edwards *et al.*, 1983b), four tRNA genes (Osinga *et al.*, 1984b; Christiansen and Rabinowitz, 1983; Miller *et al.*, 1983; Christiansen *et al.*, 1983; Frontali *et al.*, 1985; Wettstein-Edward *et al.*, 1986), and *ori* sequences (Osinga *et al.*, 1982, 1984a,b; Palleschi *et al.*, 1984a,b). Genes and the *ori* sequences possess exactly the same nonanucleotide sequence, whereas the four tRNA transcripts contain a motif with one base deviation. Nonanucleotide boxes are sometimes arranged in tandem, for example, in front of *atp9* (Edwards *et al.*, 1983b), tmm_f (Miller *et al.*, 1983b), and *tmf* (Christiansen and Rabinowitz, 1983). The initiation sequence found in front of the *tmc* gene (Bos *et al.*, 1979) does not appear to be used (Frontali *et al.*, 1982). The variation among initiation sequences may provide a possibility for regulating tRNA gene transcription. The nonanucleotide sequence is also found in front of the rRNA genes in *K. lactis* (Osinga *et al.*, 1982). It has turned out, however, that the definition of a promoter is difficult (Tabak *et al.*, 1983b; Biswas and Getz, 1986). An *in vitro* transversion (G → T) within the nonanucleotide box did not influence the efficiency of transcription, whereas substitutions in positions +10 and +13 completely abolished transcription. *In vitro* transcription assays with a complete set of single site mutations indicate that a minimal sequence of 8 nucleotides [5′ TAT/aAA/g/cGT/a/cN(+1) 3′] is the minimal sequence necessary to direct accurate initiation by mtRNA polymerase (Biswas *et al.*, 1987). Schinkel *et al.* (1987) have found in an *in vitro* transcription system that none of a series of point mutations in the conserved sequence abolishes transcription–initiation completely and that one mutation even creates a promoter up phenotype. It has been speculated that the conserved box is part of a negative control circuit comparable to a bacterial operator or an SV40 enhancer sequence (Tabak *et al.*, 1983b).

Multiple start-transcription sites have also been identified in *T. glabrata* (Sriprakash and Clark-Walker, 1983; Clark-Walker *et al.*, 1985).

In mammals, the mature transcripts originate by cleavage at tRNA sequences that flank every gene (Battey and Clayton, 1980; Ojala *et al.*, 1980, 1981). In fungal mitochondria this mechanism is also used, as shown for *N. crassa, S. cerevisiae,* and *S. pombe* in detail below.

In *N. crassa* there is evidence for a single precursor RNA that contains segments corresponding to *cob, tmc, cox1, tmr,* and *urf1* (*ndh1*) (Burger *et al.*, 1985; De Vries *et al.*, 1985b). As in HeLa cells, the two tRNAs serve as punctuation signals for the cleavage of this transcript, generating the *cox1* messenger, as well as the 3′ end of the *cob* transcript. It was believed that *Pst*I palindromes could serve as processing sites in *N. crassa* (Yin *et al.*, 1981; Macino and Morelli, 1983): since six of them are present in *cob* and *cox1,* this possibility can be ruled out. Breitenberger *et al.* (1985) studied transcription of *cox3, rns,* and *urfn1* (= *urfa1* = *ndh6*?). In all three cases, transcripts including tRNA sequences are subsequently processed to generate the mature RNAs. The endpoints of the most abundant transcripts coincide with those of tRNA sequences, and these authors reached the same conclusion that tRNA sequences act as primary signals for RNA processing in *N. crassa* mitochondria (Agsteribbe and Hartig, 1987).

In *S. pombe,* most genes are separated by tRNA sequences which were proven to act as processing signals. Only upstream of *rnl* and between *cox1* and *cox3*, no tRNA gene is found (Lang *et al.*, 1983).

Also in *S. cerevisiae*, tRNA genes are part of large primary transcripts, and their release as mature tRNAs requires processing at both their 5′ and 3′ ends (Locker and Rabinowitz, 1981; Farelly *et al.*, 1982; Frontali *et al.*, 1982; Christiansen *et al.*, 1983; Miller *et al.*, 1983b; Thalenfeld *et al.*, 1983b; Zassenhaus *et al.*, 1984). *cox1*, *atp8*, and *atp6* are cotranscribed, but no tRNA gene is contained in the primary transcript (Nagley *et al.*, 1981; Beilharz *et al.*, 1982; Cobon *et al.*, 1982).

The involvement of GC clusters in mRNA processing was discussed by Simon and Faye (1984) (see also Faye and Simon, 1983).

Nothing is known so far about the mechanism of transcript termination. For two urfs, 3′ ends of RNAs have been mapped beyond a dodacemer motif in yet unsequenced regions. The generation of 3′ ends of mature mRNAs appears to be marked by 12 nucleotides which are (almost) contained within the 3′ end of the mRNAs (Michel, 1984; Osinga *et al.*, 1984a). The dodacamer sequence is also found downstream of the intron urf in *rnl* gene in *K. thermotolerans* (Jacquier and Dujon, 1983).

In *S. pombe*, an oligo(TC) stretch is found downstream of all protein-encoding genes and also after the *rns* gene (Lang *et al.*, 1983).

In *S. cerevisiae*, 5′ termini of mature mRNAs result from RNA processing (Merten *et al.*, 1980; Locker and Rabinowitz, 1981) with the exception of *atp9* and *cox1*. However, no primary sequence similarities exist for the 5′ processing sites.

Transcript patterns have been shown to change under glucose repression (Mueller and Getz, 1986a,b) and derepression and during developmental processes such as sporulation in *S. cerevisiae*. In sporulating cells of *S. cerevisiae* circular RNAs are absent (Schroeder *et al.*, 1983) and processing of *cob* and *cox1* transcripts appears to be much more advanced in cells undergoing sporulation than in vegetative cells (Schweyen *et al.*, 1982). In sporulating cells, a novel RNA species has been described which is located 2.5 to 5 kbp in front of *cob* exon 1 (Schweyen *et al.*, 1982). Glucose repression and derepression are known to influence mt translation products (Agostinelli *et al.*, 1980; Böker-Schmitt *et al.*, 1982; Falcone *et al.*, 1983). Levels of rRNA are different in glucose-repressed and -derepressed cells (Kelly and Phillips, 1983). Transcription of protein coding and tRNA genes is strongly affected by glucose repression (Frontali *et al.*, 1982; Baldacci and Zennaro, 1982b; Baldacci *et al.*, 1983; Baldacci and Fennaro, 1987) and release from glucose repression (Zennaro *et al.*, 1985).

VI. Rearrangements of the Mitochondrial Genome

A. The *rho*$^-$ Mutation in *Saccharomyces cerevisiae*

The *rho*$^-$ mutation, which is unique to the mt genome of *S. cerevisiae* and some other yeasts, is correlated with gross alterations or complete loss of the mt genome (Faye *et al.*, 1973; Barnardi, 1979). Due to the extensive deletions, these mutants are unable to carry out mt protein synthesis. Yeast species giving rise to mutants capable of surviving with deletions or loss of the mt genome, and consequently without mt protein synthesis, are called petite positive, the remaining yeast species petite negative. *rho*$^-$ mutants provide an excellent tool for mapping mutational sites, they are interesting per se for the study of the deletion mechanism, and they provide a natural cloning system (Fukuhara and Rabinowitz, 1979).

The *petite colonie* (*rho*$^-$) mutant was first discovered in 1949 (Ephrussi *et al.*, 1949a,b) and subsequently characterized genetically and physiologically (reviewed by Dujon, 1981). This mutation occurs spontaneously at a high rate (1–2%/cell/generation), and in general, all conditions where yeast grows poorly [e.g., block of mt protein synthesis (Myers *et al.*, 1985)] promote *rho*$^-$ formation. They can be induced by a series of mutagenic agents and antibacterial antibiotics and a variety of regimens is able to modulate the induction of *rho*$^-$ mutants. In *statu nascendi, rho*$^-$ mutants are able to revert, whereas established *rho*$^-$ mutants do not revert (Rank and Person, 1970). Similarly, complementation in cytoplasmic petite mutants to form respiratory-competent cells is only possible with newly arising *rho*$^-$ mutants (Clark-Walker and Gabor-Miklos, 1975; Evans *et al.*, 1985; Evans and Clark-Walker, 1985). The fact that many nuclear genes are involved in the maintenance of the mt genome is documented by a variety of mutants that accumulate *rho*$^-$ mutants (reviewed by Dujon, 1981). The mt conditional mutant *tsm8* also accumulates *rho*$^-$ mutants when grown at the nonpermissive temperature (Handwerker *et al.*, 1973; Bandlow and Schweyen, 1975; Bandlow *et al.*, 1977; Bechmann *et al.*, 1977; Backhaus *et al.*, 1978). The mutational site maps in close proximity to the *ile* tRNA gene (Monnerot *et al.*, 1977) and the region has been sequenced in the wild type (Bandlow *et al.*, 1980). Sequence analysis has revealed a T → A change in the *tmy* gene in the mutant. This alteration strongly destabilizes the secondary structure of the extra arm in the mutant *tyr* tRNA. *In vitro* aminocylation studies have shown that their *tyr*-accepting capacity was less than 10% even at the permissive temperature. The limitation of this tRNA could explain the defect of some polypeptides and, as a consequence, *rho*⁻ formation at the high temperature (Bordonne *et al.*, 1984, 1987).

The *rho*⁻ mutation is not viable in the nuclear *op1* mutant (*pet9*) which is defective in adenyl nucleotide translocase (Kotylak and Slonimiski, 1977a). "Petite-negative" mutants have been isolated that do not grow on fermentable carbon sources while growing normally on ethanol media, in which *rho*⁻ petites are not viable (Lancashire *et al.*, 1981).

Recombination of *rho*⁻ genomes having tandemly arrayed repeat units with *rho*⁺ genomes is blocked in mutants carrying the *pif* allele (*p*etite *i*ntegration *f*requency) (Foury and Kolondynski, 1983a,b; Foury and Van Dyck, 1985; Foury and Lahaye, 1987).

Numerous studies have been carried out to characterize the mt genome from *rho*⁻ mutants of *S. cerevisiae* by buoyant density determination, RNA–DNA hybridization, DNA denaturation and reassociation, electron microscopy, restriction mapping, and DNA sequencing (reviewed by Dujon, 1981). Mutants devoid of mtDNA (*rho*°) have been described by Nagley and Linnane (1970).

In a *rho*⁻ mutant any segment of the genome can be retained and amplified. The deletion may range from 30% to greater than 99.5% of the mtDNA. *rho*⁻ mutants with very short conserved fragments have been reported (Mol *et al.*, 1974; Van Kriejl and Bos, 1977; Bos *et al.*, 1980a; Fangman and Dujon, 1984), and since *rho*⁻ mutants contain the same amount of mtDNA as the wild-type cells, the amplification factors of these segments may be considerable. The process of amplification has been studied by electron microscopy, DNA denaturation and renaturation, and restriction enzyme analysis (see Dujon, 1981). Several hypotheses have been proposed for the process of petite formation based on sequence analysis of various petites (Clark-Walker and Gabor Miklos, 1974; Perlman and Birky, 1974; Bernardi *et al.*, 1976, 1980; Slonimski and Lazowska, 1977; Fukuhara and Rabinowitz, 1979; Baldacci *et al.*, 1980; Gaillard *et al.*, 1980; Clark-Walker *et al.*, 1981a,b; de Zamaroczy *et al.*, 1983; Sor and Fukuhara, 1983; Clark-Walker, 1985).

There are several possibilities for the arrangement of the repeat units in *rho*⁻ petites: the simplest is the straight head-to-tail repeat. The other type of rearrangement is the inverted duplication of the conserved sequence. Other *rho*⁻ mutants contain mixed repeats in which the repetitions may be direct or inverted in the same molecule. In contrast to the "simple" *rho*⁻, others show internal rearrangements or "scrambled" sequences, in which originally separated sequences are made continuous. The easiest way to imagine excision is a mechanism analogous to the integration and excision of bacteriophage λ. By the excision process a circular monomer is created which is amplified until the final genome size is achieved. Excision sequences were found to be perfect direct repeats, often flanked on one or both sides by regions of imperfect homology. In spontaneous petites, excision takes place in AT spacers and GC clusters of intergenic regions (Bernardi and Bernardi, 1980). In the case of ethidium bromide-induced petites, excision sequences are also found in introns and exons of genes (Bernardi *et al.*, 1976, 1980; Bernardi and Bernardi, 1980; Bernardi, 1982a,b; Baldacci *et al.*, 1980; Gaillard *et al.*, 1980; Baldacci and Bernardi, 1983). Sor and Fukuhara (1983) proposed a model for the origin of palindromic petites, which is illustrated in Fig. 6A and B. They found that the palindromes were not symmetrical right up to the junction points. There was always an asymmetric sequence of variable length at the junction. At both ends of this sequence, short inverted sequences were found that were present in the wild-type genome. The authors proposed that at the moment of excision a single-strand cut occurs at each of these inverted repeats, in such a way that the two complementary strands are cut unequally. Thus the single-stranded sticky ends become the junction sequences between the palindromic repeating units.

Inversions in mtDNA of a petite mutant have been analyzed by Wakabayashi and Furutani (1984).

According to their behavior in crosses with *rho*$^+$ tester strain, petites are classified as neutral, suppressive (Rank, 1970a,b; Rank *et al.*, 1972a,b), or super(hyper)suppressive. The diploid progeny of a cross between a neutral *rho*$^-$ and a grande is entirely respiratory competent. In the case of suppressive petites, the portion of respiratory-competent progeny varies between 1 and 99%, and in crosses with hypersuppressive petites, the entire progeny is respiratory deficient. The phenomenon of suppressiveness in crosses between *rho*$^+$ and *rho*$^-$ cells was first described by De Margerie-Hottinguer and Roman (1955) and Ephrussi *et al.* (1955), and analyzed both genetically and biochemically (for review see Dujon, 1981). The degree of suppressiveness (Chambers and Gingold, 1986) is transmitted unchanged to the majority of subclones of a given petite. Some subclones, however, exhibit a novel degree of suppressiveness, which is again transmitted to the progeny.

Crosses between established *rho*$^-$ mutants do not yield respiratory-competent progeny. Using drug resistance markers, Michaelis *et al.* (1973) could demonstrate that recombination between petite genomes takes place, indicating that the enzymes for the recombination process are nuclearly coded (Strausberg and Birky, 1979). Recombination

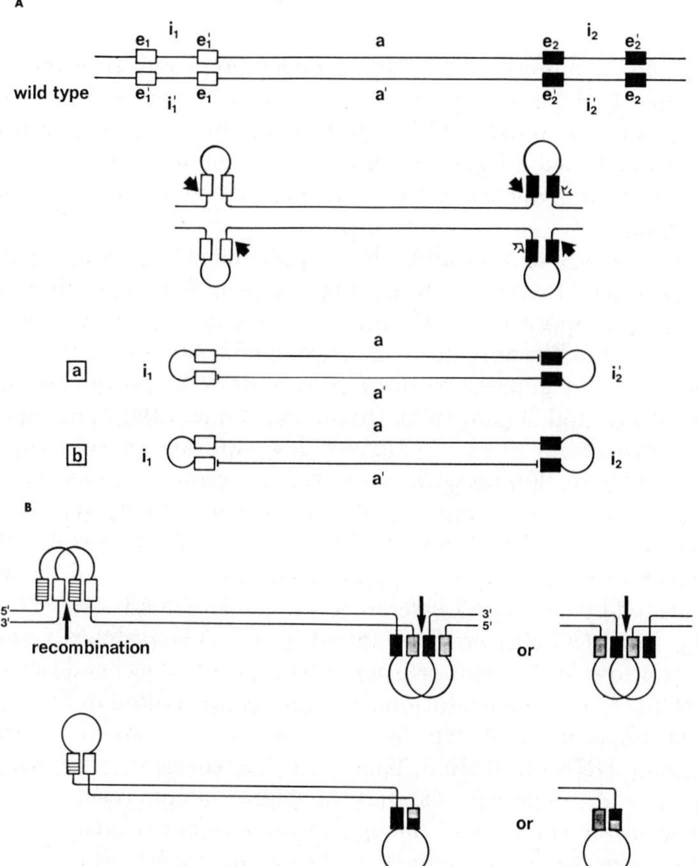

FIG. 6. (A) Region of wild-type double-stranded mtDNA from which a *rho*$^-$ DNA is to be excised. Two pairs of short inverted repeats (inversion signals) are indicated by e_1 and e_1' (white bars) and e_2 and e_2' (black bars). These pairs are separated by a segment a/a′. i_1/i_1' and i_2/i_2' are the segments between e_1 and e_1' and e_2 and e_2', respectively. In the model, the intrastrand pairings e_1/e_1' and e_2/e_2' will loop out the segments i_1, i_1', i_2, and i_2'. Then the excisions (arrows) occur at the opposite ends of each inverted pair, so that i_1 or i_1' (and i_2 or i_2') become single-stranded overhangs at the extremities of the excised segment a/a′. When both e_1/e_1' and e_2/e_2' signals are cut at the 5′ side, form a is obtained; when e_1/e_1' signals are cut at the 5′ side, but e_2/e_2' at the 3′ side (white arrows), form b is obtained. These are dimeric structures which after ligation of the gaps will form a single-stranded palindromic dimer circle. This circle will replicate to become double-stranded palindromes. (Adapted from Sor and Fukuhara, 1983b.) (B) In this model intrastrand and interstrand pairing of the inversion signals create a four-stranded situation, and can form a Holliday structure connecting the two opposite strands placed in the same orientation, through recombination at the points indicated by the arrows. This yields a palindromic dimer circle with i_1 (or i_1') and i_2 (or i_2') at the junctions. The dimer molecules will then replicate and can generate, e.g., through intermolecular recombination, a multimeric series composed of palindromic dimers. (Adapted from Sor and Fukuhara, 1983b.)

between *rho*⁻ genomes with little sequence homology resulted in an equal amount of the two parental genomes (Michaelis *et al.,* 1976). Lazowska and Slonimski (1977) could show by electron microscopy that the recombined mt genomes also form oligomeric series of circles like the parental genomes and are organized as inverted repeats of the both genomes.

The third class of mutants, the hyper(super)suppressive petites, have been studied extensively during the past few years. In crosses between hypersuppressive (*HS*) mutants, which have a genome composed of very short repeat units of roughly 400–900 bp, with the wild type, only the mt genome of the *HS* mutant is found in the diploid progeny (Blanc and Dujon, 1980; Dujon and Blanc, 1980; Goursot *et al.,* 1980; de Zamaroczy *et al.,* 1981). All *HS* mutants share a common sequence of about 300 bp with the wild-type genome, which is called *rep* or *ori*. *ori* sequences consist of distinct regions illustrated in Fig. 7 (Baldacci *et al.,* 1983, 1984; de Zamaroczy and Bernardi, 1985b; Faugeron-Fonty and Goyon, 1985): two GC clusters (A and B), which are separated by a short AT stretch, a central 200 bp AT stretch (called l), and a third GC cluster (C), flanked by the AT stretches r and r*, which contain sites for transcription initiation (Baldacci and Bernardi, 1982, 1983). A possible evolutionary origin is illustrated in Fig. 8. *ori1* and *ori5* have been shown to be used as sites for RNA-primed bidirectional DNA replication. There is a clear correlation between the *ori* sequence of the petite used and the degree of suppressiveness in a cross: the lower the overall density of *ori* sequences on the genome units, the lower the suppressivity. In a screening of *HS* mutants, Blanc and Dujon (1982) were able to identify three different *rep* (*ori*) sequences. In total, eight *ori* sequences were identified in the mt genomes of 20 wild-type strains of *S. cerevisiae* (de Zamaroczy *et al.,* 1979; Bernardi, 1982a; Baldacci *et al.,* 1983a; Faugeron-Fonty *et al.,* 1984; Mabuchi and Wakabayashi, 1984; Faugeron-Fonty and Goyon, 1985). The seven canonical *ori* sequences found in strain A (de Zamaroczy *et al.,* 1984) were also found in all 20 strains tested so far, and an additional *ori* sequence (*ori8*) was localized between the *rns* gene and *cox1* (Faugeron-Fonty *et al.,* 1984). *ori* sequences display

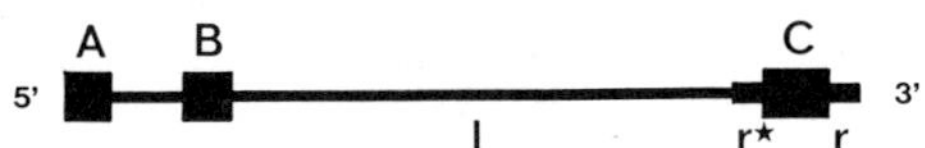

FIG. 7. Schematic presentation of an *ori* sequence. A, B, and C are GC clusters; r* and r are sites where bidirectional DNA replication is initiated by RNA primers; l is an AT stretch of approximately 200 bp (for details see the text).

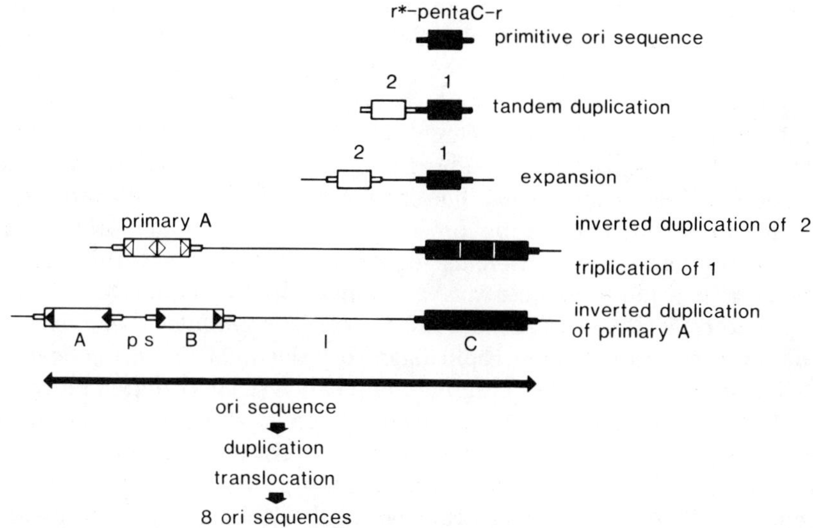

FIG. 8. Hypothetical scheme for the evolution of *ori* sequences (after de Zamaroczy and Bernardi, 1985). Formation of an *ori* sequence might have occurred in the following way: (1) duplication of the primitive *ori* sequence, composed of the r*-pentaC-r motif (see Fig. 7), followed by functional inactivation of the copy; (2) expansion, through a slippage mechanism accompanied and/or followed by internal duplication and/or intramolecular unequal crossing-over, of the sequences separating the two pentaC motifs; (3) formation of cluster A by inverted duplication of the pentaC copy; and (4) formation of the A–B region via duplication–inversion of cluster A and its flanking A + T region.

different orientations on the mt genome; *ori2* and *7* and *ori3* and *4* are close together and oriented in tandem; *ori6* and *7* have not been found at all in spontaneous petites, suggesting that these sequences may not be active *in vivo*. Partial or total deletion of the *ori* sequences as well as rearrangements affect the suppressivity of the petite. *ori*$^-$ petites, in which the *ori* sequence is partially deleted, show a decreased suppressivity. *ori*^{r} petites with an inverted orientation of the *ori* sequences within the same repeat unit exhibit a very low degree of suppressiveness (Faugeron-Fonty *et al.*, 1983; Mangin *et al.*, 1983). *ori*^{o} petites which lack the canonical *ori* sequences, and contain other *ori* sequences instead, have very low degrees of suppressiveness (Goursot *et al.*, 1982). Due to the efficient replication of *HS* petite genomes, certain types of *rho*$^-$ petites arise at an extremely high frequency in wild-type strains (*ori*^{h} petites: Marotta *et al.*, 1982) or in cultures of *mit*$^-$ mutants (*ori1:* Bingham and Nagley, 1983; Maxwell *et al.*, 1986). It is interesting in this context that spontaneous and induced (not supersuppressive) *rho*$^-$ mutants specifically delete a sequence around the *cox1* region (Fukuhara and Wesolowski, 1977; Mathews *et al.*, 1977). Mt *ori* sequences are also able to serve as origins of replication outside the mitochondria (Blanc and Dujon, 1982; Hyman *et al.*, 1983; Mabuchi and Wakabayashi, 1984; Mabuchi *et al.*, 1984; Wakabayashi and Mabuchi, 1984). Blanc (1984) has shown in a *HS* petite containing *rep2* that two modules must be present in cis to promote autonomous replication: an 11-bp consensus sequence, common to several autonomous replication sequences (*ars*) and a palindromic sequence of the mt replicator. MtDNA sequences of *C. utilis* have also been shown to support autonomous replication of plasmids in *S. cerevisiae* (Tikomirova *et al.*, 1983). Fangman and Dujon (1984) isolated mutants with reduced suppressiveness from a *HS* petite: one moderately suppressive and one nonsuppressive, which retained only 89 and 70 bp, respectively. Since these genomes, consisting only of AT, were able to replicate, it could be deduced that no single GC base pair is necessary for amplification and exhibition of suppressiveness.

B. *poky* and Related Mitochondrial Mutations in *Neurospora*

1. *General Classification*

Three years after the discovery of the petite mutant by Ephrussi and co-workers, Mitchell and Mitchell (1952) described a mutant of *Neurospora, poky* (or *mi-1*) which exhibited slow growth. Like the *rho*$^-$ mutant in yeast, it lacked cytochromes aa_3 and *b* (Mitchell *et al.*, 1953; Tissieres *et al.*, 1953; Tissieres and Mitchell, 1954), but not completely, since in older *poky* cultures the content of cytochromes aa_3 and *b* was increased. In the following years a great number of non-Mendelian slow-growing mutants have been isolated which nearly exclusively exhibited cytochrome deficiencies. Bertrand and Pittenger (1972a,b) classified these mutants into three major groups. Group I includes *poky* (*mi-1*), *SG* (slow growth), *stp* (stopper), and *exn* (extranuclear) which all show the *poky* phenotype. Group II is represented by *mi-3*, which possesses normal amounts of cytochrome *b*. Group III comprises mutants with the "stop–start" phenotype, which are called *stopper* mutants, which are deficient in cytochromes aa_3 and *b* (McDougall and Pittenger, 1966; Bertrand and Pittenger, 1969). Besides the *stopper* mutants, the *abnormal* mutants (*abn*) belong to this group (Garnjobst *et al.*, 1965; Diacumakos *et al.*, 1965). The latter mutants produce lethal segregants at a high rate. One of the most plausible explanations is that they accumulate in a similar way as chromosomal respiratory-deficient mutants in *S. cerevisiae* accumulate *rho*$^-$ mutants. This situation is in some respect reminiscent of the pseudo-wild-type (*PSW*) originating from the cross between the mutants *spg1* and *spg2* in *P. anserina*.

Complementation studies between mutants of the three classes showed that *poky* and other group I mutations predominate over the wild type in heteroplasms (Pittenger, 1956; Manella and Lambowitz, 1978; see also Puhalla and Srb, 1967, for heterokaryons with slow-growing mutants, *SG;* and Srb, 1958). The maternal inheritance of *poky* has been established by Mitchell and Mitchell (1952), and Haskins *et al.* (1953) found that cytochromes aa_3 and *b* were absent and cytochrome *c* present in great excess. It was claimed that a mt structural protein or several proteins were altered in the *poky* mutant (Woodward and Munkres, 1966; Zollinger and Woodward, 1972). In the following years, different groups have established a detailed view of the defects in electron transport and energy conservation in *poky* mutants. Rifkin and Luck (1971) found that in young *poky* cultures normal amounts of the large mt ribosomal subunit were observed, but only a few small subunits. In aging cultures, more small subunits and therefore more intact ribosomes were found, and the mutant showed elevated amounts of cytochromes aa_3 and *b*. This result was confirmed independently by Neupert *et al.* (1971). A direct proof that mt protein synthesis was impaired in *poky* mutants was provided by Sebald *et al.* (1968). Kuriyama and Luck (1974) found that the small rRNA in *poky* mutants is degraded before it is incorporated into small ribosomal subunits. Lambowitz *et al.* (1976) proposed that a specific protein of the small ribosomal subunit, S4a (now called S5: Lambowitz, 1980; Lambowitz *et al.,* 1979; La Polla and Lambowitz, 1981; Burke and RajBhandary, 1982), is the site of the primary defect in the *poky* mutation, since this protein is synthesized inside the mitochondrion, whereas all other ribosomal proteins are made in the cytosol. This is also in agreement with the finding of La Polla and Lambowitz (1977) that chloramphenicol inhibits the maturation of small ribosomal subunits. Recently, it could be shown by DNA sequencing that *poky* and other group I mutants contain a 4-bp deletion in the *rns* gene downstream from the 5′ end (Akins and Lambowitz, 1984). This deletion apparently results in synthesis of aberrant RNA which is missing 38–45 bases at the 5′ end. Akins and Lambowitz proposed that this might be the primary defect in *poky* and other group I mutants. Lemire and Nargang (1986) considered it likely that a missense mutation in the *cox1* gene is responsible for the poky phenotype.

2. *stopper* Mutants and *poky* Mutants with Large Alterations of mt DNA

The *stopper* mutants, which are grouped together with the *abnormal* (*abn*) mutants in group III, were first analyzed by McDougall and Pittenger (1966) and Bertrand and Pittenger (1969). Since these mutants are female sterile, their extrachromosomal mode of inheritance can only be shown when the mutant is used as paternal parent. Their growth pattern on solid medium is irregular and they are deficient in cytochromes aa_3 and b. They resemble the *ragged* mutation in *A. amstelodami* (see below) and, to some extent, the rho^- mutation in *S. cerevisiae.* In contrast to the *poky* mutations described in the previous section, this type of mutant is always associated with major structural changes—deletions or insertions—in the mt genome. Bertrand *et al.* (1980) proposed that the "stop–start" phenotype results from competition of defective molecules with wild-type genomes. The retained sequences of these mutants always contained both rRNA genes and most of the tRNA genes, indicating that the mechanism of deletion formation does not act randomly.

De Vries *et al.* (1980, 1981) described a stopper mutant, *E35,* with a 5-kbp deletion. De Vries *et al.* (1985, 1986) sequenced the entire region of the deletion and found that the mutant lacks the genes analogous to the human *urf2* and *urf3* (*ndh2* and *ndh3*). This mutant shows low cytochrome oxidase activity and all subunits of cytochrome aa_3 are present, but not assembled. Characterization of the deletion endpoints and of the resulting junction sequence showed that intramolecular recombination at GC-rich palindromic sequences has occurred. Gross *et al.* (1984) analyzed the mt genome of a *stopper* mutant during the stopped phase. About 20% of the genome is missing at this stage. A molecule one-third the length of the normal genome (21 kbp) is the predominant form. A complementary circle of 43 kbp appears on resumption of growth. It could be demonstrated that circles arise by reciprocal recombination at or near directly repeated *tmm* sequences. The frequency of the two genomic subsets depends on their replication rates from unique origins of replication. The two circles are also found in wild-type mt genomes. The alterations found in the *stopper* mutants are related to those found in some *poky* mutants not described in the preceding section.

In individual subcultures of *poky,* Manella *et al.* (1979) have found two variant mtDNA types (IIa and HI-10). Type IIa DNA contained an extra sequence organized as head-to-tail tandem repeats with a reiteration frequency of up to eight. This sequence appeared to be actively maintained. HI-10 DNA originated in a heteroplasmic strain with type IIa DNA as one component of the genome. This DNA was characterized by an up to 5-fold amplification of an 18-MDa segment.

Bertrand *et al.* (1980) also found deletions in the genome of three *poky* mutants. Two of them retained different sizes of the same region retained by the *stopper* mutants described above. Collins and Lambowitz (1981) identified another variant of *poky* (*exn1-548*) which contains a head-to-tail repeat of sequences with a repeat unit of 1.9 kbp in the same region as the mutant described by Manella *et al.* (1979), designed type IIb DNA. The region of insertion may contain a major replication origin. In most instances of large deletions, the remaining part of the genome is present but greatly underrepresented.

C. The *ragged* Mutation in *Aspergillus amstelodami*

The *ragged* (*rgd*) mutation in *A. amstelodami* is accompanied by the excision and head-to-tail amplification of mtDNA sequences. The mutation can originate from two regions: *rgd1* with a monomeric length of 0.9 kbp maps downstream of *rnl* (region 1); *rgd3–rgd7* are located between *cob* and *atp6* (region 2). Their sequences differ in length between 1.5 and 2.7 kbp, sharing a common sequence of 215 bp, located in *urf4* (homologous to the human *urf4* or *ndh4*). The common sequence possibly contains an origin of replication. Since region 2 DNA suppresses region 1 DNA, one can conclude that the first group of molecules possesses a more efficient *ori* sequence. No homologies have been found in the sites of excision or flanking sequences, so that the mechanism of excision is still unclear. The *ragged* mutants differ from the *rho*⁻ mutants and from senescent strains in *P. anserina,* because a wild-type copy is always present. The mutant phenotype might result from the increased dosage of amplified sequences (Lazarus and Küntzel, 1980, 1981; Lazarus *et al.,* 1980a).

D. Senescence in *Podospora anserina* and *Neurospora intermedia*

Another case of genomic rearrangement, in some respect reminiscent of the *rho*⁻ formation in *S. cerevisiae,* is observed during senescence of *P. anserina.* Strains isolated from nature become senescent and die after prolonged vegetative growth (for a review see Esser, 1974). This phenomenon was first observed and described by Rizet and collaborators (Rizet, 1953a,b). The onset of senescence depends on genetic and environmental conditions. Sexual reproduction is not affected, and senescent strains may be rejuvenated by various regimes including sexual reproduction. It is maternally transmitted and caused by an extrachromosomal infective principle. Similar senescence syndromes have been found in other organisms (Böckelmann and Esser, 1986; for further references, see Esser and Tudzynski, 1979b). In senescing cultures the cytochrome aa_3 content is gradually reduced (Belcour and Begel, 1978), indicating the involvement of a mt function in senescence. The onset of senescence can be postponed by growth in nonrepressive carbon sources, low temperature, and darkness and by treatment with different drugs. Senescence does not occur when cultures are serially passaged in liquid culture (Turker and Cummings, 1987). Senescence can be influenced by compounds which affect mt protein synthesis, DNA replication in general, or selectively that of small circular DNAs (ethidium bromide, etc.) (Esser and Tudzynski, 1977; Tudzynski and Esser, 1977; Esser and Tudzynski, 1979; Belcour and Begel, 1980; Koll *et al.*, 1984). In contrast, the life-span is decreased in all strains by cycloheximide and cholesterol (Smith and Rubenstein, 1973; Esser and Tudzynski, 1979a; Belcour and Begel, 1980). The life-span is also affected by mt mutations: in a *cap*^{r} mutant, the life-span was at least five times that of *cap*^{s} strains (Belcour and Begel, 1980). The double mutant from a cross between the mutants *spg1* and *spg2* (the abbreviation stands for impaired *sp*ore *g*ermination), called *PSW* (pseudo-wild-type) showed a great variability in longevity, which the authors interpreted as the result of a high mutation rate into a lethal and suppressive genome, similar to the suppressive *rho*$^-$ mutation in *S. cerevisiae* (Belcour and Begel, 1978). The mt mutants *mitB, mex1, mex2, mex3, mex4* (Vierny *et al.*, 1982; Belcour *et al.*, 1982), and *ex1* (Kück *et al.*, 1985) are long-lived mutants. A longevity mutant with temperature-sensitive senescence has been described by Turker *et al.* (1987). In senescent strains, circular molecules smaller than the mt genome (about 95 kbp: Cummings *et al.*, 1979a) were observed (Cummings *et al.*, 1978, 1979a–c; Stahl *et al.*, 1978, 1979). It could be demonstrated that these molecules originate from various parts of the mt genome, and are amplified during vegetative growth (Cummings *et al.*, 1979a; Jamet-Vierney *et al.*, 1980; Vierny *et al.*, 1980; Belcour *et al.*, 1981; Esser *et al.*, 1981; Kück *et al.*, 1981; Stahl *et al.*, 1981; Wright *et al.*, 1982a), originating from various parts of the mt genome. Cloning of the different molecules of senescent mycelia (senDNAs) (Cummings *et al.*, 1980) and transformation of yeast with the hybrid plasmids showed that all senDNAs and the corresponding mt sequences from young mycelia promoted extrachromosomal maintenance of the plasmids (Lazdins and Cummings, 1982).

The most frequent and most extensively studied senDNA is the α-event DNA or *plasmid-like* DNA (named *pl*DNA by Esser and collaborators). This DNA has been cloned (Cummings *et al.,* 1980; Stahl *et al.,* 1980) and sequenced (Osiewacz and Esser, 1983; Kück *et al.,* 1985b) including the region which constitute the junction sites (Cummings and Wright, 1983; Jamet-Vierny *et al.,* 1984). DNA sequencing has provided evidence that *pl*DNA represents the intron 1 of *cox1*. This group II intron, which is also present in 10 different races of *P. anserina* (Kück *et al.,* 1985a), is excised, amplified, and circularized (Esser *et al.,* 1983; Osiewacz and Esser, 1983; Kück *et al.,* 1985b). It is 2539 bp long and contains an orf. A putative autonomous replicating sequence (*ars*) has been identified within the orf. At both excision sites, inverted and direct repeats have been identified (Cummings and Wright, 1983; Osiewacz and Esser, 1984). A unique transcript of 3.2 kb hybridizes with all exon regions of the *cox1* gene, very likely representing the mature *cox1* transcript. *pl*DNA hybridizes to two different transcripts, 2.5 and 2.7 kb (Kück *et al.,* 1985) (2.4 and 2.5 kb in Wright and Cummings, 1983a). According to Wright and Cummings (1982a), these transcripts possess different 5′ ends and overlap throughout most of the length. These transcripts are found in both juvenile and senescing mycelia. Kück *et al.* (1985b), however, discuss the possibility that these transcripts represent a linear and a circular form of the same transcript, and Schmidt *et al.* (1987) were able to clearly demonstrate that these two transcripts constitute the linear and the lariat formed by the group II intron RNA.

The most intriguing finding was that, during senescence, the *pl*DNA and the β-event DNA are transposed to the nucleus and integrated into chromosomal DNA (Wright and Cummings, 1983b).

The rapidly senescing race A^+ contains a higher number of plasmid DNA sequences in the nucleus than the slower senescing race s^+. This reflects the relative proportion of plasmid versus normal DNA in the young mycelium. It is 1:20 in young s^+ mycelia, 1:1 in young A^+ mycelia, and 20:1 in senescent A^+ mycelia. Two long-lived mt deletion mutants, *mex1* (Jamet-Vierny *et al.,* 1984) and *ex1* (Kück *et al.,* 1985b) both lack *pl*DNA. In the former mutant, however, the nuclear genome should contain a single integrated copy of *pl*DNA both in young and senescent mycelia (see above), whereas a nuclear copy of *pl*DNA seems to be absent in mutant *ex1* (Kück *et al.,* 1985b). In the latter mutant, the deletion is about 25 kbp, covering the whole *cox1* gene. The latter mutant is comparable to *rho*$^-$ in yeast (Kück *et al.,* 1985b), and respiration might proceed via the salicyl hydroxamic acid-sensitive respiration as in *stopper* mutants of *N. crassa* (De Vries *et al.,* 1980). Compiling data from Belcours', Cummings', and Essers' group, Kück *et al.* (1985b) proposed the following molecular model of senescence, which they called the "mobile intron model" (see also Esser *et al.,* 1980; Tudzynski *et al.,* 1980; Esser and Tudzynski, 1980). In senescent mycelia, formation of *pl*DNA occurs via precise (or inaccurate) (Belcour and Vierny, 1986) DNA excision or reverse transcription of circularized intron RNA. Once liberated, the amplified plasmid is integrated into the mt genome or the nuclear chromosomes. Repetition of the experiment of Wright and Cummings (1983b) by Knoll (1986) lead to the conclusion that the latter event is an artifact of contaminating DNA rather than an integrated mtDNA segment. Desintegration of mtDNA finally leads to cellular death. There is evidence for "DNA splicing" via reverse transcriptase activity, since such an enzyme activity has been found by Steinhilber and Cummings (1986). The long-lived mutants could be explained as defective in functions involved the process of DNA splicing via reverse transcriptase. Sequencing of the deletion endpoints in the long-lived mutants *mex3* and *mex6* revealed the insertion of poly(A) and poly(T) of unknown origin at the recombination borders, which suggests that a reverse transcription step could be involved in the mtDNA rearrangement (Koll *et al.,* 1987; see also Benne and Tabak, 1986).

Another interesting finding is that *P. anserina* can be transformed to senescence with the *pl*DNA (Tudzynski *et al.,* 1980). Using long-lived mutants as recipients, a transformation system could be established (Tudzynski *et al.,* 1982; Stahl *et al.,* 1982), which could be important in the development of a eukaryotic cloning system in *P. anserina.* Sainsard-Chanet and Begel (1986) could show that the β-, but not the *pl*DNA sequence has autoreplicative properties in yeast.

After a large sample of natural isolates of *N. intermedia* (Rieck *et al.,* 1982) were subjected to prolonged serial subculturing, 26 cytoplasmic variants were identified which show senescence and death at a strain-specific point in the subculture series. Senescent cultures can be female fertile and genetic analysis demonstrated that senescence is induced by a cytoplasmic factor (Griffith and Bertrand, 1984; Griffiths *et al.,* 1986). It has been shown that these *kalilo* variations contain a 9-kb transposable element (*kal*DNA) (Bertrand *et al.,* 1986) that lacks nucleotide homology with mtDNA and is inserted into the mtDNA, often at sites located within orfs in the intron of the *rnl* gene. The cells die as they become deficient in functional large and small subunits of mt ribosomes.

VII. Concluding Remarks

The mt genome is a small target with a huge controller (P. P. Slonimski, personal communication). This is evident since all components of DNA replication and most of the transcriptional and translational machinery are encoded in the nucleus. The proper function of the mt-encoded components is also dependent on their transport into the mitochondrion and the assembly with the nuclear components. Research in the past has developed a fairly detailed picture of the architecture, expression, and evolution of the mt genome. Future research will concentrate on the question of interaction between organelles, and between the organelles and the nucleus. Nuclear genes involved in mt functions have and will be studied in order to understand the complex interplay. Last, not least, protein import into mitochondria and the process of assembly in the membrane will be a central problem.

Acknowledgments

We wish to thank the many colleagues who provided reprints, preprints, and unpublished results for this article and Helga Trinkl for her help in preparing the manuscript. We are grateful for the patience of the editors of *Advances in Genetics* during the long period of preparation of this review.

References

Abelson, J. (1982). Self-splicing RNA. *Nature (London)* **300,** 400–401.

Accoceberry, B., and Stahl, A. (1971). Chromatographic differences between the cytoplasmic and the mitochondrial tRNA of *Saccharomyces cerevisiae. Biochem. Biophys. Res. Commun.* **42,** 1235–1243.

Agostinelli, M., Falcone, C., and Frontali, L. (1980). Release from glucose repression and mitochondrial protein synthesis in *S. cerevisiae. In* "Organization and Expression of the Mitochondrial Genome" (A. M. Kroon and C. Saccone, eds.), pp. 347–354. Elsevier, Amsterdam.

Agsteribbe, E., and Hartog, M. (1987). Processing of precursor RNAs from mitochondria of *Neurospora crassa. Nucleic Acids Res.* **15,** 7249–7264.

Agsteribbe, E., Kroon, A. M., and Van Bruggen, E. F. J. (1972). Circular mitochondrial DNA from mitochondria of *Neurospora crassa. Biochim. Biophys. Acta* **269,** 299–303.

Agsteribbe, E., Samallo, J., de Vries, H., Hensgens, L. A. M., and Grivell, L. A. (1980). Sequence homologies between the mitochondrial DNAs of yeast and *Neurospora crassa. In* "The Organization and Expression of the Mitochondrial Genome" (A. M. Kroon and C. Saccone, eds.), pp. 51–59. Elsevier, Amsterdam.

Ahne, F., Merlos-Lange, A.-M., Lang, B. F., and Wolf, K. (1984). The mitochondrial genome of the fission yeast *Schizosaccharomyces pombe.* 5. Characterization of mitochondrial deletion mutants. *Curr. Genet.* **8,** 517–524.

Ainley, W. M., Hensley, P., and Butow, R. A. (1984). Expression of GC clusters in the yeast mitochondrial *var1* gene. Translation and secondary structure implications. *J. Biol. Chem.* **259,** 9422–9428.

Ainley, W. M., Macreadie, I. G., and Butow, R. A. (1985). *Var1* gene on the mitochondrial genome of *Torulopsis glabrata. J. Mol. Biol.* **184,** 565–576.

Akins, R. A., and Lambowitz, A. (1984). The *poky* mutant of *Neurospora* contains a 4-base-pair deletion at 5′ end of the mitochondrial small rRNA. *Proc. Natl. Acad. Sci. U.S.A.* **81,** 3791–3795.

Akins, R. A., and Lambowitz, A. M. (1987). A protein is required for splicing group I introns in *Neurospora* mitochondria is mitochondrial tyrosyl-tRNA synthetase or a derivative thereof. *Cell* **50,** 331–345.

Akins, R. A., Kelley, R. L., and Lambowitz, A. M. (1986). Mitochondrial plasmids of *Neurospora*: Integration into mitochondrial DNA and evidence for reverse transcription in mitochondria. *Cell* **47,** 505–516.

Alexander, N. J., Vincent, R. D., Perlman, P. S., Miller, D. H., Hanson, D. K., and Mahler, H. R. (1979). Regulatory interactions between mitochondrial genes. I. Genetic and biochemical characterization of some mutant types affecting apocytochrome *b* and cytochrome oxidase. *J. Biol. Chem.* **254,** 2471–2479.

Alexander, N. J., Perlman, P. S., Hanson, D. K., and Mahler, H. R. (1980). Mosaic organization of a mitochondrial gene: Evidence from double mutants in the cytochrome *b* region of *Saccharomyces cerevisiae*. *Cell* **20,** 199–206.

Algeri, A., Marmiroli, N., Viola, A., and Puglisi, P. P. (1977). Dependence of cytoplasmic on mitochondrial protein synthesis in *K. lactis* CBS 2360. II. Genetic studies. *Mol. Gen. Genet.* **150,** 141–145.

Altman, S., Baer, M., Guerrier-Takada, C., and Vioque, A. (1986). Enzymatic cleavage of RNA by RNA. *Trends Biochem.* **11,** 515–517.

Anderson, S., Bankier, A. T., Barell, B. G., de Bruijn, M. H. L., Coulson, A. R., Drouin, J., Eperon, I. C., Nierlich, D. P., Roe, B. A., Sanger, F., Schreier, P. H., Smith, A. J. H., Staden, R., and Young, I. G. (1981). Sequence and organization of the human mitochondrial genome. *Nature (London)* **290,** 457–464.

Anderson, S., de Bruijn, M. H. L., Coulson, A. R., Eperon, I. C., Sanger, F., and Young, G. (1982a). Complete sequence of bovine mitochondrial DNA: Conserved features of the mammalian mitochondrial genome. *J. Mol. Biol.* **157,** 683–717.

Anderson, S., Bankier, A. T., Barell, B. G., de Bruijn, M. H. L., Coulson, A. R., Drouin, J., Eperon, I. C., Nierlich, D. P., Roe, B. A., Sanger, F., Schreier, P. H., Smith, A. J. H., Staden, R., and Young, I. G. (1982b). Comparison of the human and bovine mitochondrial genomes. *In* "Mitochondrial Genes" (P. Slonimski, P. Borst, and G. Attardi, eds.), pp. 5–43. Cold Spring Harbor Laboratory, Cold Spring Harbor, New York.

Anziano, P. Q., Hanson, D. K., Mahler, H. R., and Perlman, P. S. (1982). Functional domains in introns: Trans-acting and cis-acting regions of intron 4 of the *cob* gene. *Cell* **30,** 925–932.

Anziano, P. Q., Perlman, P. S., Lang, B. F., and Wolf, K. (1983). The mitochondrial genome of the fission yeast *Schizosaccharomyces pombe*. I. Isolation and physical mapping of mitochondrial DNA. *Curr. Genet.* **7,** 273–284.

Arlett, C. F. (1957). Induction of cytoplasmic mutants in *Aspergillus nidulans*. *Nature (London)* **179,** 1250–1251.

Arnberg, A. C., Van Ommen, G.-J. B., Grivell, L. A., Van Bruggen, E. F. J., and Borst, P. (1980). Some yeast mitochondrial RNAs are circular. *Cell* **19,** 313–319.

Arnberg, A. C., Van der Horst, G., and Tabak, H. F. (1986). Formation of lariats and circles in self-splicing of the precursors to the large ribosomal RNA of yeast mitochondria. *Cell* **44,** 235–242.

Atchison, B. A., Choo, K.-B., Devenish, R. J., Linnane, A. W., and Nagley, P. (1979). Biogenesis of mitochondria. 53. Physical map of genetic loci in the 21S ribosomal RNA region of mitochondrial DNA in *Saccharomyces cerevisiae*. *Mol. Gen. Genet.* **174,** 307–316.

Attardi, G. (1981a). Organization and expression of the mammalian mitochondrial genome: A lesson in economy (Part I). *Trends Biochem. Sci. (Pers. Ed.)* **6,** 86–89.

Attardi, G. (1981b). Organization and expression of the mammalian mitochondrial genome: A lesson in economy (Part II). *Trends Biochem. Sci. (Pers. Ed.)* **6,** 100–103.

Attardi, G., Cantatore, P., Ching, E., Crews, S., Gelfund, R., Merkel, C., Montoya, J., and Ojala, D. (1980). The remarkable features of gene organization and expression of human mitochondrial DNA. *In* "The Organization and Expression of the Mitochondrial Genome" (A. M. Kroon and C. Saccone, eds.), pp. 103–119. Elsevier, Amsterdam.

Avner, P. R., Coen, D., Dujon, B., and Slonimski, P. P. (1973). Mitochondrial genetics. IV. Allelism and mapping studies of oligomycin resistant mutants in *S. cerevisiae*. *Mol. Gen. Genet.* **125,** 9–52.

Backhaus, B., Schweyen, R. J., and Kaudewitz, F. (1978). On the formation of ρ^--petites in yeast. III. Effects of temperature on transmission and recombination of mitochondrial markers and on ρ^--cell formation in temperature sensitive mutants of *Saccharomyces cerevisiae. Mol. Gen. Genet.* **161,** 153–173.

Baldacci, G., and Bernardi, G. (1982). Replication origins are associated with transcription initiation sequences in the mitochondrial genome of yeast. *EMBO J.* **1,** 987–994.

Baldacci, G., and Bernardi, G. (1983). The replication of yeast mitochondrial DNA. *In* "Mitochondria 1983" (R. J. Schweyen, K. Wolf, and F. Kaudewitz, eds.), pp. 57–63. De Gruyter, Berlin.

Baldacci, G., and Zennaro, E. (1982). Mitochondrial transcripts in glucose-repressed cells of *Saccharomyces cerevisiae. Eur. J. Biochem.* **127,** 411–416.

Baldacci, G., Falcone, C., Frontali, L., Macino, G., and Palleschi, C. (1976a). tRNA in mitochondria from *Saccharomyces cerevisiae* grown in different physiological conditions: Hybridization on mitochondrial DNA fragments and import of cytoplasmic species. *In* "The Genetic Function of Mitochondrial DNA" (C. Saccone and A. M. Kroon, eds.), pp. 305–312. Elsevier, Amsterdam.

Baldacci, G., Falcone, C., Frontali, L., Macino, G., and Palleschi, C. (1976b). Isoaccepting tRNA*ser* in mitochondria from *Saccharomyces cervisiae:* Mitochondrially encoded and cytoplasmic species. *In* "Genetics and Biogenesis of Chloroplasts and Mitochondria" (T. Bücher, W. Neupert, W. Sebald, and S. Werner, eds.), pp. 759–762. Elsevier, Amsterdam.

Baldacci, G., Falcone, C., Frontali, L., Macino, G., Palleschi, C. (1977). Differences in mitochondrial isoaccepting tRNAs from *Saccharomyces cerevisiae* as a function of growth conditions. *In* "Mitochondria 1977" (W. Bandlow, R. J. Schweyeu, K. Wolf, and F. Kaudewitz, eds.), pp. 571–578.

Baldacci, G., Cundari, E., Francisci, S., and Palleschi, C. (1978). Variability of mitochondrial alanyl tRNA isoaccepting species in *Saccharomyces cerevisiae. Bull. Mol. Biol. Med.* **3,** 243–249.

Baldacci, G., Falcone, C., Francisci, S., Frontali, L., and Palleschi, C. (1979). Mitochondrial protein-synthesizing machinery in *Saccharomyces cerevisiae.* Growth in different metabolic conditions. Variability of seryl-tRNA and alanyl-tRNA isoacceptor patterns. *Eur. J. Biochem.* **98,** 181–186.

Baldacci, G., de Zamaroczy, M., and Bernardi, G. (1980). Excision sites in the GC clusters of the mitochondrial genome of yeast. *FEBS Lett.* **114,** 234–236.

Baldacci, G., Colin, Y., Faugeron-Fonty, G., Goursot, R., Huyard, A., Levankim, C., Mangin, M., Marotta, R., de Zamaroczy, M. (1983a). The origins of replication of the mitochondrial genome of yeast. *In* "Manipulation and Expression of Genes in Eukaryotes" (P. Nagley *et al.*, eds.), pp. 279–289. Academic Press, Sidney.

Baldacci, G., Francisci, S., Palleschi, C., Zennaro, E., and Frontali, L. (1983b). Expression of mitochondrial genes in *Saccharomyces cerevisiae:* Release from glucose repression and transcription of the tRNA genes. *In* "Mitochondria 1983" (R. J. Schweyen *et al.,* eds.), pp. 165–178. De Gruyter, Berlin.

Baldacci, G., Cherif-Zakar, B., and Bernardi, G. (1984). The initiation of DNA replication in the mitochondrial genome of yeast. *EMBO J.* **3,** 2115–2120.

Bandlow, W., and Schweyen, R. J. (1975). On the mechanism of petite genesis in yeast. IV. Biochemical characterization of conditional cytoplasmic mutant producing petites at restrictive temperature. *Biochem. Biophys. Res. Commun.* **67,** 1078–1085.

Bandlow, W., Metzke, R., Klein, A., Kotzias, K., Doxiadis, I., Bechmann, H., Schweyen, R. J., and Kaudewitz, F. (1977). Macromolecular synthesis and energy level in a mitochondrial conditional mutant *tsm-8. Eur. J. Biochem.* **76,** 373–382.

Bandlow, W., Baumann, U., and Schnittchen, P. (1980). The nucleotide sequence of the *tsm-8* region of yeast mitochondrial DNA. *In* "The Organization and Expression of the Mitochondrial Genome" (A. M. Kroon and C. Saccone, eds.), pp. 207–210.

Banroques, J., Delahodde, A., and Jacq, C. (1986). A mitochondrial RNA maturase gene transferred to the yeast nucleus can control mitochondrial mRNA splicing. *Cell* **46,** 837–844.

Banroques, J., Perea, J., and Jacq, C. (1987). Efficient splicing of two yeast mitochondrial introns controlled by nuclear-encoded maturase. *EMBO J.* **6,** 1085–1091.

Baranowska, H., Szczesniak, B., Eijchart, A., Kruszewska, A., and Claisse, M. (1983). Recombinational analysis of *oxi2* mutants and preliminary analysis of their translation products in *S. cerevisiae. Curr. Genet.* **7,** 225–233.

Barell, B. G., Bankier, A. T., and Drouin, J. (1979). A different genetic code in human mitochondria. *Nature (London)* **282,** 189–194.

Bartik, E., Pieniazek, J., and Stepien, P. P. (1979). Molecular cloning of the 4.2 Md *EcoRI* fragment of *Aspergillus nidulans* mitochondrial DNA. *Mol. Gen. Genet.* **171,** 75–78.

Bartik, E., Biderman, A., Hahn, U., Küntzel, H., and Stepien, P. P. (1981). The cloning of *Aspergillus nidulans* mitochondrial DNA in *Escherichia coli* on plasmid *pBR322. Mol. Gen. Genet.* **182,** 332–335.

Bass, B. L., and Cech, T. R. (1984). Specific interaction between the selfsplicing RNA of *Tetrahymena* and its guanosine substrate: Implications for biological catalysis by RNA. *Nature (London)* **308,** 820–826.

Battey, J., and Clayton, D. A. (1980). The transcription map of human mitochondrial DNA implicates transfer RNA excision as a major processing event. *J. Biol. Chem.* **255,** 11599–11606.

Bechmann, H., Krüger, M., Böker, E., Bandlow, W., Schweyen, R. J., and Kaudewitz, F. (1977). On the formation of ρ^--petites in yeast. II. Effects of mutation *tsm-8* on mitochondrial functions and ρ-factor stability in *Saccharomyces cerevisiae. Mol. Gene. Genet.* **155,** 41–51.

Bechmann, H., Haid, A., Schmelzer, C., Schweyen, R. J., and Kaudewitz, F. (1980). Processing of the mRNA for apocytochrome *b* in yeast depends on a product encoded by an intervening sequence. *In* "The Organization of the Mitochondrial Genome" (A. M. Kroon and C. Saccone, eds.), pp. 173–178. Elsevier, Amsterdam.

Bechmann, H., Haid, A., Schweyen, R. J., Mathews, S., Kaudewitz, F. (1981). Expression of the "split gene" *COB* in yeast mitochondrial DNA. *J. Biol. Chem.* **256,** 3525–3531.

Been, M. D., and Cech, T. R. (1985). Sites of circularization of the *Tetrahymena* rRNA IVS are determined by sequence and influenced by position and secondary structure. *Nucleic Acids Res.* **13,** 8389–8407.

Been, M. D., and Cech, T. R. (1986). One binding site determines sequence specificity of *Tetrahymena* pre-rRNA self-splicing, trans-splicing and RNA enzyme activity. *Cell* **47,** 207–216.

Been, M. D., and Cech, T. R. (1987). Selection of circularization sites in a group I IVS RNA requires multiple alignments of an internal template-like sequence. *Cell* **50,** 951–961.

Beilharz, M. W., Cobon, G. S., and Nagley, P. (1982). Physiological alteration of the pattern of transcription of the *oli2* region of yeast mitochondrial DNA. *FEBS Lett.* **147,** 235–238.

Belcour, L. (1975). Cytoplasmic mutations isolated from protoplasts of *Podospora anserina. Genet. Res.* **25,** 155–161.

Belcour, L. (1981). Mitochondrial DNA and senescence in *Podospora anserina. Curr. Genet.* **4,** 81–82.

Belcour, L., and Begel, O. (1977). Mitochondrial genes in *Podospora anserina:* Recombination and linkage. *Mol. Gen. Genet.* **153,** 11–21.

Belcour, L., and Begel, O. (1978). Lethal mitochondrial genotypes in *Podospora anserina*: A model for senescence. *Mol. Gen. Genet.* **163,** 113–123.

Belcour, L., and Begel, O. (1980). Life-span and senescence in *Podospora anserina*: Effect of mitochondrial genes and functions. *J. Gen. Microbiol.* **119,** 505–515.

Belcour, L., and Vierny, C. (1986). Variable DNA splicing sites of a mitochondrial intron: Relationship to the senescence process in *Podospora. EMBO J.* **5,** 609–614.

Belcour, L., Begel, O., Mosse, M.-O., and Vierny, C. (1981). Mitochondrial DNA amplified in senescent cultures of *Podospora anserina*: Variability between the retained, amplified sequences. *Curr. Genet.* **3,** 13–21.

Belcour, L., Begel, O., Keller, A.-M., and Vierny, C. (1982). Does senescence in *Podospora anserina* result from instability of the mitochondrial genome? *In* "Mitochondrial Genes" (P. Slonimski, P. Borst, and G. Attardi, eds.), pp. 415–421. Cold Spring Harbor Laboratory, Cold Spring Harbor, New York.

Benne, R., and Tabak, H. F. (1986). Senescence comes of age. *Trends Genet.* **June,** 147–148.

Berlani, R. E., Bonitz, S. G., Coruzzi, G., Nobrega, M., and Tzagoloff, A. (1980a). Transfer RNA genes in the *cap–oxi1* region of yeast mitochondrial DNA. *Nucleic Acids Res.* **8,** 5017–5031.

Berlani, R. E., Pentella, C., Macino, G., and Tzagoloff, A. (1980b). Assembly of the mitochondrial membrane system: Isolation of mitochondrial transfer ribonucleic acid mutants and characterization of transfer ribonucleic acid genes of *Saccharomyces cerevisiae. J. Bacteriol.* **141,** 1086–1097.

Bernard, U., and Küntzel, H. (1976). Physical mapping of mitochondrial DNA from *Neurospora crassa. In* "The Genetic Function of Mitochondrial DNA" (C. Saccone and A. M. Kroon, eds.), pp. 105–109. Elsevier, Ansterdam.

Bernard, U., Bade, E., and Küntzel, H. (1975a). Specific fragmentation of mitochondrial DNA from *Neurospora crassa* by restriction endonuclease *EcoRI. Biochem. Biophys. Res. Commun.* **64,** 783–789.

Bernard, U., Pühler, A., and Küntzel, H. (1975b). Physical map of circular mitochondrial DNA from *Neurospora crassa. FEBS Lett.* **60,** 119–121.

Bernard, U., Pühler, A., Mayer, F., and Küntzel, H. (1975c). Denaturation map of the circular mitochondrial genome of *Neurospora crassa*. *Biochim. Biophys. Acta* **402,** 270–278.

Bernard, U., Goldthwaite, C., Küntzel, H. (1976). Physical map of *Neurospora crassa* mitochondrial DNA and its transcription unit for ribosomal RNA. *Nucleic Acids Res.* **3,** 3101–3108.

Bernardi, G. (1976). Organization and evolution of the mitochondrial genome of yeast. *J. Mol. Evol.* **9,** 25–35.

Bernardi, G. (1978). Intervening sequences in the mitochondrial genome. *Nature (London)* **276,** 558–559.

Bernardi, G. (1979). The petite mutation in yeast. *Trends Biochem. Sci. (Pers. Ed.)* **4,** 187–201.

Bernardi, G. (1982a). The origins of replication of the mitochondrial genome in yeast. *Trends Biochem. Sci. (Pers. Ed.)* **7,** 404–408.

Bernardi, G. (1982b). Evolutionary origin and the biological function of non-coding sequences in the mitochondrial genome of yeast. *In* "Mitochondrial Genes" (P. Slonimski, P. Borst, and G. Attardi, eds.), pp. 269–278. Cold Spring Harbor Laboratory, Cold Spring Harbor, New York.

Bernardi, G., and Bernardi, G. (1980). Repeated sequences in the mitochondrial genome of yeast. *FEBS Lett.* **115,** 159–162.

Bernardi, G., Piperno, G., and Fonty, G. (1972). The mitochondrial genome of wild-type yeast cells. I. Preparation and heterogeneity of mitochondrial DNA. *J. Mol. Biol.* **65,** 173–189.

Bernardi, G., Prunell, A., Fonty, G., Kopecka, H., and Strauss, F. (1976). The mitochondrial genome of yeast: Organization, evolution and the petite mutation. *In* "The Genetic Function of Mitochondrial DNA" (C. Saccone and A. M. Kroon, eds.), pp. 185–198. Elsevier, Amsterdam.

Bernardi, G., Baldacci, G., Bernardi, G., Faugeron-Fonty, G., Gaillard, C., Goursot, R., Hujard, A., Mangin, M., Marotta, R., and de Zamaroczy, M. (1980). The petite mutation: Excision sequences, replication origins and suppressivity. *In* "The Organization and Expression of the Mitochondrial Genome" (A. M. Kroon and C. Saccone, eds.), pp. 21–32. Elsevier, Amsterdam.

Bertrand, H., and Pittenger, T. H. (1969). Cytoplasmic mutants selected from continuously growing cultures of *Neurospora crassa*. *Genetics* **61,** 643–659.

Bertrand, H., and Pittenger, T. H. (1972a). Isolation and classification of extranuclear mutants of *Neurospora crassa*. *Genetics* **71,** 521–533.

Bertrand, H., and Pittenger, T. H. (1972b). Complementation among cytoplasmic mutants in *Neurospora crassa*. *Mol. Gen. Genet.* **117,** 82–90.

Bertrand, H., Collins, R. A., Stohl, L. L., Goewert, R. R., and Lambowitz, A. M. (1980). Deletion mutants of *Neurospora crassa* mitochondrial DNA and their relationship to the "stop-start" growth phenotype. *Proc. Natl. Acad. Sci. U.S.A.* **77,** 6032–6036.

Bertrand, H., Griffiths, A. J. F., Court, D. A., and Cheng, C. K. (1986). An extrachromosomal plasmid is the etiological precursor of kalDNA insertion sequences in the mitochondrial chromosome of senescent *Neurospora*. *Cell* **47,** 827–829.

Bibb, M. J., Van Etten, R. A., Wright, C. T., Walberg, M. W., and Clayton, D. A. (1981). Sequence and gene organization of mouse mitochondrial DNA. *Cell* **26,** 167–180.

Bingham, C. G., and Nagley, P. (1983a). Occurrence of a super-abundant petite genome in different *mit*⁻ mutants of yeast. *In* "Manipulation and Expression of Genes in Eukaryotes" (P. Nagley *et al.,* eds.), pp. 291–298. Academic Press, Sidney.

Bingham, C. G., and Nagley, P. (1983b). A petite mitochondrial DNA segment arising in exceptionally high frequency in a *mit*⁻ mutant of *Saccharomyces cerevisiae*. *Biochim. Biophys. Acta* **740,** 88–98.

Birky, C. W., Jr. (1978). Transmission genetics of mitochondria and chloroplasts. *Annu. Rev. Genet.* **12,** 471–512.

Birky, C. W., Jr. (1983). Relaxed cellular controls and organelle heredity. *Science* **222,** 468–475.

Biswas, T. K., and Getz, G. S. (1986). Nucleotides flanking the promoter sequence influence the transcription of the yeast mitochondrial gene coding for ATPase subunit 9. *Proc. Natl. Acad. Sci. U.S.A.* **83,** 270–274.

Biswas, T. K., Ticho, B., and Getz, G. S. (1987). In vitro characterization of the yeast mitochondrial promoter using single-base substitution mutants. *J. Biol. Chem.* **262,** 13690–13696.

Blanc, H. (1984). Two modules from the hypersuppressive *rho*$^-$ mitochondrial DNA are required for plasmid replication in yeast. *Gene* **30,** 47–61.

Blanc, H., and Dujon, B. (1980). Replicator regions of the yeast mitochondrial DNA responsible for suppressiveness. *Proc. Natl. Acad. Sci. U.S.A.* **77,** 3942–3946.

Blanc, H., and Dujon, B. (1982). Replicator regions of the yeast mitochondrial DNA active *in vivo* and in yeast transformants. *In* "Mitochondrial Genes" (P. Slonimski, P. Borst, and G. Attardi, eds.), pp. 279–294. Cold Spring Harbor Laboratory, Cold Spring Harbor, New York.

Blanc, H., Adams, C. W., and Wallace, D. C. (1981). Different nucleotide changes in the large rRNA gene of the mitochondrial DNA confer chloramphenicol resistance on two human cell lines. *Nucleic Acids Res.* **9,** 5785–5795.

Böckelmann, B., and Esser, K. (1986). Plasmids of mitochondrial origin in senescent mycelia of *Podospora curviolla. Curr. Genet.* **10,** 803–810.

Böker-Schmitt, E., Francisci, S., and Schweyen, R. (1982). Mutations releasing mitochondrial biogenesis from glucose repression in *Saccharomyces cerevisiae. J. Bacteriol.* **151,** 303–310.

Bolden, A., Noy, G. P., Weissback, A. (1977). DNA polymerase of mitochondria is a gamma-polymerase. *J. Biol. Chem.* **252,** 3351–3356.

Bolotin, M., Coen, D., Deutsch, J., Dujon, B., Netter, P., Petrochilo, E., and Slonimski, P. P. (1971). La recombinaison des mitochondries chez *Saccharomyces cerevisiae. Bull. Institut Pasteur* **69,** 215–239.

Bolotin-Fukuhara, M. (1979). Mitochondrial and nuclear mutations that affect the biogenesis of the mitochondrial ribosomes of yeast. *Mol. Gen. Genet.* **177,** 39–46.

Bolotin-Fukahara, M., Faye, G., and Fukuhara, H. (1976). Localization of some mitochondrial mutations in relation to transfer and RNA genes in *Saccharomyces cerevisiae. In* "The Genetic Function of Mitochondrial DNA" (C. Saccone and A. M. Kroon, eds.), pp. 243–250. Elsevier, Amsterdam.

Bolotin-Fukuhara, M., Faye, G., and Fukuhara, H. (1977). Temperature-sensitive respiratory-deficient mitochondrial mutations: Isolation and genetic mapping. *Mol. Gen. Genet.* **152,** 295–305.

Bolotin-Fukuhara, M., Faye, G., and Fukuhara, H. (1978). Localization of some mitochondrial mutations in relation to transfer and ribosomal RNA genes in *Saccharomyces cerevisiae. In* "The Genetic Function of Mitochondrial DNA" (C. Saccone and A. M. Kroon, eds.), pp. 243–250. Elsevier, Amsterdam.

Bolotin-Fukuhara, M., Sor, F., and Fukuhara, H. (1983). Mitochondrial ribosomal RNA mutations and their nuclear suppressors. *In* "Mitochondria 1983" (R. J. Schweyen, K. Wolf, and F. Kaudewitz, eds.), pp. 455–467. De Gruyter, Berlin.

Bonitz, S. G., and Tzagoloff, A. (1980). Assembly of the mitochondrial membrane system. Sequences of yeast mitochondrial tRNA genes. *J. Biol. Chem.* **255,** 9075–9081.

Bonitz, S. G., Bertani, R., Coruzzi, G., Li, M., Macino, G., Nobrega, F. G., Nobrega, M. P.,

Thalenfeld, B. E., and Tzagoloff, A. (1980a). Codon recognition rules in yeast mitochondria. *Proc. Natl. Acad. Sci. U.S.A.* **77,** 3167–3170.

Bonitz, S. D., Coruzzi, G., Thalenfeld, B. E., and Tzagoloff, A. (1980b). Assembly of the mitochondrial membrane system. Structure and nucleotide sequence of the gene coding for subunit 1 of yeast cytochrome oxidase. *J. Biol. Chem.* **255,** 11927–11941.

Bonitz, S. G., Coruzzi, G., Thalenfeld, B. E., Tzagoloff, A., and Macino, G. (1980c). Assembly of the mitochondrial membrane system. Physical map of the *Oxi3* locus of yeast mitochondrial DNA. *J. Biol. Chem.* **255,** 11922–11926.

Bonitz, S. G., Homison, G., Thalenfeld, B. E., Tzagoloff, A., and Nobrega, F. G. (1982). Assembly of the mitochondrial membrane system. Processing of the apocytochrome *b* precursor RNAs in *Saccharomyces cerevisiae* D273–10B. *J. Biol. Chem.* **237,** 6268–6274.

Bonjardin, C. A., and Nobrega, F. G. (1984). Nucleotide substitutions in yeast mitochondrial cis-acting mutant located in the last intron of the apocytochrome *b* gene. *FEBS Lett.* **169,** 73–78.

Bordonne, R., Bandlow, W., Dirheimer, G., and Martin, R. P. (1984). Identification of the *tsm8* mutation in the structural portion of the yeast mitochondrial tRNA Tyr gene. *Int. Conf. Yeast Genet. Mol. Biol., 12th, Edinburgh* Abstract.

Bordonne, R., Bandlow, W., Dirheimer, G., and Martin, R. P. (1987a). A single base change in the extra-arm of yeast mitochondrial tyrosine tRNA affects its conformational stability and impairs aminoacylation. *Mol. Gen. Genet.* **206,** 498–504.

Bordonne, R., Dirheimer, G., and Martin, R. P. (1987b). Transcription initiation and RNA processing of a yeast mitochondrial tRNA gene cluster. *Nucleic Acids Res.* **15,** 7381–7394.

Borst, P. (1978). The genes for ribosomal RNA in yeast mitochondria. *In* "Specific Eukaryontic Genes. Structural Organization and Function" (J. Engberg *et al.,* eds.), pp. 244–255. Muncksgaard, Copenhagen.

Borst, P. (1980). The optional introns in yeast mitochondrial DNA. *In* "Biological Chemistry of Organelle Formation" (T. Bücher, W. Neupert, W. Sebald, and S. Werner, eds.), pp. 27–42. Springer-Verlag, Berlin.

Borst, P. (1981). Control of mitochondrial biosynthesis. *In* "Cellular Controls in Differentiation" (C.W. Lloyd and D. A. Ress, eds.), pp. 231–254. Academic Press, New York.

Borst, P., and Grivell, L. A. (1978). The mitochondrial genome of yeast *Cell* **15,** 705–723.

Borst, P., and Grivell, L. A. (1981a). One gene's intron is another gene's exon. *Nature (London)* **289,** 439–440.

Borst, P., and Grivell, L. A. (1981b). Small is beautiful—portrait of a mitochondrial genome. *Nature (London)* **290,** 443–444.

Borst, P., Bos, J. L., Grivell, L. A., Groot, G. S. P., Heyting, C., Moorman, A. F. M., Sanders, J. P. M., Talen, J. L., Van Kreijl, C. F., and Van Ommen, G. J. B. (1977). The physical map of yeast mitochondrial DNA anno 1977. *In* "Mitochondria 1977" (W. Bandlow, R. J. Schweyen, K. Wolf, and F. Kaudewitz, eds.), pp. 213–254. De Gruyter, Berlin.

Borst, P., Sanders, J. P. M., and Heyting, C. (1979). Biochemical methods to locate genes on the physical map of yeast mitochondrial DNA. *In* "Methods in Enzymology" (S. Fleischer and L. Packer, eds.), Vol. 56, pp 182–197. Academic Press, New York.

Borst, P., Grivell, L. A., and Groot, G. S. P. (1984). Organelle DNA. *Trends Biochem. Sci. (Pers. Ed.)* **7,** 128–130.

Bos, J. L., Heyting, C., Borst, P., Arnberg, A. C., and Van Bruggen, E. F. J. (1978). An insert in the single gene for the large ribosomal RNA in yeast. *Nature (London)* **275,** 336–338.

Bos, J. L., Osinga, K. A., Van der Horst, G., and Borst, P. (1979). Nucleotide sequence of the mitochondrial structural genes for cysteine-tRNA and histidine-tRNA of yeast. *Nucleic Acids Res.* **6,** 3255–3266.

Bos, J. L., Heyting, C., Van der Horst, G., and Borst, P. (1980a). The organization of repeating units in mitochondrial DNA from yeast petite mutants. *Curr. Genet.* **1,** 233–239.

Bos, J. L., Osinga, K. A., Van der Horst, G., Hecht, N. B., Tabak, H. F., Van Ommen, G.-J. B., and Borst, P. (1980b). Splice point sequence and transcripts of the intervening sequence of the mitochondrial 21 S ribosomal RNA gene of yeast. *Cell* **20,** 207–214.

Brehm, S. L., and Cech, T. R. (1983). Fate of an intervening sequence ribonucleic acid: Excision and cyclization of the *Tetrahymena* ribosomal ribonucleic acid intervening sequence *in vivo. Biochemistry* **22,** 2390–2397.

Breitenberger, C. A., and RajBhandary, U. L. (1985). Some highlights of mitochondrial research based on analysis of *Neurospora crassa* mitochondrial DNA. *Trends Biochem. Sci. (Pers. Ed.)* **Dec.** 478–483.

Breitenberger, C. A., Browning, K. S., Alzner-DeWeerd, B., and RajBhandary, U. L. (1985). RNA processing in *Neurospora crassa* mitochondria: Use of transfer RNA sequence as signals. *EMBO J.* **4,** 185–196.

Brown, T. A. Davies, R. W., Ray, J. A., Waring, R. B., and Scazzocchio, C. (1983a). The mitochondrial genome of *Asperqillus nidulans* contains reading frames homologous to the human URFs 1 and 4. *EMBO J.* **2,** 427–435.

Brown, T. A., Davies, R. W., Waring, R. B., Ray, J. A., and Scazzocchio, C. (1983b). DNA duplication has resulted in transfer of an amino-terminal peptide between two mitochondrial proteins. *Nature (London)* **302,** 721–723.

Brown, T. A., Ray, A. J., Waring, R. B., Scazzocchio, C., and Davies, R. W. (1984). A mitochondrial reading frame which may code for a second form of ATPase subunit 9 in *Aspergillus nidulans. Curr. Genet.* **8,** 489–492.

Brown, T. A., Waring, R. B., Scazzocchio, C., and Davies, R. W. (1985). The *Aspergillus nidulans* mitochondrial genome. *Curr. Genet.* **9,** 113–117.

Browning, K. S., and RajBhandary, U. L. (1982). Cytochrome oxidase subunit III gene in *Neurospora* mitochondria. Location and sequence. *J. Biol. Chem.* **257,** 5253–5256.

Brunner, A., Tuena de Cobos, A., and Griffith, D. E. (1977). The isolation and genetic characterization of extrachromosomal chloramphenicol- and oligomycin-resistant mutants from the petite-negative yeast *Kluyveromyces lactis. Mol. Gen. Genet.* **152,** 183–191.

Bulder, C. J. E. A. (1964a). Induction of petite mutation and inhibition of synthesis of respiratory enzymes in various yeasts. *Antonie v. Leeuwenhoek* **30,** 1–9.

Bulder, C. J. E. A. (1964b). Lethality of the petite mutation in petite negative yeast. *Antonie v. Leeuwenhoek* **30,** 442–454.

Bunn, C. L., Mitchell, C. H., Lukins, H. B., and Linnane, A. W. (1970). Biogenesis of mitochondria. XVIII. A new class of cytoplasmically determined antibiotic resistant mutants in *Saccharomyces cerevisiae. Proc. Natl. Acad. Sci. U.S.A.* **67,** 1233–1240.

Burger, G. (1984). Cytochrome *b* of *cob* revertants in yeast. 1. Isolation and characterization of revertants derived from *cob* exon mutants of *Saccharomyces cerevisiae. Mol. Gen. Genet.* **196,** 158–166.

Burger, G. (1985). Unassigned reading frames in the mitochondrial DNA of *Neurospora crassa. In* "Achievements and Perspectives of Mitochondrial Research" (E. Quagliariello, E. C. Slater, F. Palmieri, C. Saccone, and A. M. Kroon, eds.), Vol. 2, pp. 305–316. Elsevier, Amsterdam.

Burger, G., and Hofner, E. (1984). Cytochrome *b* of *cob* revertants in yeast. Bioenergetic characterization of revertants with reduced content and shifted maximum absorption wavelength of cytochrome *b*. *Eur. J. Biochem.* **142,** 299–304.

Burger, G., and Werner, S. (1983). Nucleotide sequence and transcript mapping of a mit DNA segment comprising *CO1, tRNAarg,* and several unidentified reading frames in *Neurospora crassa. In* "Mitochondria 1983" (R. J. Schweyen, K. Wolf, and F. Kaudewitz, eds.), pp. 331–342. De Gruyter, Berlin.

Burger, G., and Werner, S. (1985). The mitochondrial *URF1* gene in *Neurospora crassa* has an intron that contains a novel type of URF. *J. Mol. Biol.* **186,** 231–242.

Burger, G., and Werner, S. (1986). Mitochondrial gene URF N of *Neurospora crassa* codes for a long polypeptide with highly repetitive structure. *J. Mol. Biol.* **191,** 589–600.

Burger, G., and Wolf, K. (1981). Mitochondrially inherited resistance to antimycin and diuron in the petite negative yeast *Schizosaccharomyces pombe. Mol. Gen. Genet.* **181,** 134–139.

Burger, G., Lang, B., Bandlow, W., Schweyen, R. J., Backhaus, B., and Kaudewitz, F. (1976a). Antimycin resistance in *Saccharomyces cerevisiae:* A new mutation on the mitDNA conferring antimycin resistance on the mitochondrial respiratory chain. *Biochem. Biophys. Res. Commun.* **72,** 1201–1208.

Burger, G., Lang, B., Wolf, K., Bandlow, W., and Kaudewitz, F. (1976b). The inhibitory effect of HQNO compared with extrareduction and binding in the wild-type and in an antimycin resistant mutant $ANT^{R}8$, in *Schizosaccharomyces pombe. In* "Genetics, Biogenesis and Bioenergetics of Mitochondria" (W. Bandlow, R. J. Schweyen, D. Y. Thomas, K. Wolf, and F. Kaudewitz, eds.), pp. 399–407, De Gruyter, Berlin and New York.

Burger, G., Lang, B., Backhaus, B., Wolf, K., Bandlow, W., and Kaudewitz, F. (1977). Mutations to drug-resistance in the *cob* region of the mitochondrial genome. *In* "Mitochondria 1977" (W. Bandlow, R. J. Schweyen, K. Wolf, and F. Kaudewitz, eds.), pp. 205–212. De Gruyter, Berlin.

Burger, G., Scriven, C., Machleidt, W., and Werner, S. (1982). Subunit 1 of cytochrome oxidase from *Neurospora crassa:* Nucleotide sequence of the coding gene and partial amino acid sequence of the protein. *EMBO J.* **1,** 1385–1391.

Burger, G., Helmer-Citterich, M., Nelson, M. A., Werner, S., and Macino, G. (1985), RNA processing in *Neurospora crassa* mitochondria: Transfer RNAs punctuate a large precursor transcript. *EMBO J.* **4,** 197–204.

Burke, J. M., and RajBhandary, U. L. (1982). Intron within the large rRNA gene of *N. crassa* mitochondria: A long open reading frame and a consensus sequence possibly important in splicing. *Cell* **31,** 509–520.

Burke, J. M., Breitenberger, C., Heckman, J. E., Dujon, B., and RajBhandary, U. (1984). Cytochrome *b* gene of *Neurospora crassa* mitochondria. Partial sequence and location of introns at sites different from those in *Saccharomyces cerevisiae* and *Aspergillus nidulans. J. Biol. Chem.* **259,** 504–511.

Burke, J. M., Belfort, M., Cech, T. R., Davies, R. W., Schweyen, R. J., Shuts, D. A., Szostak, J. W., Tabak, H. T. (1987). Structural conventions for group I introns. *Nucleic Acids Res.* **15,** 7217–7222.

Burke, J. M., Irvine, K. D., Kaneko, K. J., Kerker, B. J., Oettgen, A. B., Tierney, W. M., Williamson, C. L., Zaug, A. J., and Cech, T. R. (1986). Role of conserved sequence elements 9L and 2 in self-splicing of the *Tetrahymena* ribosomal RNA precursor. *Cell* **45,** 167–176.

Butow, R. A. (1985). Nonreciprocal exchanges in the yeast mitochondrial genome. *Trends Biochem. Sci. (Pers. Ed.)* **1,** 81–84.

Butow, R. A., Vincent, R. D., Strausberg, R. L., Zanders, E., and Perlman, P. S. (1977). Genetic, physical and biochemical analysis of a mitochondrial gene. *In* "Mitochondria 1977" (W. Bandlow, R. J. Schweyen, K. Wolf, and F. Kaudewitz, eds.), pp. 317–336. De Gruyter, Berlin.

Butow, R. A., Terpstra, P., and Strausberg, R. L. (1979). Genetic and biochemical analysis of *var1*. *In* "Extrachromosomal DNA" (D. J. Cummings, P. Borst, I. B. Dawid, S. M. Weissman, and C. F. Fox, eds.), pp. 269–286. Academic Press, New York.

Butow, R. A., Lopez, I. C., Change, H.-P., and Farelly, F. (1980). The specification of *var1* polypeptide by the *var1* determinant. *In* "The Organization and Expression of the Mitochondrial Genome" (A. M. Kroon and C. Saccone, eds.), pp. 195–206. Elsevier, Amsterdam.

Butow, R. A., Farelly, F., Zassenhaus, H. P., Hudspeth, M. E. S., Grossman, L. I., and Perlman, P. S. (1982). *Var1* determinant region of yeast mitochondrial DNA. *In* "Mitochondrial Genes" (P. Slonimski, P. Borst, and G. Attardi, eds.), pp. 241–254. Cold Spring Harbor Laboratory, Cold Spring Harbor, New York.

Butow, R. A., Perlman, P. S., and Grossman, L. I. (1985). Organization and evolution of the unusual *var1*-gene of yeast mitochondrial DNA. *Science* **228,** 1496–1501.

Cabral, F., Solioz, M., Deters, D., Rudin, Y., Schatz, G., Clavilier, L., Groudinski, O., and Slonimski, P. P. (1977). Effect of mitochondrial mutation on cytochrome *c* oxidase in yeast. *In* "Mitochondria 1977" (W. Bandlow, R. J. Schweyen, K. Wolf, and F. Kaudewitz, eds.), pp. 401–414. De Gruyter, Berlin.

Cabral, F., Solioz. M., Rudin, Y., Schatz, G., Clavilier, L., and Slonimski, P. P. (1978). Identification of the structural gene for yeast cytochrome *c* oxidase subunit II on mitochondrial DNA. *J. Biol. Chem.* **253,** 287–304.

Canaday, J., Dirheimer, G., and Martin, R. P. (1980). Yeast mitochondrial methionine initiator tRNA: Characterization and nucleotide sequence. *Nucleic Acids Res.* **8,** 1445–1457.

Carignani, G., Dujardin, G., and Slonimski, P. P. (1979). Petite deletion map of the mitochondrial *oxi3* region in *Saccharomyces cerevisiae. Mol. Gen. Genet.* **167,** 301–308.

Carignani, G., Groudinsky, O., Frezza, D., Schiavon, E., Bergantino, E., and Slonimski, P. P. (1983). A mRNA maturase is encoded by the first intron of the mitochondrial gene for the subunit I of cytochrome oxidase in *S. cerevisiae. Cell* **35,** 733–742.

Cariganani, G., Netter, P., Bergantino, E., and Robineau, S. (1986). Expression of the mitochondrial split gene coding for cytochrome oxidase subunit 1 in *S. cerevisiae:* RNA splicing pathway. *Curr. Genet.* **11,** 55–65.

Caron, F., Jacq, C., and Ranviere-Yaniv, J. (1979). Characterization of a histone like protein extracted from yeast mitochondria. *Proc. Natl. Acad. Sci U.S.A.* **76,** 4265–4269.

Casey, J., Cohen, M., Rabinowitz, M., Fukuhara, H., and Getz, G. S. (1972). Hybridization of mitochondrial transfer RNAs with mitochondrial and nuclear DNA of grande (wild type) yeast. *J. Mol. Biol.* **63,** 431–440.

Casey, J. W., Hsu, H.-J., Getz, G. S., and Rabinowitz, M. (1974a). Transfer RNA genes in mitochondrial DNA of grande (wild type) yeast. *J. Mol. Biol.* **88,** 735–747.

Casey, J. W., Hsu, H.-J., Rabinowitz, M., and Getz, G. (1974b). Transfer RNA genes in the mitochondrial DNA of cytoplasmic petite mutants of *Saccharomyces cerevisiae. J. Mol. Biol.* **88,** 717–733.

Casey, J., Gordon, P., and Rabinowitz, M. (1974c). Characterization of mitochondrial DNA from grande and petite yeasts by renaturation and denaturation analysis and by tRNA hybridization: Evidence for internal repetition or heterogeneity in mitochondrial DNA populations. *Biochemistry* **13,** 1059–1067.

Cech, T. R. (1983). RNA splicing: Three themes with variations. *Cell* **34,** 713–716.

Cech, T. R. (1986a). The generality of self-splicing RNA: Relationship to nuclear mRNA splicing. *Cell* **44,** 207–210.

Cech, T. R. (1986b). A model for the RNA-catalyzed replication of RNA. *Proc. Natl. Acad. Sci. U.S.A.* **83,** 4360–4363.

Cech, T. R. (1986c). RNA as an enzyme. *Sci. Am.* **255,** 76–84.

Cech, T. R., and Bass, B. L. (1986). Biological catalysis by RNA. *Annu. Rev. Biochem.* **55,** 599–630.

Cech, T. R., Tanner, N. K., Tinoco, I., J., Weir, B. R., Zuker, M., and Perlman, P. S. (1983). Secondary structure of the *Tetrahymena* ribosomal RNA intervening sequences. *Proc. Natl. Acad. Sci. U.S.A.* **80,** 3903–3907.

Celis, E., Mas, J., and Brunner, A. (1975). Nuclear and cytoplasmic cross-resistance and correlated sensitivity to DNA intercalating drugs in a petite-negative yeast. *Genet. Res.* **25,** 59–69.

Chambers, P., and Gingold, E. (1986). A direct study of the relative synthesis of petite and grande mitochondrial DNA in zygotes from crosses involving suppressive petite mutants of *Saccharomyces cerevisiae. Curr. Genet.* **10,** 565–572.

Chang, D. D., Wong, T. W., Hixson, J. E., and Clayton, D. A. (1985). Regulatory sequences for mammalian mitochondrial transcription and replication. *In* "Achievements and Perspectives of Mitochondrial Research" (E. Quagliariello, E. C. Slater, F. Palmieri, C. Saccone, and A. M. Kroon, eds.), Vol. 2, pp. 135–144. Elsevier, Amsterdam.

Chang, L. T., Tuveson, R. W., and Munroi, M. H. (1969). Nonnuclear inheritance of UV sensitivity in *Neurospora crassa. Can. J. Genet. Cytol.* **10,** 920–927.

Chevilotte-Brivet, P., and Meunier-Lemesle, D. (1980). Cytochrome *b-565* in *Saccharomyces cerevisiae:* Use of mutants in the *cob–box* region of the mitochondrial DNA to study the functional role of this spectral species of cytochrome *b*. 2. Relation between energetic data and cytochrome *b-565* content. *Eur. J. Biochem.* **111,** 161–169.

Chevillotte-Brivet, P., Meunier-Lemesle, D., Forget, N., and Pajot P. (1983). Is cytochrome *b* really the antimycin-binding component of the cytochrome bc_1 complex of yeast mitochondria? *Eur. J. Biochem.* **129,** 653–661.

Chevillotte-Brivet, P., Salon, G., and Meunier-Lemesle, D. (1987). Missense exonic mitochondrial mutation in cytochrome b gene of *Saccharomyces cerevisiae* resulting in core protein deficiency in complex III of the respiratory chain. *Curr. Genet.* **12,** 111–118.

Chomyn, A., Mariottini, P., Cleeter, M. W. J., Ragan, C. I., Doolittle, R. F., Matsuno-Yagi, A., Hatefi, Y., and Attardi, G. (1985a). Functional assignment of the products of the unidentified reading frames of human mitochondrial DNA. *In* "Achievements and Perspectives of Mitochondrial Research" (E. Quagliariello, E. C. Slater, F. Palmieri, C. Saccone, and A. M. Kroon, eds.), Vol. 2, pp. 259–276. Elsevier, Amsterdam.

Chomyn, A., Mariottini, P., Cleeter, M. W., J., Ragan, C. I., Matsuno-Yagi, A., Hatefi, Y., Doolittle, R., and Attardi, G. (1985b). Six unidentified reading frames of human mitochondrial DNA encode components of the respiratory-chain NADH dehydrogenase. *Nature (London)* **314,** 592–597.

Chomyn, A., Cleeter, M. W. J., Ragan, C. I., Riley, M., Doolittle, R. F., and Attardi, G. (1986). *URF 6,* last unidentified reading frame of human mtDNA, codes for an NADH dehydrogenase subunit. *Science* **234,** 614–618.

Choo, K. B., Nagley, P., Lukins, H. B., and Linnane, A. W. (1977). Biogenesis of mitochondria. 47. Refined physical map of mitochondrial genome of *Saccharomyces cerevisiae* determined by analysis of an extended library of genetically and molecularly defined petite mutants. *Mol. Gen. Genet.* **153,** 279–288.

Christiansen, G., and Christiansen, C. (1976). Comparison of the fine structure of mitochondrial DNA from *Saccharomyces cerevisiae* and *S. carlsbergensis*. Electron microscopy of partially denatured molecules. *Nucleic Acids Res.* **3,** 465–476.

Christiansen, T., and Rabinowitz, M. (1983). Identification of multiple transcriptional initiation sites on the yeast mitochondrial genome by *in vitro* capping with guanylyltransferase. *J. Biol. Chem.* **258,** 14025–14033.

Christiansen, T., Edwards, J., Levens, D. Locker, J., and Rabinowitz, M. (1982). Transcription initiation and processing of the small ribosomal RNA of yeast mitochondria. *J. Biol. Chem.* **257,** 6494–6500.

Christiansen, T., Edwards, J. C., Mueller, D. M., and Rabinowitz, M. (1983). Identification of a single transcriptional site for the glutamic tRNA and *COB* genes in yeast mitochondria. *Proc. Natl. Acad. Sci. U.S.A.* **80,** 5564–5568.

Church, G. M., and Gilbert, W. (1980). Yeast mitochondrial intron products required in *trans* for RNA splicing. *In* "Mobilization and Reassembly of Genetic Information" (D. R. Joseph *et al.,* eds.), pp. 379–394. Academic Press, New York.

Church, G. M., Slonimski, P. P., and Gilbert, W. (1979). Pleiotropic mutations within two yeast mitochondrial cytochrome genes block mRNA processing. *Cell* **18,** 1209–1215.

Claisse, M. L., Spyridakis, A., and Slonimski, P. P. (1977). Mutation at anyone of three unlinked mitochondrial genetic loci, *BOX1, BOX4* and *BOX6,* modify the structure of cytochrome *b* polypeptide(s). *In* "Mitochondria 1977" (W. Bandlow, R. J. Schweyen, K. Wolf, and F. Kaudewitz, eds.), pp. 337–344. De Gruyter, Berlin.

Claisse, M. L., Spyridakis, A., Wambier-Kluppel, M. L., Pajot, P., and Slonimski, P. P. (1978). Mosaic organization and expression of the mitochondrial DNA region controlling cytochrome *c* reductase and oxidase. II. Analysis of proteins translated from the *box* region. *In* "Biochemistry and Genetics of Yeast" (M. Bacila *et al.,* eds.), pp. 369–390. Academic Press, New York.

Claisse, M., Slonimski, P. P., Johnson, J., and Mahler, H. R. (1980). Mutations within an intron and its flanking sites: Patterns of novel polypeptides generated by mutants in one segment of the *cob–box* region of yeast mitochondrial DNA. *Mol. Gen. Genet.* **177,** 375–387.

Clark-Walker, G. D. (1985). Basis of diversity in mitochondrial DNAs. *In* "The Evolution of Genome Size" (T. Cavalier-Smith, *ed.*), pp. 277–297. Wiley, Sussex.

Clark-Walker, G. D., Evans, R. J., Hoeben, P., and McArthur, C. R. (1985a). The basis of diversity in yeast mitochondria. *In* "Achievements and Perspectives of Mitochondrial Research" (E. Quagliarello, E. C. Slater, F. Palmieri, C. Saccone, and A. M. Kroon, eds.), Vol. 2, pp. 71–78. Elsevier, Amsterdam.

Clark-Walker, G. D., and Gabor Miklos, G. L. (1974). Mitochondrial genetics, circular DNA and the mechanism of the petite mutation in yeast. *Genet. Res.* **24,** 43–57.

Clark-Walker, G. D., and Gabor-Miklos, G. L. (1975). Complementation in cytoplasmic petite mutants of yeast to form respiratory competent cells. *Proc. Natl. Acad. Sci. U.S.A.* **72,** 372–375.

Clark-Walker, G. D., and Linnane, A. W. (1966). *In vivo* differentiation of yeast cytoplasmic and mitochondrial protein synthesis with antibiotics. *Biochem. Biophys. Res. Commun.* **25,** 8–13.

Clark-Walker, G.D., and McArthur, C. R. (1978). Structural and functional relationships of mitochondrial DNAs from various yeasts. *In* "Biochemistry and Genetics of Yeast" (M. Bacila *et al.,* eds.), pp. 255–272. Academic Press, New York.

Clark-Walker, G. D., and Sriprakash, K. S. (1981). Sequence rearrangements between mitochondrial DNAs of *Torulopsis glabrata* and *Kloeckera africana* identified by hybridization with six polypeptide encoding regions from *Saccharomyces cerevisiae* mitochondrial DNA. *J. Mol. Biol.* **151,** 367–387.

Clark-Walker, G. D., and Sriprakash, K. S. (1982). Size diversity and sequence rearrangements in mitochondrial DNAs from yeasts. *In* "Mitochondrial Genes" (P. Slonimski, P. Borst, and G. Attardi, eds.), pp. 349–354. Cold Spring Harbor Laboratory, Cold Spring Harbor, New York.

Clark-Walker, G. D., and Sriprakash, K. S. (1983a). Analysis of a five gene cluster and a unique orientation of large genic sequences in *Torulopsis glabrata* mitochondrial DNA. *J. Mol. Evol.* **19,** 342–351.

Clark-Walker, G. D., and Sriprakash, K. S. (1983b). Map location of transcripts from *Torulopsis glabrata* mitochondrial DNA. *EMBO J.* **2,** 1465–1472.

Clark-Walker, G. D., Sriprakash, K. S., McArthur, C. R., and Azad, A. A. (1980). Mapping of mitochondrial DNA from *Torulopsis glabrata.* Location of ribosomal and transfer RNA genes. *Curr. Genet.* **1,** 209–217.

Clark-Walker, G. D., McArthur, C. R., and Daley, D. J. (1981a). Does mitochondrial DNA length influence the frequency of spontaneous petite mutants in yeasts? *Curr. Genet.* **4,** 7–12.

Clark-Walker, G. D., McArthur, C. R., and Sriprakash, K. S. (1981b). Partial duplication of the large ribosomal RNA sequence in an inverted repeat in circular mitochondrial DNA from *Kloeckera africana.* Implications for the mechanism of the petite mutation. *J. Mol. Biol.* **147,** 399–415.

Clark-Walker, G. D., McArthur, C. R., and Sriprakash, K. S. (1983). Order and orientation of genic sequences in circular mitochondrial DNA from *Saccharomyces exiguus:* Implications for evolution of yeast mt DNAs. *J. Mol. Evol.* **19,** 333–341.

Clark-Walker, G. D., McArthur, C. R., and Sriprakash, K. S. (1985b). Location of transcriptional control signals and transfer RNA sequences in *Torulopsis glabrata* mitochondrial DNA. *EMBO J.* **4,** 465–473.

Clavilier, L. (1976). Mitochondrial genetics. XII. An oligomycin-resistant mutant localized at a new mitochondrial locus in *Saccharomyces cerevisiae. Genetics* **83,** 227–243.

Clayton, D. A. (1982). Replication of animal mitochondrial DNA. *Cell* **28,** 693–705.

Clayton, D. A. (1984). Transcription of the mammalian mitochondrial genome. *Annu. Rev. Biochem.* **53,** 573–594.

Clayton, D. A., and Brambl, R. M. (1972). Detection of circular DNA from mitochondria of *Neurospora crassa. Biochem. Biophys. Res. Commun.* **46,** 1277–1482.

Clejan, L., Sidhu, A., and Beattie, D. S. (1983). Studies on the function and biogenesis of cytochrome *b* in mutants of *Saccharomyces cerevisiae* resistant to 3-(3,4-dichlorophenyl)-1,1-dimethyl-urea. *Biochemistry* **22,** 52–57.

Cobon, G. S., Groot-Obbink, D. J., Hall, R. M., Maxwell, R., Murphy, M., Rytka, J., and Linnane, A. W. (1976). Mitochondrial genes determining cytochrome *b* (complex III) and cytochrome oxidase function. *In* "Genetics and Biogenesis of Chloroplasts and Mitochondria" (T. Bücher, W. Neupert, W. Sebald, and S. Werner, eds.), pp. 453–460. Elsevier, Amsterdam.

Cobon, G. S., Beilharz, M. W., Linnane, A. W., and Nagley, P. (1982). Biogenesis of mitochondria: Mapping of transcripts from the *oli2* region of mitochondrial DNA in two grande strains of *Saccharomyces cerevisiae. Curr. Genet.* **5,** 97–107.

Coen, D., Deutsch, J., Netter P., Petrochilo, E., and Slonimski, P. P. (1970). Mitochondrial genetics—methodology and phenomenology. "Control of Organelle Development," pp. 449–496. Cambridge Univ. Press, London.

Cohen, M., and Rabinowitz, M. (1972). Analysis of grande and petite yeast mitochondrial DNA by tRNA hybridization. *Biochim. Biophys. Acta* **281,** 192–201.

Colin, Y., Baldacci, G., and Bernardi, G. (1985). A new putative gene in the mitochondrial genome of *Saccharomyces cerevisiae. Gene* **36,** 1–13.

Coletti, E., Frontali, L., Palleschi, C., Wesolowski, M., and Fukuhara, H. (1979). Two isoaccepting seryl tRNAs coded by separate mitochondrial genes in yeast. *Mol. Gen. Genet.* **175,** 1–4.

Colleaux, L., d'Auriol, L., Betermier, M., Cottarel, G., Jacquier, A., Galibert, F., and Dujon, B. (1986). Universal code equivalent of a yeast mitochondrial intron reading frame is expressed in *E. coli* as a specific double strand endonuclease. *Cell* **44,** 521–533.

Collins, R. A., and Lambowitz, A. M. (1981). Characterization of a variant *Neurospora crassa* mitochondrial DNA which contains tandem reiterations of 1.9 kb sequence. *Curr. Genet.* **4,** 131–133.

Collins, R. A., and Lambowitz, A. M. (1983). Structural variations and optional introns in the mitochondrial DNAs of *Neurospora* strains isolated from nature. *Plasmid* **9,** 53–70.

Collins, R. A., Bertrand, H., La Polla, R. J., and Lambowitz, A. M. (1981a). A novel extranuclear mutant of *Neurospora* with a temperature-sensitive defect in mitochondrial protein synthesis and mitochondrial ATPase. *Mol. Gen. Genet.* **181,** 13–19.

Collins, R. A., Stohl, L. L., Cole, M. D., Lambowitz, A. M. (1981b). Characterization of a novel plasmid DNA found in mitochondria of *N. crassa. Cell* **24,** 443–452.

Colson, A.-M., and Slonimski, P. P. (1977). Mapping of drug-resistant loci in the coenzyme QH_2:cytochrome *c* reductase region of the mitochondrial DNA map in *Saccharomyces cerevisiae. In* "Mitochondria 1977" (W. Bandlow, R. J. Schweyen, K. Wolf, and F. Kaudewitz, eds.), pp. 185–198. De Gruyter, Berlin.

Colson, A.-M., and Slonimski, P. P. (1979). Genetic localization of diuron- and mucidin-resistant mutants, relative to a group of loci of the mitochondrial DNA controlling coenzyme QH_2–cytochrome *c* reductase in *Saccharomyces cerevisiae. Mol. Gen. Genet.* **167,** 287–298.

Colson, A.-M., and Wouters, L. (1980). Selection of a new class of cytoplasmic diuron-resistant mutations in *Saccharomyces cerevisiae:* Tentative explanation for unexpected genetic and phenotypic properties of the mitochondrial cytochrome *b* split gene in these mutants. *In* "The Organization and Expression of the Mitochondrial Genome" (A. M. Kroon and C. Saccone, eds.), pp. 71–74. Elsevier, Amsterdam.

Colson, A.-M., Labaille, F., and Goffeau, A. (1976). A cytoplasmic gene for partial suppression of a nuclear pleiotropic respiratory deficient mutant in the petite negative yeast *Schizosaccharomyces pombe. Mol. Gen. Genet.* **149,** 101–109.

Colson, A.-M., The Van, L., Convent, B., Briquet, M., and Goffeau, A. (1977). Mitochondrial heredity of resistance to 2-(3,4-dichlorophenyl)-1,1-dimethylurea, an inhibitor of cytochrome *b* oxidation, in *Saccharomyces cerevisiae. Eur. J. Biochem.* **74,** 521–526.

Connerton, I. F., Ray, M. K., Lancashire, W. E., and Griffith, D. E. (1984). Genetics of oxidative phosphorylation—petite deletion mapping in the *oli2* region of the mitochondrial genome of *Saccharomyces cerevisiae. Mol. Gen. Genet.* **193,** 149–152.

Contamine, V., and Bolotin-Fukuhara, M. (1984). A mitochondrial ribosomal RNA mutation and its nuclear suppressor. *Mol. Gen. Genet.* **193,** 280–287.

Convent, B., and Briquet, M. (1975). Diuron and related herbicides, inhibitors of the oxidation of mitochondrial cytochrome *b* in *Saccharomyces cerevisiae. Arch. Int. Physiol. Biochem.* **83,** 358–359.

Convent, B., Briquet, and M., Goffeau, A. (1978). Kinetic evidence for two sites in the inhibition by diuron of the electron transport in the bc_1 segment of the respiratory chain in *Saccharomyces cerevisiae. Eur. J. Biochem.* **92,** 137–145.

Coruzzi, G., and Tzagoloff, A. (1979). Assembly of the mitochondrial membrane system. DNA sequence of subunit 2 of yeast cytochrome oxidase. *J. Biol. Chem.* **254,** 9324–9330.

Coruzzi, G., Trembath, M. K., and Tzagoloff, A. (1978). Assemby of the mitochondrial membrane system: Mutations in the *pho2* locus of the mitochondrial genome of *Saccharomyces cerevisiae. Eur. J. Biochem.* **92,** 279–287.

Coruzzi, G., Trembath, M. K., and Tzagoloff, A. (1979). The isolation of mitochondrial and nuclear mutants of *Saccharomyces cerevisiae* with specific defects in mitochondrial functions. *In* "Methods in Enzymology" (S. Fleischer and L. Packer, eds.), Vol. 56, pp. 95–107. Academic Press, New York.

Coruzzi, G., Bonitz, S. G., Thalenfeld, B. E., and Tzagoloff, A. (1981). Assembly of the mitochondrial membrane system. Analysis of the nucleotide sequence and transcripts of the *oxi1* region of yeast mitochondrial DNA. *J. Biol. Chem.* **256,** 12780–12787.

Cosson, J., and Tzagoloff, A. (1979). Sequence homologies of (guanosine + cytosine)-rich regions of mitochondrial DNA of *Saccharomyces cerevisiae. J. Biol. Chem.* **254,** 42–43.

Cottrell, S. F. (1981). Mitochondrial DNA synthesis in synchronous cultures of the yeast, *Saccharomyces cerevisiae. Exp. Cell Res.* **132,** 89–98.

Cottrell, S. F., and Lee, L. H. (1981). Evidence for the synchronous replication of mitochondrial DNA during the yeast cell cycle. *Biochem. Biophys. Res. Commun.* **101,** 1350–1356.

Cottrell, S. F. Rabinowitz, M., and Getz, G. S. (1973). Mitochondrial DNA synthesis in a temperature sensitive mutant of DNA replication of *Saccharomyces cerevisiae. Biochemistry* **12,** 4373–4378.

Croft, J. H., Dales, R. B. G., Turner, G., and Earl, A. (1979). The transfer of mitochondria between species of *Aspergillus. In* "Advances in Protoplast Research" (L. Ferency and G. L. Farkas, eds.), pp. 85–92. Pergamon, Oxford.

Cummings, D. J., and Wright, R. M. (1983). DNA sequence of the excision sites of a mitochondrial plasmid from senescent *Podospora anserina. Nucleic Acids Res.* **11,** 2111–2119.

Cummings, D. J., Belcour, L., and Grandchamp, C. (1978). Etude au microscope éléctronique du DNA mitochondrial de *Podospora anserina* et présence d'une série multimerique de molecules circulaires de DNA dans des cultures sénéscentes. *C. R. Acad. Sci. Paris* **287,** 157–162.

Cummings, D. J., Belcour, L., and Grandchamp, C. (1979a). Mitochondrial DNA and senescence in *Podospora anserina. In* "Extrachromosomal DNA" (D. J. Cummings, P. Borst, I. B. Dawid, J. M. Weissman, and C. F. Fox, eds.), pp. 549–559. Academic Press, New York.

Cummings, D. J., Belcour, L., and Grandchamp, C. (1979b). Mitochondrial DNA from *Podospora anserina.* 1. Isolation and characterization. *Mol. Gen. Genet.* **171,** 229–238.

Cummings, D. J., Belcour, L., and Grandchamp, C. (1979c). Mitochondrial DNA from *Podospora anserina*. 2. Properties of mutant DNA and multimeric circular DNA from senescent cultures. *Mol. Gen. Genet.* **171,** 239–250.

Cummings, D. J., Laping, J. L., and Nolan, P. (1980). Cloning of senescent mitochondrial DNA from *Podospora anserina:* A beginning. *In* "The Organization and Expression of the Mitochondrial Genome" (A. M. Kroon and C. Saccone, eds.), pp. 97–102. Elsevier, Amsterdam.

Cummings, D. J., MacNeil, A. I., Domenico, J., and Matsuura, E. T. (1985). Excision-amplification of mitochondrial DNA during senescence in *Podospora anserina.* DNA sequence analysis of three unique plasmids. *J. Mol. Biol.* **185,** 659–680.

Darlison, M. G., and Lancashire, W. E. (1980). Genetics of oxidative phosphorylation—Allelism studies of mitochondrial loci in the *PHO1-OLI2* region of the genome. *Mol. Gen. Genet.* **180,** 227–229.

Davies, R. W., Waring, R. B., Ray, J. A., Brown, T. A., and Scazzocchio, C. (1982). Making ends meet: A model for RNA splicing in fungal mitochondria. *Nature (London)* **300,** 719–724.

Davies, R. W., Waring, R. B., Brown, T. A., and Scazzocchio, C. (1983). Implications of fungal mitochondrial intron RNA structure for the mechanism of RNA splicing. *In* "Mitochondria 1983" (R. J. Schweyen, K. Wolf, and F. Kaudewitz, eds.), pp. 17–189. De Gruyter, Berlin.

Dawes, I. W., and Carter, B. L. A. (1974a). Nitrosoguanidine mutagenesis during nuclear and mitochondrial gene replication. *Nature (London)* **250,** 709–712.

Dawes, I. W., and Carter, B. L. A. (1974b). The timing of mitochondrial deoxyribonucleic acid replication during the cell cycle in *Saccharomyces cerevisiae. Biochem. Soc. Trans.* **2,** 224–227.

Dawid, I. B., and Rastl, E. (1979). Structure and evolution of animal mitochondrial DNA. *In* "Extrachromosomal DNA" (D. J. Cummings, P. Borst, I. B. Dawid, S. M. Weissman, and C. F. Fox, eds.), pp. 395–407. Academic Press, New York.

De Jonge, J., and de Vries, H. (1983). The structure of the gene for subunit I of cytochrome *c* oxidase in *Neurospora crassa* mitochondria. *Curr. Genet.* **7,** 21–28.

Delahodde, A., Banroques, J., Becam, A. M., Goguel, V., Perea, J., Schroeder, R., and Jacq, C. (1985). Purification of yeast bI4 mRNA maturase from *E. coli* and from yeast. Nucleic acids binding properties. *In* "Achievements and Perspectives of Mitochondrial Research" (E. Quagliariello, E. C. Slater, F. Palmieri, C. Saccone, and A. M. Kroon, eds.), Vol. 2, pp. 79–88. Elsevier, Amsterdam.

De La Salle, H., Jacq, C., and Slonimski, P. P. (1982). Critical sequences within mitochondrial introns: Pleiotropic mRNA maturase and cis-dominant signals of the *box* intron controlling reductase and oxidase. *Cell* **28,** 721–732.

Del Giudice, L. (1981). Cloning of mitochondrial DNA from the petite-negative yeast *Schizosaccharomyces pombe* in the bacterial plasmid *pBR322*. *Mol. Gen. Genet.* **184,** 465–470.

Del Giudice, L., and Brunner, A. (1977). Chromosomal and extrachromosomal inheritance of erythromycin-resistance in the "petite-negative" yeast *Kluyveromyces lactis. Mol. Gen. Genet.* **152,** 325–329.

Del Giudice, L., and Wolf, K. (1980). Evidence for a joint control of nuclear and mitochondrial DNA synthesis in the petite negative yeast *Schizosaccharomyces pombe. In* "Endosymbiosis and Cell Research" (W. Schwemmler and H. E. A. Schenk, eds.), pp. 775–781. De Gruyter, Berlin.

Del Giudice, L., Wolf, K., Seitz, G., Burger, G., Lang, B., and Kaudewitz, F. (1977).

Extrachromosomal inheritance in *Schizosaccharomyces pombe*. III. Isolation and characterization of paromomycin-resistant mutants. *Mol. Gen. Genet.* **152,** 319–324.

Del Giudice, L., Wolf, K., Buono, C., and Manna, F. (1981a). Nucleo–cytoplasmic interactions in the petite-negative yeast *Schizosaccharomyces pombe*. Inhibition of nuclear and mitochondrial DNA syntheses in the absence of cytoplasmic protein synthesis. *Mol. Gen. Genet.* **181,** 306–308.

Del Giudice, L., Wolf, K., Manna, F., and Pagliuca, N. (1981b). Synthesis of mitochondrial DNA during the cell cycle of the petite negative yeast *Schizosaccharomyces pombe*. *Mol. Gen. Genet.* **182,** 252–254.

De Margerie-Hottinguer, H., and Roman, H. (1955). Suppressiveness: A new factor in genetic determination of the synthesis of respiratory enzymes in yeast. *Proc. Natl. Acad. Sci. U.S.A.* **41,** 1065–1071.

De Ronde, A., Van Loon, A. P. G. M., Grivell, L. A., and Kohli, J. (1980). *In vitro* suppression of UGA codons in a mitochondrial mRNA. *Nature (London)* **287,** 361–363.

Deutsch, J., Dujon, B., Netter, P., Petrochilo, E., Slonimski, P. P., Bolotin-Fukuhara, M., and Coen, D. (1974) Mitochondrial genetics. VI. The petite mutation in *Saccharomyces cerevisiae:* Interrelations between the loss of the *rho*$^+$ factor and the loss of the drug mitochondrial genetic markers. *Genetics* **76,** 195–219.

Devenish, R. J., English, K. J., Hall, R. M., Linnane, A. W., and Lukins, H. B. (1978). Biogenesis of mitochondria. 49. Identification and mapping of a new mitochondrial locus (*tsr1*) which maps in the polar region of the yeast mitochondrial genome. *Mol. Gen. Genet.* **161,** 251–259.

Devenish, R. J., Hall, R. M., Linnane, A: W., and Lukins, H. B. (1979). Biogenesis of mitochondria. 52. Deletions in petite strains occurring in the mitochondrial gene for the 21S ribosomal RNA, that affect the properties of mitochondrial recombination. *Mol. Gen. Genet.* **174,** 297–305.

De Vries, H., De Jonge, J. C., Schneller, J.-M., Martin, R. P., Dirheimer, G., and Stahl, A. J. C. (1978). *Neurospora crassa* mitochondrial transfer RNAs. *Biochim. Biophys. Acta* **520,** 419–427.

De Vries, H., De Jonge, J. C., Bakker, H., Meurs, H., and Kroon, A. (1979). The anatomy of the tRNA–rRNA region of *Neurospora crassa* mitochondrial DNA. *Nucleic Acids Res.* **6,** 1791–1803.

De Vries, H., De Jonge, J. C., and Van't Sant, P. (1980). Characterization of a mitochondrial "stopper" mutant of *Neurospora crassa:* Deletions and rearrangements in the mitochondrial DNA result in disturbed assembly of respiratory chain components. *In* "The Organization and Expression of the Mitochondrial Genome" (A. M. Kroon and C. Saccone, eds.), pp. 333–342. Elsevier, Amsterdam.

De Vries, H., De Jonge, J. C., Van't Sant, P., Agsteribbe, E., and Arnberg, A. (1981). A "stopper" mutant of *Neurospora crassa* containing two populations of aberrant mitochondrial DNA. *Curr. Genet.* **3,** 205–211.

De Vries, H., de Jonge, C. J., Arnberg, A., and Peijneburg, A. A. C. M. (1983). The expression of the mitochondrial genes for subunit 1 of cytochrome *c* oxidase and for an ATPase proteolipid in *Neurospora crassa:* Nucleotide sequences and transcript analysis. *In* "Mitochondria 1983" (R. J. Schweyen, K. Wolf, and F. Kaudewitz, eds.), pp. 343–356. De Gruyter, Berlin.

De Vries, H., De Jonge, J. C., and Schrage, C. (1985a). The *Neurospora* mitochondrial "stopper" mutant. [*E35*], lacks two protein genes indispensable for the formation of complexes I, III and IV. *In* "Achievements and Perspectives of Mitochondrial Research" (E. Quagliariello, E. C. Slater, F. Palmieri, C. Saccone, and A. M. Kroon, eds.), Vol. 2, pp. 285–292. Elsevier, Amsterdam.

De Vries, H., Haima, P., Brinker, M., and De Jonge, J. C. (1985b). The *Neurospora* mitochondrial genome: The region coding for the polycistronic cytochrome oxidase subunit I transcript is preceded by a transfer RNA gene. *FEBS Lett.* **179,** 337–342.

De Vries, H., Alzner-De Weerd, B., Breitenberger, C. A., Chang, D. D., de Jonge, J. C., RajBhandary, U. L. (1986). The [*E 35*] stopper mutant of *Neurospora crassa:* Precise localization of deletion endpoints in mitochondrial DNA and evidence that the deleted DNA codes for a subunit of NADH dehydrogenase. *EMBO J.* **5,** 779–785.

de Zamaroczy, M., and Bernardi, G. (1985a). Sequence organization of the mitochondrial genome of yeast—a review. *Gene* **37,** 1–17.

de Zamaroczy, M., and Bernardi, G. (1985b). The organization and evolution of the mitochondrial genome of yeast: *ori* sequences, genes and intergenic sequences. *In* "Achievements and Perspectives of Mitochondrial Research" (E. Quagliariello, E. C. Slater, F. Palmieri, C. Saccone, and A. M. Kroon, eds.), Vol. 2, pp. 89–98. Elsevier, Amsterdam.

de Zamaroczy, M., and Bernardi, G. (1986a). The GC clusters of the mitochondrial genome of yeast and their evolutionary origin. *Gene* **41,** 1–22.

de Zamaroczy, M., and Bernardi, G. (1986b). The primary structure of the mitochondrial genome of *Saccharomyces cerevisiae*—a review. *Gene* **47,** 155–177.

de Zamaroczy, M., and Bernardi, G. (1987). The AT spacers and the *var 1* genes from the mitochondrial genomes of *Saccharomyces cerevisiae* and *Torulopsis glabrata:* Evolutionary origin and mechanism of formation. *Gene* **54,** 1–22.

de Zamaroczy, M., Baldacci, G., and Bernardi, G. (1979). Putative origins of replication in the mitochondrial genome of yeast. *FEBS Lett.* **108,** 429–432.

de Zamaroczy, M., Marotta, R., Faugeron-Fonty, G., Goursot, R., Mangin, M., Baldacci, G., and Bernardi, G. (1981). The origins of replication of the yeast mitochondrial genome and the phenomenon of suppressivity. *Nature (London)* **292,** 75–78.

de Zamaroczy, M., Faugeron-Fonty, G., and Bernardi, G. (1983). Excision sequences in the mitochondrial genome of yeast. *Gene* **21,** 193–202.

de Zamaroczy, M., Faugeron-Fonty, G., Baldacci, G., Goursot, R., and Bernardi G. (1984). The *ori* sequences of the mitochondrial genome of a wild-type yeast strain: Number, location, orientation and structure. *Gene* **32,** 439–457.

Dhawale, S., Hanson, D. K., Alexander, N. J., Perlman, P. S., and Mahler, H. R. (1981). Regulatory interactions between mitochondrial genes: Interactions between two mosaic genes. *Proc. Natl. Acad. Sci. U.S.A.* **78,** 1778–1782.

Diacumakos, E. G., Garnjobst, L., and Tatum, E. L. (1965). A cytoplasmic character in *Neurospora.* The role of nuclei and mitochondria. *J. Cell Biol.* **26,** 427–443.

Dieckmann, C. L., Bonitz, S. G., Hill, J., Homison, G., McGraw, P., Pape, L., Thalenfeld, B. E., and Tzagoloff, A. (1982). Structure of the apocytochrome-*b* gene and processing of apocytochrome-*b* transcripts in *Saccharomyces cerevisiae. In* "Mitochondrial Genes" (P. Slonimski, P. Borst, and G. Attardi, eds.), pp. 213–223. Cold Spring Harbor Laboratory, Cold Spring Harbor, New York.

di Rago, J.-P., Perea, X., and Colson, A.-M. (1986). DNA sequence analysis of diuron-resistant mutations in the mitochondrial cytochrome b gene of *Saccharomyces cerevisiae. FEBS Lett.* **208,** 208–210.

Dobres, M., Gerbl-Rieger, S., Schmelzer, C., Müller, M. W., and Schweyen, R. J. (1985). Deletions in the *cob* gene of yeast mtDNA and their phenotypic effect. *Curr. Genet.* **10,** 283–290.

Douglas, M. G., and Butown, R. A. (1976). Variant forms of mitochondrial translation products in yeast—evidence for location of determinants on mitochondrial DNA. *Proc. Natl. Acad. Sci. U.S.A.* **71,** 1083–1086.

Douglas, M. G., Kendrick, E., Boulikas, P., Perlman, P. S., and Butow, R. A. (1976a).

Electrophoretic variants of yeast mitochondrial translation products: Genetic and biochemical analysis. *In* "The Genetic Function of Mitochondrial DNA" (C. Saccone and A. M. Kroon, eds.), pp. 199–203. Elsevier, Amsterdam.

Douglas, M. G., Strausberg, R. L., Perlman, P. S., and Butow, R. A. (1976b). Genetic analysis of mitochondrial polymorphic proteins in yeast. *In* "Genetics and Biogenesis of Chloroplasts and Mitochondria" (T. Bücher, W. Neupert, W. Sebald, and S. Werner, eds.), pp. 435–442. Elsevier, Amsterdam.

Dujardin, G., Groudinsky, O., Kruczewska, A., Pajot, P., and Slonimski, P. P. (1980a). Cytochrome *b* messenger RNA maturase encoded in an intron regulates the expression of the split gene. III. Genetic and phenotypic suppression of intron mutations. *In* "The Organization and Expression of the Mitochondrial Genome" (A. M. Kroon and C. Saccone, eds.), pp. 157–160. Elsevier, Amsterdam.

Dujardin, G., Pajot, P., Groudinsky, O., and Slonimski, P. P. (1980b). Long range control circuits within mitochondria and between nucleus and mitochondria. I. Methodology and phenomenology of suppressors. *Mol. Gen. Genet.* **179,** 469–482.

Dujardin, G., Jacq, C., and Slonimski, P. P. (1982). Single base substitutions in an intron of oxidase gene compensates splicing defects of the cytochrome *b* gene. *Nature (London)* **298,** 628–632.

Dujardin, G., Lund, P., and Slonimski, P. P. (1984). The effect of paromomycin and *psi* on the suppression of mitochondrial mutations in *Saccharomyces cerevisiae. Curr. Genet.* **9,** 21–30.

Dujon, B. (1974). Recombination of mitochondrial genes in yeast. *In* "Mechanisms in Recombination" (R. F. Grell, ed.), pp. 307–372. Plenum, New York.

Dujon, B. (1977). Transmission, recombination and segregation of mitochondrial genes in *Saccharomyces cerevisiae. In* "Genetics, Biogenesis and Bioenergetics of Mitochondria" (W. Bandlow, R. J. Schweyen, K. Wolf, and F. Kaudewitz, eds.), pp. 1–6. De Gruyter, Berlin.

Dujon, B. (1979). Mutants in a mosaic gene reveal functions for intron. *Nature (London)* **282,** 777–778.

Dujon, B. (1980). Sequence of the intron and flanking exons of the mitochondrial 21S rRNA gene of yeast strains having different alleles at the ω and *rib-1* loci. *Cell* **20,** 185–197.

Dujon, B. (1981). Mitochondrial genetics and functions. *In* "Molecular Biology of the Yeast *Saccharomyces cerevisiae:* Life Cycle and Inheritance" (J. N. Strathern, E. W. Jones, and J. R. Broach, eds.), pp. 505–635. Cold Spring Harbor Laboratory, Cold Spring Harbor, New York.

Dujon, B., and Blanc, H. (1980). Yeast mitochondrial minilysates and their use to screen a collection of hypersensitive *rho*$^-$ mutants. *In* "The Organization and Expression of the Mitochondrial Genome" (A. M. Kroon and C. Saccone, eds.). Elsevier, Amsterdam.

Dujon, B., Colleaux, L., Jacquier, A., Michel, F., and Monteilhet, C. (1986). Mitochondrial introns as mobile genetic elements: The role of intron-encoded proteins. *In* "Extrachromosomal Elements in Lower Eukaryotes" (R. B. Wickner, A. Hinnebusch, A. M. Lambowitz, I. C. Gunsalus, and A. Hollaender, eds.). Plenum, New York.

Dujon, B., and Jacquier, A. (1983). Organization of the mitochondrial 21S rRNA gene in *Saccharomyces cerevisiae:* Mutants of the peptidyl transferase centre and nature of the *omega* locus. *In* "Mitochondria 1983" (R. J. Schweyen, K. Wolf, and F. Kaudewitz, eds.), pp. 389–403. De Gruyter, Berlin.

Dujon, B., and Michel, F. (1976). Genetic and physical characterization of a segment of the mitochondrial DNA involved in the control of genetic recombination. *In* "The Genetic Function of Mitochondrial DNA" (C. Saccone and A. M. Kroon, eds.), pp. 175–180. Elsevier, Amsterdam.

Dujon, B., and Slonimski, P. P. (1976). Mechanisms and rules for transmission, recombination and segregation of mitochondrial genes in *Saccharomyces cerevisiae*. *In* "Genetics and Biogenesis of Chloroplasts and Mitochondria" (T. Bücher, W. Neupert, W. Sebald, and S. Werner, eds.), pp. 393–403. Elsevier, Amsterdam.

Dujon, B., Bolotin-Fukuhara, M., Coen, D., Deutsch, J., Netter, P. Slonimski, P. P., and Weill, L. (1976). Mitochondrial genetics. XI. Mutations at the mitochondrial locus ω affecting the recombination of mitochondrial genes in *Saccharomyces cerevisiae*. *Mol. Gen. Genet.* **143,** 131–165.

Dujon, B., Colson, A. M., and Slonimski, P. P (1977). The mitochondrial genetic map of *Saccharomyces cerevisiae:* Compilation of mutations, genes, genetic and physical maps. *In* "Mitochondria 1977" (W. Bandlow *et al.,* eds.), pp. 579–669. De Gruyter, Berlin.

Dujon, B., Cottarel, G., Colleaux, L., Betermier, M., Jacquier, A., D'Auriol, L., and Galibert, F. (1985). Mechanism of integration of an intron within a mitochondrial gene: A double strand break and the transposase function of an intron encoded protein as revealed by *in vivo* and *in vitro* assays. *In* "Achievements and Perspectives of Mitochondrial Research" (E. Quagliariello, E. C. Slater, F. Palmieri, C. Saccone, and A. M. Kroon, eds.), Vol. 2, pp. 215–226. Elsevier, Amsterdam.

Earl, A. J., Turner, G., Croft, J. H., Dales, R. B. G., Lazarus, C. M., Lünsdorf, H., and Küntzel, H. (1981). High frequency transfer of species specific mitochondrial DNA sequences between members of the Aspergillaceae. *Curr. Genet.* **3,** 221–228.

Eccleshall, T. R., Needleman, R. B., Storm, E. M., Buchferer, B., and Marmur, J. (1978). A temperature-sensitive yeast mitochondrial mutant with altered cytochrome *c* oxidase subunit. *Nature (London)* **273,** 67–70.

Edelmann, M., Verma, I., and Littauer, U. (1970). Mitochondrial ribosomal RNA from *Aspergillus nidulans:* Characterization of a novel molecular species. *J. Mol. Biol.* **49,** 67–83.

Edelmann, M., Herzog, R., Galun, E., and Littauer, U. (1971). Physicochemical properties of mitochondrial ribosomal RNA from fungi. *Eur. J. Biochem.* **19,** 372–378.

Edwards, J. C., Levens, D., and Rabinowitz, M. (1982). Analysis of transcriptional initiation of yeast mitochondrial DNA in a homologous *in vitro* transcription system. *Cell* **31,** 337–346.

Edwards, J. C., Christianson, T., Mueller, D., Biswas, T. K., Levens, D., Li, D., Wettstein, J., and Rabinowitz, M. (1983a). Initiation and transcription of yeast mitochondrial RNA. *In* "Mitochondria 1983" (R. J. Schweyen, K. Wolf, and F. Kaudewitz, eds.), pp. 69–78. De Gruyter, Berlin.

Edwards, J. C., Osinga, K. A., Christiansen, T., Hensgens, L. A. M., Janssens, P. M., Rabinowitz, M., and Tabak, H. F. (1983b). Initiation of transcription of the yeast mitochondrial gene coding for ATPase subunit 9. *Nucleic Acids Res.* **11,** 8269–8282.

Egel, R., Kohli, J., Thuriaux, P., and Wolf, K. (1980). Genetics of the fission yeast *Schizosaccharomyces pombe*. *Annu. Rev. Genet.* **14,** 77–108.

Ephrussi, B., Hottinguer, H., and Chimenes, A. M. (1949a). Action de l'acriflavine sur les lévures. I. La mutation "petite colonie." *Ann. Inst. Pasteur* **76,** 351–367.

Ephrussi, B., Hottinguer, H., and Tavlitzki, J. (1949b). Action de l'acriflavine sur les lévures. II. Étude génétique du mutant "petite colonie." *Ann. Inst. Pasteur* **76,** 419–450.

Ephrussi, B., de Margerie-Hottinguer, H., and Roman, H. (1955). Suppressiveness: A new factor in the genetic determinism of the synthesis of respiratory enzymes in yeast. *Proc. Natl. Acad. Sci. U.S.A.* **41,** 1065–1070.

Epler, J. L. (1969). The mitochondrial and cytoplasmic transfer ribonucleic acids of *Neurospora crassa. Biochemistry* **8,** 2285–2290.

Epler, J. L., Shugart, L. R., and Barnett, W. E. (1970). *N*-Formylmethionyl transfer ribonucleic acid in mitochondria from *Neurospora. Biochemistry* **9,** 3575–3579.

Esser, K. (1966). Incompatibility. *In* "The Fungi" (G. C. Ainsworth and A. S. Sussman, eds.), Vol. 2, pp. 661–676. Academic Press, New York.

Esser K. (1974). *Podospora anserina. In* "Handbook of Genetics" (R. C. King, ed.), pp. 532–553. Plenum, New York.

Esser, K., and Raper, J. R., eds. (1965). "Incompatibility in Fungi." Springer-Verlag, Berlin.

Esser, K., and Tudzynski, P. (1977). The prevention of senescence in the ascomycete *Podospora anserina* by the antibiotic Tiamulin. *Nature (London)* **265,** 454–456.

Esser, K., and Tudzynski, P. (1979a). Genes and mitochondrial inhibitors preventing senescence in the ascomycete *Podospora anserina. Gerontology* **25,** 163–168.

Esser, K., and Tudzynski, P. (1979b). Genetic control and expression of senescence in *Podospora anserina.* "Viruses and Plasmids in Fungi," pp. 595–602. Dekker, New York.

Esser, K., and Tudzynski, P. (1980). Senescence in fungi. *In* "Senescence in Plants" (K. V. Thimann, ed.), Vol. 2, pp. 67–83. CRC Press, Boca Raton, Florida.

Esser, K., Tudzynski, B., Stahl, U., and Kück, U. (1980). A model to explain senescence in the filamentous fungus *Podospora anserina* by the action of plasmid-like DNA. *Mol. Gen. Genet.* **178,** 213–216.

Esser, K., Kück, U., Stahl, U., and Tudzynski, P. (1981). Mitochondrial DNA and senescence in *Podospora anserina. Curr. Genet.* **4,** 83–84.

Esser, K., Osiewacz, H. D., Stahl, U., and Tudzynski, P. (1983). Nuclear–mitochondrial interactions cause senescence in the filamentous fungus *Podospora anserina. In* "Mitochondria 1983" (R. J. Schweyen, K. Wolf, and F. Kaudewitz, eds.), pp. 251–258. De Gruyter, Berlin.

Evans, R. J., and Clark-Walker, G. D. (1985). Elevated levels of petite formation in strains of *Saccharomyces cerevisiae* restored to respiratory competence. II. Organization of mitochondrial genomes in strains having high and moderate frequencies of petite mutant formation *Genetics* **111,** 403–432.

Evans, R. J., Oakley, K. M., and Clark-Walker, G. D. (1985). Elevated levels of petite formation in strains of *Saccharomyces cerevisiae* restored to respiratory competence. I. Association of both high and moderate frequencies of petite mutant formation with the presence of aberrant mitochondrial DNA. *Genetics* **111,** 389–402.

Falcone, C. (1984). The mitochondrial DNA of the yeast *Hansenula petersonii:* Genome organization and mosaic genes. *Curr. Genet.* **8,** 449–455.

Falcone, C., Agostinelli, M., and Frontali, L. (1983). Mitochondrial translation products during release from glucose repression in *Saccharomyces cerevisiae. J. Bacteriol.* **153,** 1125–1132.

Fangman, W. L., and Dujon, B. (1984). Yeast mitochondrial genomes consisting of only A-T base pairs replicate and exhibit suppressiveness. *Proc. Natl. Acad. Sci. U.S.A.* **81,** 7156–7160.

Farelly, F., and Butow, R. A. (1983). Rearranged mitochondrial genes in the yeast nuclear genome. *Nature (London)* **301,** 296–301.

Farelly, F., Zassenhaus, H. P., and Butow, R. A. (1982). Characterization of transcripts from the *var1* region on mitochondrial DNA of *Saccharomyces cerevisiae*. *J. Biol. Chem.* **257,** 6581–6587.

Faugeron-Fonty, G., and Goyon, C. (1985). Polymorphic variations in the *ori* sequences from the mitochondrial genomes of different wild-type yeast strains. *Curr. Genet.* **10,** 269–282.

Faugeron-Fonty, G., Mangin, M., Huyard, A., and Bernardi, G. (1983). The mitochondrial genomes of spontaneous *ori*^{r} petite mutants have rearranged repeat units organized as inverted tandem dimers. *Gene* **24,** 61–71.

Faugeron-Fonty, G., Le Van Kim, C., de Zamaroczy, M., Goursot, R., and Bernardi, G. (1984). A comparative study of the *ori* sequences from the mitochondrial genomes of 20 wild-type yeast strains. *Gene* **32,** 459–473.

Faye, G., and Simon, M. (1983). Processing of the *oxi3* pre-messenger RNA in yeast. *In* "Mitochondria 1983" (R. J. Schweyen, K. Wolf, and F. Kaudewitz, eds.), pp. 434–439. De Gruyter, Berlin.

Faye, G., Fukuhara, H., Grandchamp, C., Lazowska, J., Michel, F., Casey, J., Getz, G. S., Locker, J., Rabinowitz, M., Bolotin-Fukuhara, M., Coen, D., Deutsch, H., Dujon, B., Netter, P., and Slonimski, P. P. (1973). Mitochondrial nucleic acids in the petite colonie mutants: Deletion and repetition of genes. *Biochimie* **55,** 779–792.

Faye, G., Kujawa, C., and Fukuhara, H. (1974). Physical and genetic organization of petite and grande yeast mitochondrial DNA and localization of 23S ribosomal RNA in petite mutants of *Saccharomyces cerevisiae*. *J. Biol. Chem.* **88,** 185–203.

Faye, G., Kujawa, C., Dujon, B., Bolotin-Fakuhara, M., Wolf, K., Fukuhara, H., and Slonimski, P. P. (1975). Localization of the gene coding for the 16S ribosomal mitochondrial RNA using *rho*$^-$ mutants of *Saccharomyces cerevisiae*. *J. Mol. Biol.* **99,** 203–217.

Faye, G., Bolotin-Fukuhara, M., and Fukuhara, H. (1976a). Mitochondrial mutations that affect mitochondrial transfer ribonucleic acid in *Saccharomyces cerevisiae*. *In* "Genetics and Biogenesis of Chloroplasts and Mitochondria" (T. Bücher, W. Neupert, W. Sebald, and S. Werner, eds.), pp. 547–555. Elsevier, Amsterdam.

Faye, G., Kujawa, C., Fukuhara, H., and Rabinowitz, M. (1976b). Mapping of the mitochondrial 16S ribosomal RNA gene and its expression in the cytoplasmic petite mutants of *Saccharomyces cerevisiae*. *Biochem. Biophys. Res. Commun.* **68,** 476–482.

Faye, G., Dennebouy, N., Kujawa, C., and Jacq, C. (1979). Inserted sequence in the mitochondrial 23S ribosomal RNA gene of the yeast *Saccharomyces cerevisiae*. *Mol. Gen. Genet.* **168,** 101–109.

Flavell, A. (1985). Introns continue to amaze. *Nature* (*London*) **316,** 574–575.

Flury, U., Mahler, H. R., and Feldman, F. (1974). A novel respiration deficient mutant of *Saccharomyces cerevisiae*. I. Preliminary characterization of phenotype and mitochondrial inheritance. *J. Biol. Chem.* **249,** 6130–6137.

Fonty, G., Goursot, R., Wilkie, D., and Bernardi, G. (1978). The mitochondrial genome of wild-type yeast cells. *J. Mol. Biol.* **119,** 213–235.

Foury, F., and Kolodynski, J. (1983). *Pif* mutation blocks recombination between mitochondrial ρ^+ amd ρ^- genomes having tandemly arrayed repeat units in *Saccharomyces cerevisiae*. *Proc. Natl. Acad. Sci. U.S.A.* **80,** 5345–5349.

Foury, F., and Lahaye, A. (1987). Cloning and sequencing of the *PIF* gene involved in repair and recombination of yeast mitochondrial DNA. *EMBO J.* **6,** 1441–1449.

Foury, F., and Tzagoloff, A. (1976a). Localization on mitochondrial DNA of mutations leading to a loss of rutamycin-sensitive adenosine triphosphatase. *Eur. J. Biochem.* **68,** 113–119.

Foury, F., and Tzagoloff, A. (1976b). Assembly of the mitochondrial membrane system XIX. Genetic characterization of *mit*⁻ mutants with deficiencies in cytochrome oxidase and coenzyme QH_2–cytochrome *c* reductase. *Mol. Gen. Genet.* **149,** 43–50.

Foury, F., and Van Dyck, E. (1985). A *PIF*-dependent recombinogenic signal in the mitochondrial DNA of yeast. *EMBO J.* **4,** 3525–3530.

Fox, T. D. (1979a). Genetic and physical analysis of the mitochondrial gene for subunit II of yeast cytochrome *c* oxidase. *J. Mol. Biol.* **130,** 63–82.

Fox, T. D. (1979b). 5 TGA stop codons occur within the translated sequence of the yeast mitochondrial gene for cytochrome *c* oxidase subunit II. *Proc. Natl. Acad. Sci. U.S.A.* **76,** 6534–6538.

Fox, T. S., and Boerner, P. (1980). Transcripts of the *oxi-1* locus are asymmetric and may be spliced. *In* "The Organization and Expression of the Mitochondrial Genome" (A. M. Kroon and C. Saccone, eds.), pp. 191–194. Elsevier, Amsterdam.

Fox, T. D., and Staempfli, S. (1982). Suppressor of yeast mitochondrial ochre mutations that maps in or near the 15S ribosomal gene of mtDNA. *Proc. Natl. Acad. Sci. U.S.A.* **79,** 1583–1587.

Fox, T. D., and Weiss-Brummer, B. (1980). Leaky +1 and −1 frameshift mutations at the same site in a yeast mitochondrial gene. *Nature (London)* **288,** 60–63.

Francisci, S., Palleschi, C., Ragnini, A., and Frontali, L. (1987). Analysis of transcripts of the major cluster of tRNA genes in the mitochondrial genome of *S. cerevisiae. Nucleic Acids Res.* **15,** 6387–6404.

Frontali, L., Agostinelli, M., Baldacci, G., Falcone, C., and Zennaro, E. (1982a). Expression of the mitochondrial genes in *Saccharomyces cerevisiae:* Analysis of translation and transcription products in repressed and derepressed cells. *In* "Mitochondrial Genes" (P. Slonimski, P. Borst, and G. Attardi, eds.), pp. 327–331. Cold Spring Harbor Laboratory, Cold Spring Harbor, New York.

Frontali, L., Palleschi, C., and Francisci, S. (1982b). Transcripts of mitochondrial tRNA genes in *Saccharomyces cerevisiae. Nucleic Acids Res.* **10,** 7283–7293.

Frontali, L., Francisci, S., Palleschi, C., Stifani, S., and Zennaro, E. (1985). Transcription, initiation, and processing sites in the tRNA region of the yeast mitochondrial genome. *In* "Achievements and Perspectives of Mitochondrial Research" (E. Quagliariello, E. C. Slater, F. Palmieri, C. Saccone, and A. M. Kroon, eds.), Vol. 2, pp. 203–214. Elsevier, Amsterdam.

Fukuhara, H., and Rabinowitz, M. (1979). Yeast petite mutants for DNA amplification. *In* "Methods in Enzymology" (S. Fleischer and L. Packer, eds.), Vol. 56, pp. 154–163. Academic Press, New York.

Fukuhara, H., and Wesolowski, M. (1977). Preferential loss of a specific region of mitochondrial DNA by *rho*⁻ mutation. *In* "Mitochondria 1977" (W. Bandlow, R. J. Schweyen, K. Wolf, and F. Kaudewitz, eds.), pp. 123–131. De Gruyter, Berlin.

Fukuhara, H., Bolotin-Fukuhara, M., Hsu, H. J., and Rabinowitz, M. (1976a). Deletion mapping of mitochondrial transfer RNA genes in *Saccharomyces cerevisiae* by means of cytoplasmic petite mutants. *Mol. Gen. Genet.* **145,** 7–17.

Fukuhara, H., Faye, G., Bolotin-Fukuhara, M., Hsu, H. J., Martin, N., and Rabinowitz, M. (1976b). Mapping of transfer and ribosomal RNA genes of yeast mitochondria. *In* "Genetics, Biogenesis and Bioenergetics of Mitochondria" (W. Bandlow, R. J. Schweyen, D. Y. Thomas, K. Wolf, and F. Kaudewitz, eds.), pp. 57–68. De Gruyter, Berlin.

Gaillard, C., Strauss, F., and Bernadi, G. (1980). Excision sequences in the mitochondrial genome of yeast. *Nature (London)* **283,** 218–220.

Gampel, A., and Tzagoloff, A. (1987). *In vitro* splicing of the terminal intervening sequence of *Saccharomyces cerevisiae* cytochrome b pre-mRNA. *Mol. Cell. Biol.* **7,** 2545–2551.

Garber, R. C., and Yoder, O. C. (1984). Mitochondrial DNA of the filamentous ascomycete *Cochliobolus heterostrophus*. Characterization of the mitochondrial chromosome and population genetics of a restriction enzyme polymorphism. *Curr. Genet.* **8,** 621–628.

Gargouri, A., Lazowska, J., and Slonimski, P. P. (1983). DNA-splicing of introns in the gene: A general way of reverting intron mutations. *In* "Mitochondria 1983" (R. J. Schweyen, K. Wolf, I. Kaudewitz, eds.), pp. 259–268. De Gruyter, Berlin.

Garnjobst, L., Wilson, J. F., and Tatum, E. L. (1965). Studies on a cytoplasmic character in *Neurospora. J. Cell Biol.* **26,** 413–426.

Garriga, G., and Lambowitz, A. M. (1984). RNA splicing in *Neurospora* mitochondria: Self-splicing of a mitochondrial intron *in vitro. Cell* **38,** 631–641.

Garriga, G., and Lambowitz, A. M. (1986). Protein-dependent splicing of the group I intron in ribonucleoprotein particles and soluble fractions. *Cell* **46,** 669–680.

Garriga, G., Lambowitz, A. M., Inoue, T., and Cech, T. R. (1986). Mechanism of recognition of the 5′ splice site in self-splicing group I introns. *Nature* (*London*) **322,** 86–88.

Glotz, C., Zwieb, C., Brimacombe, R., Edwards, K., and Kössel, H. (1981). Secondary structure of the large subunit ribosomal RNA from *Escherichia coli, Zea mays* chloroplasts and human and mouse mitochondrial ribosomes. *Nucleic Acids Res.* **9,** 3287–3306.

Goffeau, A., Boutry, M., and Dufour, J. P. (1977). Structure of the mitochondrial and plasma membrane ATPases of the yeast *Schizosaccharomyces pombe. In* "Mitochondria 1977" (W. Bandlow *et al.,* eds.), pp. 451–462. De Gruyter, Berlin.

Goursot, R., de Zamaroczy, M., Baldacci, G., and Bernardi, G. (1980). Supersuppressive "petite" mutants of yeast. *Curr. Genet.* **1,** 173–176.

Goursot, R., Mangin, M., and Bernardi, G. (1982). Surrogate origins of replication in the mitochondrial genomes of *ori*° petite mutants of yeast. *EMBO J.* **1,** 705–711.

Grabowski, P. J., Zaug, A. J., and Cech, T. R. (1981). The intervening sequence of the ribosomal RNA precursor is converted to a circular RNA in isolated nuclei of *Tetrahymena. Cell* **23,** 467–476.

Grant, D. M., and Lambowitz, A. M. (1981). Mitochondrial ribosomal RNA genes. *In* "The Cell Nucleus" (X. H. Busch and L. Rothblum, eds.), pp. 387–408. Academic Press, New York.

Gray, M. W., Sankoff, D., and Cedergren, R. J. (1984). On the evolutionary descent of organisms and organelles: A global phylogeny based on a highly conserved structural core in small ribosomal RNA. *Nucleic Acids Res.* **12,** 5837–5852.

Green, M. R., Grimm, M. F., Goewert, R. R., Collins, R. A. Cole, M. D., Lambowitz, A. M., Heckman, J. E., Yin, S., and RajBhandary, U. L. (1981). Transcripts and processing patterns for the ribosomal RNA and transfer RNA region of *Neurospora crassa* mitochondrial DNA. *J. Biol. Chem.* **256,** 2027–2034.

Griffith, A. J. F., and Bertrand, H. (1984). Unstable cytoplasm in Hawaiian strains of *Neurospora intermedia. Curr. Genet.* **8,** 387–398.

Griffith, D. E., and Houghton, R. L. (1974). Studies on energy-linked reactions: Modified mitochondrial ATPase of oligomycin-resistant mutants of *Saccharomyces cerevisiae. Eur. J. Biochem.* **46,** 157–167.

Griffith, D. E., Houghton, R. L., Lancashire, W. E., and Meadows, P. A. (1975). Studies on energy-linked reactions: Isolation and properties of mitochondrial venturicidin-resistant mutants of *Saccharomyces cerevisiae. Eur. J. Biochem.* **51,** 393–402.

Griffith, A. J. F., Kraus, S., and Bertrand, H. (1986). Expression of senescence in *Neurospora intermedia*. *Can. J. Genet. Cytol.* **28,** 459–467.

Grimes, G. W., Mahler, H. R., and Perlman, P. S. (1974). Nuclear gene dosage effects on mitochondrial mass and DNA. *J. Cell Biol.* **61,** 565–574.

Grimm, M. F., Cole, M. D., and Lambowitz, A. M. (1981). Ribonucleic acid splicing in *Neurospora* mitochondria: Secondary structure of the 35S ribosomal precursor ribonucleic acid investigated by digestion with ribonuclease III and by electron microscopy. *Biochemistry* **20,** 2836–2842.

Grisi, E., Brown, T. A., Waring, R. B., Scazzocchio, C., and Davies, R. W. (1982). Nucleotide sequence of a region of the mitochondrial genome of *Aspergillus nidulans* including the genes for ATPase subunit 6. *Nucleic Acids Res.* **10,** 3531–3539.

Grivell, L. A. (1982). Restriction and genetic maps of yeast mitochondrial DNA. *In* "Genetic Maps: A Compilation of Linkage and Restriction Maps of Genetically Studied Organisms" (S. J. O'Brien, ed.). National Institutes of Health, Bethesda, Maryland.

Grivell, L. A. (1983). Mitochondrial DNA. *Sci. Am.* **248,** 73–78.

Grivell, L. A., and Borst, P. (1982). Mitochondrial mosaics—maturases on the move. *Nature (London)* **298,** 703–704.

Grivell, L. A., and Moorman, A. F. M. (1977). A structural analysis of the *oxi3* region of yeast mitDNA. *In* "Mitochondria 1977" (W. Bandlow, R. J. Schweyen, K. Wolf, and F. Kaudewitz, eds.), pp. 371–384. De Gruyter, Berlin.

Grivell, L. A., Netter, P., Borst, P., and Slonimski, P. P. (1973). Mitochondrial antibiotic resistance in yeast: Ribosomal mutants resistant to chloramphenicol, erythromycin and spiramycin. *Biochim. Biophys. Acta* **312,** 358–367.

Grivell, L. A., Arnberg, A. C., Boer, P. H., Borst, P., Bos, J. L., Van Bruggen, E. F. J., Groot, G. S. P., Hecht, N. B., Hensgens, L. A. M., Van Ommen, G. J. B., and Tabak, H. F. (1979). Transcripts of yeast mitochondrial DNA and their processing. *In* "Extrachromosomal DNA" (D. J. Cummings, P. Borst, I. B. Dawid, J. M. Weissman, and C. F. Fox, eds.), pp. 305–324. Academic Press, New York.

Grivell, L. A., Arnberg, A. C., Hensgens, L. A. M., Roosendaal, E., Van Ommen, G. J. B., and Van Bruggen, E. J. F. (1980). Split genes on yeast mitochondrial DNA: Organization and expression. *In* "The Organization and Expression of the Mitochondrial Genome" (A. M. Kroon and C. Saccone, eds.), pp. 37–49. Elsevier, Amsterdam.

Grivell, L. A., Hensgens, L. A. M., Osinga, K. A., Tabak, H. F., Boer, P. H., Crusius, J. B. A., Van der Laan, J. C., de Haan, M., Van der Horst, G., Evers, R. F., and Arnberg, A. C. (1982). RNA processing in yeast mitochondria. *In* "Mitochondrial Genes" (P. Slonimski, P. Borst, and G. Attardi, eds.), pp. 225–240. Cold Spring Harbor Laboratory, Cold Spring Harbor, New York.

Groot, G. S. P., and Van Harten-Loosbroek, N. (1980). The organization of the genes for ribosomal RNA on mitochondrial DNA of *Kluyveromyces lactis*. *Curr. Genet.* **1,** 133–135.

Groot, G. S. P., Obbink, D. J., Spithill, T. W., Maxwell, R. J., and Linnane, A. W. (1977). Biogenesis of mitochondria 48. Mikamycin resistance in *Saccharomyces cerevisiae*—A mitochondrial mutation conferring resistance to an antimycin A-like contaminant in mikamycin. *Mol. Gen. Genet.* **151,** 127–136.

Groot, G. S. P., Mason, T. L., and Van Harten-Loosbroek, N. (1979). *Var 1* is associated with the small ribosomal subunit of mitochondrial ribosomes in yeast. *Mol. Gen. Genet.* **174,** 339–342.

Grosch, G., Schmelzer, C., and Mathews, S. (1981). Physical map of the *COB* region in mitochondrial DNA of *Saccharomyces cerevisiae*. *Curr. Genet* **3,** 65–71.

Gross, S. R., Hsiah, J.-S., and Levine, P. H. (1984). Intramolecular recombination as a source of mitochondrial chromosome heteromorphism in *Neurospora*. *Cell* **38,** 233–239.

Grossman, L. I., and Hudspeth, M. E. S. (1986). Fungal mitochondrial genomes. *In* "Gene Manipulations in Fungi" (J. W. Bennet and L. Lasure, eds.). Academic Press, New York.

Grossman, L. I., Goldring, E. S, and Marmur, J. (1969). Preferential synthesis of yeast mitochondrial DNA in the absence of protein synthesis. *J. Mol. Biol.* **46,** 367–376.

Groudinsky, O., Carignani, G., Schiavon, E., Frezza, D., Bergantino, E., and Slonimski, P. P. (1983). The first intron of the gene *oxi3* in yeast mitochondria encodes a mRNA maturase. *In* "Mitochondria 1983" (R. J. Schweyen, K. Wolf, and F. Kaudewitz, eds.), pp. 227–232. De Gruyter, Berlin.

Guerrini, A. M., Luciano, B., and Cavaliere, F. (1977). Cytoplasmic and mitochondrial protein synthesis in the "petite-negative" yeast *Kluyveromyces lactis*. Cytoplasmic inheritance of resistance to tetracyclin. *Mol. Gen. Genet.* **154,** 181–184.

Guiso, N., Dreyfus, M., Siffert, O., Dauchin, A., Spyridakis, A., Gargouri, A., Claisse, M., and Slonimski, P. P. (1984). Antibodies against synthetic oligopeptides allow the identification of the mRNA-maturase encoded by the second intron of the yeast *cob–box* gene. *EMBO J.* **3,** 1769–1772.

Gunatilleke, I. A. U. N., Scazzocchio, C., and Arst, H. N., Jr. (1975). Cytoplasmic and nuclear mutations to chloramphenicol resistance in *Aspergillus nidulans*. *Mol. Gen. Genet.* **137,** 269–276.

Haid, A., Schweyen, R. J., Bechmann, H., Kaudewitz, F., Solioz, M., and Schatz, G. (1979). The mitochondrial *COB* region in yeast codes for apocytochrome *b* and is mosaic. *Eur. J. Biochem.* **94,** 451–464.

Haid, A., Grosch, G., Schmelzer, C., Schweyen, R. J., and Kaudewitz, F. (1980). Expression of the split gene *COB* in yeast mtDNA. Mutational arrest in the pathway of transcript splicing. *Curr. Genet.* **1,** 155–161.

Halbreich, A., Rabinowitz, M. (1971). Isolation of *Saccharomyces cerevisiae* mitochondrial formyl-tetrahydrofolic acid: methionyl tRNA transformylase and the hybridization of mitochondrial fMet-tRNA with mitochondrial DNA. *Proc. Natl. Acad. Sci. U.S.A.* **68,** 294–298.

Halbreich, A., Pajot, P., Foucher, M., Grandchamp, C., and Slonimski, P. (1980). A pathway of cytochrome *b* mRNA processing in yeast mitochondria: Specific splicing steps and an intron-derived circular RNA. *Cell* **19,** 321–329.

Halbreich, A., Grandchamp, C., and Foucher, M. (1981). A low-molecular-weight RNA species in yeast mitochondria arising from a 3′ end trimming of cytochrome *b* pre-mRNA. *Biosci. Rep.* **1,** 533–538.

Halbreich, A., Grandchamp, C., and Foucher, M. (1984). Yeast mitochondria contain a linear RNA strand complementary to the circular intronic bI1 RNA of cytochrome b. *Eur. J. Biochem.* **144,** 261–269.

Hall, R. M., Mattick, J. S., Marzuki, S., and Linnane, A. W. (1975). Evidence for a functional association of DNA synthesis with the membrane in mitochondria of *Saccharomyces cerevisiae*. *Mol. Biol. Rep.* **2,** 101–106.

Handwerker, A., Schweyen, R. J., Wolf, K., and Kaudewitz, F. (1973). Evidence for an extrakaryotic mutation affecting the maintenance of the *rho* factor in yeast. *J. Bacteriol.* **113,** 1307–1310.

Hanson, D. K., Miller, D. H., Mahler, H. R., Alexander, N. J., and Perlman, P. S. (1979). Regulatory interaction between mitochondrial genes. II. Detailed characterization of novel mutants mapping within one cluster in the *cob2* region. *J. Biol. Chem.* **254,** 2480–2490.

Hanson, D. K., Lamb, M. R., Mahler, H. R., and Perlman, P. S. (1982). Evidence for translated intervening sequences in the mitochondrial genome of *Saccharomyces cerevisiae*. *J. Biol. Chem.* **257,** 3218–3224.

Haskins, F., Tissieres, A., Mitchell, H., and Mitchell, M. (1953). Cytochromes and the succinic acid oxidase system of *poky* strains of *Neurospora*. *J. Biol. Chem.* **200,** 819–826.

Heckman, J. E., and RajBhandary, U. L. (1979). Organization of tRNA and rRNA genes in *N. crassa* mitochondria: Intervening sequence in the large rRNA gene and strand distribution of the RNA genes. *Cell* **17,** 583–595.

Heckman, J. E., Hecker, L. I., Schwartzbach, S. D., Barnett, W. E., Baumstark, B., and RajBhandary, U. L. (1978). Structure and function of initiator methionine tRNA from the mitochondria of *Neurospora crassa*. *Cell* **13,** 83–95.

Heckman, J. E., Alzner-De Weerd, B., and RajBhandary, U. L. (1979a). Interesting and unusual features in the sequence of *Neurospora crassa* mitochondrial tyrosine transfer RNA. *Proc. Natl. Acad. Sci. U.S.A.* **76,** 717–721.

Heckman, J. E., Yin, S., Alzner-De Weerd, B., and RajBhandary, U. L. (1979b). Mapping and cloning of *Neurospora crassa* mitochondrial transfer RNA genes. *J. Biol. Chem.* **254,** 12694–12700.

Heckman, J. E., Sarnoff, B., Alzner-De Weerd, B., Yin, S., and RajBhandary, U. L. (1980). Novel features in the genetic code and codon reading patterns in *Neurospora crassa* mitochondria based on sequences of six mitochondral tRNAs. *Proc. Natl. Acad. Sci. U.S.A.* **77,** 3159–3163.

Hefta, L. J. F., Lewin, A. S., Daignan-Fornier, B., and Bolotin-Fukuhara, M. (1987). Nuclear and mitochondrial revertants of a mitochondrial mutant with a defect in the ATP synthetase complex. *Mol. Gen. Genet.* **207,** 106–113.

Helmer-Citterich, M., Morelli, G., and Macino, G. (1983a). Nucleotide sequence and intron structure of the apocytochrome *b* gene of *Neurospora crassa* mitochondria. *EMBO J.* **2,** 1235–1242.

Helmer-Citterich, M., Morelli, G., Nelson, M. A., and Macino, G. (1983b). Expression of split genes of the *Neurospora crassa* mitochondrial genome. *In* "Mitochondria 1983" (R. J. Schweyen, K. Wolf, F. Kaudewitz, eds.), pp. 357–369. De Gruyter, Berlin.

Hensgens, L. A. M., Grivell L. A., Borst, P., and Bos, J. L. (1979). Nucleotide sequence of the mitochondrial structural gene for subunit 9 of the yeast ATPase complex. *Proc. Natl. Acad. Sci. U.S.A.* **76,** 1663–1667.

Hensgens, L. A. M., Arnberg, A. C., Roosendaal, E., Van der Horst, G., Van der Veen, R., Van Ommen, G.-J. B., and Grivell, L. A. (1983a). Variation, transcription and circular RNAs of the mitochondrial gene for subunit I of cytochrome *c* oxidase. *J. Mol. Biol.* **164,** 35–58.

Hensgens, L. A. M., Bonen, L., de Haan, M., Van der Horst, G., and Grivell, L. A. (1983b). Two intron sequences in yeast mitochondrial *cox1* gene: Homology among URF-containing introns and strain-dependent variation in flanking exons. *Cell* **32,** 379–389.

Hensgens, L. A. M., Van der Horst, G., and Grivell, L. A. (1984a). Interaction between mitochondrial genes in yeast: Evidence of novel *box* effect(s). *Plasmid* **12,** 41–51.

Hensgens, L. A. M., Van der Horst, G., Vos, H. L., and Grivell, L. A. (1984b). RNA processing in yeast mitochondria: Characterization of *mit*$^-$ mutants disturbed in the synthesis of subunit I of cytochrome *c* oxidase. *Curr. Genet.* **8,** 457–465.

Heyting, C., and Meijlink, F. C. P. W. (1979). Detailed restriction enzyme fragment map of the C-region of the mitochondrial DNA from *Saccharomyces cerevisiae,* strain JS1-3D. *Mol. Gen. Genet.* **168,** 247–250.

Heyting, C., and Menke, H. H. (1979). Fine structure of the 21S ribosomal RNA region on yeast mitochondrial DNA. 3. Physical location of mitochondrial genetic markers and the molecular nature of *omega*. *Mol. Gen. Genet.* **168,** 279–291.

Heyting, C., and Sanders, J. P. M. (1976). The physical map of some genetic markers in the 21S ribosomal region of the mitochondrial DNA of yeast. *In* "The Genetic Function of Mitochondrial DNA" (C. Saccone and A. M. Kroon, eds.), pp. 273–280. Elsevier, Amsterdam.

Heyting, C., Meijlink, F. C. P. W., Verbeet, M. P., Sanders, J. P. M., Bos, J. L., and Borst, P. (1979a). Fine structure of the 21S ribosomal RNA region on yeast mitochondrial DNA. 1. Construction of the physical map and location of the cistron for the mitochondrial ribosomal RNA. *Mol. Gen. Genet.* **168,** 231–246.

Heyting, C., Talen, J. L., Weijers, P. J., and Borst, P. (1979b). Fine structure of the 21S ribosomal RNA region on yeast mitochondrial DNA. 2. Organization of sequences in petite mitochondria carrying markers from the 21S region. *Mol. Gen. Genet.* **168,** 251–277.

Hickey, D. A., and Benkel, B. (1986). Introns as relict retrotransposons: Implications for the evolutionary origin of eukaryotic mRNA splicing mechanisms. *J. Theor. Biol.* **121,** 283–292.

Hoeben, P., and Clark-Walker, G. D. (1986). An approach to yeast classification by mapping mitochondrial DNA from *Dekkera/Brettanomyces* and *Eniella* genera. *Curr. Genet.* **10,** 371–279.

Holl, J., Rödel, G., and Schweyen, R. J. (1985a). Suppressor mutations identify *box9* as the central nucleotide sequence in the highly ordered structure of intron RNA in yeast mitochondria. *EMBO J.* **4,** 2081–2085.

Holl, J., Schmidt, C., and Schweyen, R. J. (1985b). *Cob* intron 3 in yeast mtDNA: Nucleotide sequence and mutations in a novel RNA domain. *In* "Achievements and Perspectives of Mitochondrial Research" (E. Quagliariello, E. C. Slater, F. Palmieri, C. Saccone, and A. M. Kroon, eds.), Vol. 2, pp. 227–236. Elsevier, Amsterdam.

Hollenberg, C. P., Borst, P., and Van Bruggen, E. F. J. (1970). Mitochondrial DNA. 25 μm closed circular duplex DNA molecule in wild type yeast mitochondria. Structure and genetic complexity. *Biochim. Biophys. Acta* **209,** 1–15.

Hollingsworth, M. J., and Martin, N. C. (1986). RNase P activity in the mitochondria of *Saccharomyces cerevisiae* depends on both mitochondrion- and nucleus-encoded components. *Mol. Cell. Biol.* **6,** 1058–1064.

Howell, N., Trembath, M. K., Linnane, A. W., and Lukins, H. B. (1973). Biogenesis of mitochondria. 30. An analysis of polarity of mitochondrial gene recombination and transmission. *Mol. Gen. Genet.* **122,** 37–51.

Hudspeth, M. E. S., Ainley, W. M., Shumard, D. S., Butow, R. A., and Grossman, L. (1982). Location and structure of the *var1* gene on yeast mitochondrial DNA: Nucleotide sequence of the 40.0 allele. *Cell* **30,** 617–626.

Hudspeth, M. E. S., Vincent, R. D., Perlman, P. S. Shumard, D. S. Treisman, L. O., and Grossman, L. I. (1984). Expandable *var1* gene of yeast mitochondrial DNA: In-frame insertions can explain the strain-specific protein size polymorphism. *Proc. Natl. Acad. Sci. U.S.A.* **81,** 3148–3152.

Hyman, B. C., Cramer, J. H., and Rownd, R. H. (1983). The mitochondrial genome of *Saccharomyces cerevisiae* contains numerous densely spaced autonomous replicating sequences. *Gene* **26,** 223–230.

Inoue, T., Sullivan, F. X., and Cech, T. R. (1986). New reactions of the ribosomal RNA precursor of *Tetrahymena* and the mechanism of self-splicing. *J. Mol. Biol.* **189,** 143–165.

Ise, W., Haiker, H., and Weiss, H. (1985a). Mitochondrial translation of subunits of the rotenone-sensitive NADH-ubiquinone reductase in *Neurospora crassa. EMBO J.* **4,** 2075–2080.

Ise, W., Haiker, H., and Weiss, H. (1985b). Mitochondrial translation of subunits of the rotenone-sensitive NADH: ubiquinone reductase in *Neurospora crassa. In* "Achievements and Perspectives of Mitochondrial Research" (E. Quagliariello, E. C. Slater, F. Palmieri, C. Saccone, and A. M. Kroon, eds.), Vol. 2, pp. 317–324. Elsevier, Amsterdam.

Iwashima, A., and Rabinowitz, M. (1969). Partial purification of mitochondrial and supernatant DNA polymerase from *Saccharomyces cerevisiae. Biochim. Biophys. Acta* **178,** 283–293.

Jacq, C., Kujawa, C., Grandchamp, C., and Netter, P. (1977). Physical characterization of the difference between yeast mitDNA alleles *omega*$^+$ and *omega*$^-$. *In* "Mitochondria 1977" (W. Bandlow, R. J. Schweyen, K. Wolf, and F. Kaudewitz, eds.), pp. 255–270. De Gruyter, Berlin.

Jacq, C., Lazowska, J., and Slonimski, P. P. (1980a). Cytochrome *b* messenger RNA maturase encoded in an intron regulates th expression of the split gene: I. Physical location and base sequence of intron mutations. *In* "The Organization and Expression of the Mitochondrial Genome" (C. Saccone and A. M. Kroon, eds.), pp. 139–152. Elsevier, Amsterdam.

Jacq, C., Lazowska, J., and Slonimski, P. P. (1980b). Sur un nouveau mecanisme de la régulation de l'expression génétique. *C. R. Acad. Sci. Paris* **290,** 89–92.

Jacq C., Pajot, P., Lazowska, J., Dujardin, G., Claisse, M., Groudinsky, O., de la Salle, H., Grandchamp, C., Labouesse, M., Gargouri, A., Guiard, B. Spyridakis, A., Dreyfus, M., and Slonimski, P. P. (1982). Role of introns in the yeast cytochrome-*b* gene: Cis- and trans-acting signals, intron manipulation, expression, and intergenic communications. *In* "Mitochondrial Genes" (P. Slonimski, P. Borst, and G. Attardi, eds.), pp. 155–183. Cold Spring Harbor Laboratory, Cold Spring Harbor, New York.

Jacq, C., Banroques, J., Becam, A. M., Slonimski, P. P., Guiso, N., and Danchin, A. (1984). Antibodies against a fused "*lacZ*–yeast mitochondrial intron" gene product allow identification of the mRNA maturase encoded by the fourth intron of the yeast *cob–box* gene. *EMBO J.* **3,** 1567–1572.

Jacquier, A., and Dujon, B. (1983). The intron of the mitochondrial 21S rRNA gene: Distribution in different yeast species and sequence comparison between *Kluyveromyces thermotholerans* and *Saccharomyces cerevisiae. Mol. Gen. Genet.* **192,** 487–499.

Jacquier, A., and Dujon, B. (1985). An intron-encoded protein is active in a gene conversion process that spread an intron into a mitochondrial gene. *Cell* **41,** 383–394.

Jacquier, A., and Michel, F. (1987). Multiple exon-binding sites in class II self-splicing introns. *Cell* **50,** 17–29.

Jacquier, A., and Rosbash, M. (1986). Efficient trans-splicing of a yeast mitochondrial group II intron implicates a strong 5′-exon–intron interaction. *Science* **234,** 1099–1104.

Jakovcic, S., Casey, J., and Rabinowitz, M. (1975). Sequence homology of the mitochondrial leucyl-tRNA cistron in different organisms. *Biochemistry* **14,** 2037–2042.

Jamet-Vierny, C., Begel, O., and Belcour, L. (1980). Senescence in *Podospora anserina:* Amplification of a mitochondrial DNA sequence. *Cell* **21,** 189–194.

Jamet-Vierny, C., Begel, O., and Belcour, L. (1984). A 20×10^3-base mosaic gene identified on the mitochondrial chromosome of *Podospora anserina. Eur. J. Biochem.* **143,** 389–394.

Jean-Francois, M. J. B., Lukins, H. B., and Marzuki, S. (1986). Post-transcriptional defects in the synthesis of the mitochondrial H^+-ATPase subunit 6 in yeast mutants with lesions in the subunit 9 strucutral gene. *Biochim. Biophys. Acta* **868,** 178.

John, U. P., and Nagley, P. (1986). Amino acid substitutions in mitochondrial ATPase subunit 6 of *Saccharomyces cerevisiae* leading to oligomycin resistance. *FEBS Lett.* **207,** 79–83.

John, U. P., Willson, T. A., Linnane, A. W., and Nagley, P. (1986). Biogenesis of mitochondria: DNA sequence analysis of mit^- mutations in the mitochondrial *oli 2* gene coding for mitochondrial ATPase subunit 6 in *Saccharomyces cerevisiae. Nucleic Acids Res.* **14,** 7437–7452.

Joulou, C., and Bolotin-Fukuhara, M. (1982). Genetics of mitochondrial ribosome of yeast: Mitochondrial lethality of a double mutant carrying two mutations of the 21S ribosomal RNA gene. *Mol. Gen. Genet.* **188,** 256–260.

Joulou, C., Contamine, V., Sor, F., and Bolotin-Fukuhara, M. (1984). Mitochondrial ribosomal RNA genes of yeast—their mutations and a common nuclear suppressor. *Mol. Gen. Genet.* **193,** 275–279.

Kaldma, J. A. (1975). A new type of mitochondrial mutation in yeast: temperature dependent antibiotic resistant mitochondrial mutations. *Genetika* **5,** 151–183.

Kan, N. C., and Gall, J. G. (1982). The intervening sequence of the ribosomal RNA gene is highly conserved between two *Tetrahymena species. Nucleic Acids Res.* **10,** 2809–2822.

Karsch, T., Kück, U., and Esser, K. (1987). Mitochondrial group I introns from the filamentous fungus *Podospora anserina* code for polypeptides related to maturases. *Nucleic Acids Res.* **15,** 6743–6744.

Kay, P. S., and Inoue, T. (1987). Catalysis of splicing-related reactions between dinucleotides by a ribozyme. *Nature* (*London*) **327,** 343–346.

Kearsey, S. E., and Craig, I. W. (1981). Altered ribosomal RNA genes in mitochondria from mammalian cells with chloramphenicol resistance. *Nature* (*London*) **290,** 607–608.

Kelly, R., and Phillips, S. L. (1983). Comparison of the levels of the 21S mitochondrial rRNA in derepressed and glucose-repressed *Saccharomyces cerevisiae. Mol. Cell. Biol.* **3,** 1949–1957.

Keyhani, E. (1979). Identification of the structural gene for yeast mitochondrial cytochrome *c* oxidase subunit I on mitochondrial DNA. *Biochem. Biophys. Res. Commun.* **89,** 1212–1216.

Keyhani, E., and Keyhani, J. (1982). Biochemical characterization of the *OXI* mutants of the yeast *Saccharomyces cerevisiae. Biochim. Biophys. Acta* **717,** 355–364.

Kleese, R. A., Grotbeck, R. C., and Snyder, J. R. (1972). Two cytoplasmically inherited chloramphenicol resistance loci in yeast (*S. cerevisiae*). *Can. J. Genet. Cytol.* **14,** 713–715.

Knight, J. A. (1980). New antibiotic resistance loci in the ribosomal region of yeast mitochondrial DNA. *Genetics* **94,** 69–92.

Knight, J. A., Courey, A. J., and Stebbins, B. (1982). Second-site antibiotic resistance mutations in the ribosomal region of yeast mitochondrial DNA. *Curr. Genet.* **5,** 21–27.

Kochko, A., Colson, A. M., and Slonimski, P. P. (1979). Expression èn cis, lors de la complementation, des exons du gène mosaique mitochondrial controlant de cytochrome *b* chez *Saccharomyces cerevisiae. Arch. Int. Physiol. Biochem.* **87,** 619–620.

Köchel, H. G., and Küntzel, H. (1981). Nucleotide sequence of the *Aspergillus nidulans* mitochondrial gene coding for the small ribosomal subunit RNA: Homology to *E. coli* 16S rRNA. *Nucleic Acids Res.* **9,** 5689–5696.

Köchel, H. G., and Küntzel, H. G. (1982). Mitochondrial L-rRNA from *Aspergillus nidulans:* potential secondary structure and evolution. *Nucleic Acids Res.* **10,** 4795–4801.

Köchel, H. G., Lazarus, C. M., Basak, N., Küntzel, H. (1981). Mitochondrial tRNA gene cluster in *Aspergillus nidulans:* Organization and nucleotide sequence. *Cell* **23,** 625–633.

Kohli, J. (1987). Genetic nomenclature and gene list of the fission yeast *Schizosaccharomyces pombe L. Curr. Genet.* **11,** 575–590.

Koike, K., Taira, M., Kuchino, Y., Katsuyuki, Y., Sekiguchi, T., and Kobayashi, M. (1983). Mutations in the rat mitochondrial genome. *In* "Mitochondria 1983" (R. J. Schweyen, K. Wolf, and F. Kaudewitz, eds.), pp. 371–387. De Gruyter, Berlin.

Kojo, H. (1976). Estimation of the molecular length of petite negative yeast mitochondrial DNA. *FEBS Lett.* **67,** 134–136.

Koll, F. (1986). Does nuclear integration of mitochondrial sequences occur during senescence in *Podospora? Nature* (*London*) **324,** 597–599.

Koll, F., Begel, O., Keller, A.-M., Vierny, C., and Belcour, L. (1984). Ethidium bromide rejuvenation of senescent cultures of *Podospora anserina:* Loss of senescence-specific DNA and recovery of normal mitochondrial DNA. *Curr. Genet.* **8,** 127–134.

Koll, F., Begel, O., and Belcour, L. (1987). Insertion of short poly *d*(*A*)*d*(*T*) sequences at recombination junctions in mitochondrial DNA of *Podospora. Mol. Gen. Genet.* **209,** 630–632.

Kostriken, R., Strathern, J. N., Klar, A. J. S., Hicks, B., and Heffron, F. (1983). A site-specific endonuclease essential for mating-type switching in *Saccharomyces cerevisiae. Cell* **35,** 167–174.

Kotylak, Z., and Slonimski, P. P. (1976). Joint control of cytochrome *a* and *b* by a unique mitochondrial DNA region comprising four genetic loci. *In* "The Genetic Functions of Mitochondrial DNA" (C. Saccone and A. M. Kroon, eds.), pp. 143–154. Elsevier, Amsterdam.

Kotylak, Z., and Slonimski, P. P. (1977a). Mitochondrial mutants isolated by a new screening method based upon the use of the nuclear mutations *op1*. *In* "Mitochondria 1977" (W. Bandlow, R. J. Schweyen, K. Wolf, and F. Kaudewitz, eds.), pp. 83–90. De Gruyter, Berlin.

Kotylak, Z., and Slonimski, P. P. (1977b). Fine structure genetic map of the mitochondrial DNA region controlling coenzyme QH_2:cytochrome *c* reductase. *In* "Mitochondria 1977" (W. Bandlow, R. J. Schweyen, K. Wolf, and F. Kaudewitz, eds.), pp. 161–172. De Gruyter, Berlin.

Kotylak, Z., Lazowska, J., and Slonimski, P. P. (1985). Intron encoded proteins of mitochondria: Key elements of gene expression and genomic evolution. *In* "Achievements and Perspectives of Mitochondrial Research" (E. Quagliariello, E. C. Slater, F. Palmieri, C. Saccone, and A. M. Kroon, eds.), Vol. 2, pp. 1–21. Elsevier, Amsterdam.

Kovac, L. (1974). Biochemical mutants: An approach to the mitochondrial energy coupling. *Biochim. Biophys. Acta* **346,** 101–135.

Kovac, L., Lazowska, J., and Slonimski, P. P. (1984). A yeast with linear molecules of mitochondrial DNA. *Mol. Gen. Genet.* **197,** 420–424.

Kozlowski, M., Bartnik, E., and Stepien, P. P. (1982). Restriction enzyme analysis of mitochondrial DNA of members of the genus *Aspergillus* as an aid in taxonomy. *J. Gen. Microbiol.* **128,** 471–476.

Kreike, J., Bechmann, H., van Hemert, F. J., Schweyen, R. J., de Boer, P. H., Kaudewitz, F., and Groot, G. S. P. (1979). The identification of apocytochrome *b* as a mitochondrial gene product and immunological evidence for altered apocytochrome *b* in yeast strains having mutations in the *COB* region of mitochondrial DNA. *Eur. J. Biochem.* **101,** 607–617.

Kroon, A. M., Terpstra, P., Holtrop, M., de Vries, H., van den Boogaart, C., de Jonge, J., and Agsteribbe, E. (1976). The mitochondrial RNAs of *Neurospora crassa:* Their function in translation and their relation to the mitochondrial genome. *In* "Genetics and Biogenesis of Chloroplasts and Mitochondria" (T. Bücher, W. Neupert, W. Sebald, and S. Werner, eds.), pp. 685–696. Elsevier, Amsterdam.

Kruger, K., Grabowski, P. J., Zaug, A. J., Sand, J., Gottschling, D. E., and Cech, T. R. (1982). Self-splicing RNA: Autoexcision and autocyclization of the ribosomal RNA intervening sequence of *Tetrahymena. Cell* **31,** 147–157.

Kruszewska, A. (1982). Nuclear and mitochondrial informational suppressors of *box3* intron mutations in *Saccharomyces cerevisiae. In* "Mitochondrial Genes" (P. Slonimski, P. Borst, and G. Attardi, eds.), pp. 323–326. Cold Spring Harbor Laboratory, Cold Spring Harbor, New York.

Kruszewska, A., and Slonimski, P. P. (1984a). Mitochondrial and nuclear mitoribosomal suppressors that enable misreading of ochre codons in yeast mitochondria. I. Isolation, localization and allelism of suppressors. *Curr. Genet.* **9,** 1–10.

Kruszewska, A., and Slonimski, P. P. (1984b). Mitochondrial and nuclear mitoribosomal suppressors that enable misreading of ochre codons in yeast mitochondria. II. Specificity and extent of suppressor action. *Curr. Genet.* **9,** 11–19.

Kruszewska, A., Szczesniak, B., and Claisse, M. (1980). Recombinational analysis of *OXI1* mutants and preliminary analysis of their translation products in *S. cerevisiae. Curr. Genet.* **2,** 45–51.

Kück, U., and Esser, K. (1982). Genetic map of mitochondrial DNA in *Podospora anserina. Curr. Genet.* **5,** 143–147.

Kück, U., Stahl, U., Lhermitte, A., and Esser, K. (1980). Isolation and characterization of mitochondrial DNA from the alkane yeast *Saccharomycopsis lipolytica. Curr. Genet.* **2,** 97–101.

Kück, U. Stahl, U., and Esser, K. (1981). Plasmid-like DNA is part of mitochondrial DNA in *Podospora anserina. Curr. Genet.* **3,** 151–156.

Kück, U., Kappelhoff, B., and Esser, K. (1985a). Despite mtDNA polymorphism the mobile intron (*p1DNA*) of the *COI* gene is present in ten different races of *Podospora anserina. Curr. Genet.* **10,** 59–68.

Kück, U., Osiewacz, H. D., Schmidt, U., Kappelhoff, B., Schulte, E., Stahl, U., and Esser, K. (1985b). The onset of senescence is affected by DNA rearrangement of a discontinuous mitochondrial gene in *Podospora anserina. Curr. Genet.* **9,** 373–382.

Küntzel, H., and Köchel, H. G. (1981). Evolution of rRNA and origin of mitochondria. *Nature (London)* **293,** 751–755.

Küntzel, H., and Schäfer, K. P. (1971). Mitochondrial RNA polymerase from *Neurospora crassa. Nature (London) New Biol.* **231,** 265–269.

Küntzel, H., Pühler, A., and Bernard, U. (1976). Physical map of circular mitochondrial DNA from *Neurospora crassa. FEBS Lett.* **60,** 119–121.

Küntzel, H., Basak, N., Imim, G., Köchel, H., Lazarus, C. M., Lünsdorf, H., Bartnik, E., Bidermann, A., and Stepien, P. P. (1980). The mitochondrial genome of *Aspergillus nidulans. In* "The Organization and Expression of the Mitochondrial Genome" (A. M. Kroon and C. Saccone, eds.), pp. 79–86. Elsevier, Amsterdam.

Küntzel, H., Köchel, H. G., Lazarus, C. M., and Lunsdorf, H. (1982). Mitochondrial genes in *Aspergillus. In* "Mitochondrial Genes" (P. Slonimski, P. Borst, and G. Attardi, eds.), pp. 391–403. Cold Spring Harbor Laboratory, Cold Spring Harbor, New York.

Küenzi, M. T., and Roth R. (1974). Timing of mtDNA synthesis during meiosis in *Saccharomyces cerevisiae*. *Exp. Cell Res.* **85,** 377–382.

Kuriyama, Y., and Luck, D. J. L. (1974). Methylation and processing of mitochondrial ribosomal RNAs in *poky* and wild-type *Neurospora crassa*. *J. Mol. Biol.* **83,** 253–266.

Kutzleb, R., Schweyen, R. J., and Kaudewitz, F. (1973). Extrachromosomal inheritance of paromomycin resistance in *Saccharomyces cerevisiae*. Genetic and biochemical characterization of mutants. *Mol. Gen. Genet.* **125,** 91–98.

Labaille, F., Colson, A.-M., Petit, L., and Goffeau, A. (1977). Properties of a mitochondrial suppressor mutation restoring oxidative phosphorylation in a nuclear mutant of the yeast *Schizosaccharomyces pombe*. *J. Biol. Chem.* **252,** 5716–5723.

Labouesse, M., and Slonimski, P. P. (1983). Construction of novel cytochrome *b* genes in yeast mitochondria by subtraction or addition of introns. *EMBO J.* **2,** 269–276.

Labouesse, M., Netter, P., and Schroeder, R. (1984). Molecular basis of the "*box*-effect." A maturase deficiency leading to the absence of splicing of two introns located in two split genes of yeast mitochondrial DNA. *Eur. J. Biochem.* **144,** 85–93.

Lamb, A. J., Clark-Walker, G. D., and Linnane, A. W. (1968). The biogenesis of mitochondria. IV. The differentiation of mitochondrial and cytoplasmic protein synthesizing system *in vitro* by antibiotics. *Biochim. Biophys. Acta* **161,** 415–427.

Lamb, M. R., Anziano, P. Q., Glaus, K. R., Hanson, D. K., Klapper, H. J., Perlman, P. S., and Mahler, H. R. (1983). Functional domains in introns. RNA processing intermediates in cis- and trans-acting mutants in the penultimate intron of the mitochondrial gene for cytochrome *b*. *J. Biol. Chem.* **258,** 1991–1999.

Lambowitz, A. M. (1980). Mitochondrial ribosome assembly and RNA splicing in *Neurospora crassa*. *In* "The Organization and Expression of the Mitochondrial Genome" (A. M. Kroon and C. Saccone, eds.), pp. 291–300. Elsevier, Amsterdam.

Lambowitz, A. M., Chua, H.-H., and Luck, D. J. L. (1976). Mitochondrial ribosome assembly in *Neurospora:* Preparation of mit ribosomal precursor particles, site of synthesis of mit ribosomal proteins and studies on the *poky* mutant. *J. Mol. Biol.* **107,** 223–253.

Lambowitz, A. M., La Polla, R. J., and Collins, R. A. (1979). Mitochondrial assembly in *Neurospora*. Two-dimensional gel electrophoretic analysis of mitochondrial ribosomal proteins. *J. Cell Biol.* **82,** 17–31.

Lambowitz, A. M., Akins, R. A., Garriga, G., Henderson, M., Kubelik A. R., and Maloney, K. A. (1985). Mitochondrial introns and mitochondrial plasmids of *Neurospora*. *In* "Achievements and Perspectives of Mitochondrial Research" (E. Quagliariello, E. C. Slater, F. Palmieri, C. Saccone, and A. M. Kroon, eds.), Vol. 2, pp. 237–248. Elsevier, Amsterdam.

Lamouroux, A., Pajot, P., Kochko, A., Halbreich, A., and Slonimski, P. P. (1980). Cytochrome *b* messenger RNA maturase encoded in an intron regulates the expression of the split gene. II. Trans- and cis acting mechanisms of mRNA splicing. *In* "The Organization and Expression of the Mitochondrial Genome" (A. M. Kroon and C. Saccone, eds.), pp. 153–156. Elsevier, Amsterdam.

Lancashire, W. E., and Griffith, D. E. (1975). Studies on energy-linked reactions: Genetic analysis of venturicidin-resistant mutants. *Eur. J. Biochem.* **51,** 403–413.

Lancashire, W. E., and Mattoon, J. R. (1974). Genetics of oxidative phosphorylation. Mitochondrial loci determining ossamycin resistance, venturicidin resistance and oligomycin resistance in yeast. *Mol. Gen. Genet.* **176,** 255–264.

Lancashire, W. E., Houghton, R: L., and Griffiths, D. E. (1974). Two mitochondrial genes specifying venturicidin resistance in yeast. *Biochem. Soc. Trans.* **2,** 213–215.

Lancashire, W. E., Poyton, M. A., Webber, M. J., and Hartley, B. S. (1981) Petite-negative mutants of *Saccharomyces cerevisiae*. *Mol. Gen. Genet.* **181,** 409–410.

Lang, B. F. (1984). The mitochondrial genome of the fission yeast *Schizosaccharomyces pombe:* Highly homologous introns are inserted at the same position of the otherwise less conserved *cox1* genes in *Schizosaccharomyces pombe* and *Aspergillus nidulans. EMBO J.* **3,** 2129–2136.

Lang, B. F., and Wolf, K. (1984). The mitochondrial genome of the fission yeast *Schizosaccharomyces pombe.* 2. Localization of genes by interspecific hybridization in strain *ade7-50h*$^-$ and cloning of the genome in small fragments. *Mol. Gen. Genet.* **196,** 465–472.

Lang, B., Burger, G., Wolf, K., Bandlow, W., and Kaudewitz, F. (1975). Studies on the mechanism of electron transport in the bc_1-segment of the respiratory chain in yeast. III. Isolation and characterization of an antimycin-resistant mutant *ANT8* in *Schizosaccharomyces pombe. Mol. Gen. Genet.* **137,** 353–363.

Lang, B., Burger, G., Bandlow, W., Kaudewitz, F., and Schweyen, R. J. (1976a). Antimycin- and funiculosin-resistant mutants in *Saccharomyces cerevisiae:* New markers on the mitochondrial DNA. *In* "Genetics and Biogenesis of Chloroplasts and Mitochondria" (T. Bücher, W. Neupert, W. Sebald, and S. Werner, eds.), pp. 461–466. Elsevier, Amsterdam.

Lang, B., Burger, G., Wolf, K., Bandlow, W., and Kaudewitz, F. (1976b). Characterization of respiratory-deficient and antimycin-resistant mutants of *Schizosaccharomyces pombe* with extrachromosomal inheritance. *In* "Genetics, Biogenesis, and Bioenergetics of Mitochondria" (W. Bandlow, R. J. Schweyen, D. Y. Thomas, K. Wolf, and F. Kaudewitz, eds.), pp. 379–387. De Gruyter, Berlin.

Lang, B. F., Ahne, F., Distler, S., Trinkl, H., Kaudewitz, F., and Wolf, K. (1983). Sequence of the mitochondrial DNA, arrangement of genes and processing of their transcripts in *Schizosaccharomyces pombe. In* "Mitochondria 1983" (R. J. Schweyen, K. Wolf, and F. Kaudewitz, eds.), pp. 313–329. De Gruyter, Berlin.

Lang, B. F., Ahne, F., and Bonen, L. (1985). The mitochondrial genome of the fission yeast *Schizosaccharomyces pombe:* The cytochrome *b* gene has an intron closely related to the first two introns in the *Saccharomyces cerevisiae cox 1* gene. *J. Mol. Biol.* **184,** 353–366.

Lang, B. F., Cedergren, R., and Gray, M. W. (1987). The mitochondrial genome of the fission yeast, *Schizosaccharomyces pombe.* Sequence of the large subunit ribosomal RNA gene, comparison of potential secondary structure in fungal mitochondrial LSU rRNAs, and evolutionary considerations. *Eur. J. Biochem.* **169,** 527–537.

La Polla, R. J., and Lambowitz, A. M. (1977). Mitochondrial ribosome assembly in *Neurospora crassa.* Chloramphenicol inhibits the maturation of small ribosomal subunits. *J. Mol. Biol.* **116,** 189–205.

La Polla, R. J., and Lambowitz, A. M. (1979). The binding of mitochondrial ribosomal proteins to a mitochondrial ribosomal precursor RNA containing a 2.3 kb intron. *J. Biol. Chem.* **254,** 11746–11750.

La Polla, R. J., and Lambowitz, A. M. (1981). Mitochondrial ribosome assembly in *Neurospora crassa.* Purification of the mitochondrially synthesized ribosomal protein, *S-5. J. Biol. Chem.* **256,** 7064–7067.

Lawson, J. E., and Deters, D. W. (1985). Identification and isolation of the cytochrome oxidase subunit II gene in mitochondria of the yeast *Hansenula saturnus. Curr. Genet.* **9,** 345–350.

Lazdins, J. B., and Cummings, D. J. (1982). Autonomously replicating sequences in young and senescent mitochondrial DNA from *Podospora anserina. Curr. Genet.* **6,** 173–178.

Lazarus, C. M., and Turner, G. (1977). Extranuclear recombination in *Aspergillus nidulans:* Closely linked multiple chloramphenicol and oligomycin resistance loci. *Mol. Gen. Genet.* **156,** 303–311.

Lazarus, C. M., and Küntzel, H. (1980). Amplification of a common mitochondrial DNA sequence in three new ragged mutants of *Aspergillus amstelodami. In* "The Organization and Expression of the Mitochondrial Genome" (A. M. Kroon and C. Saccone eds.), pp. 87–90. Elsevier, Amsterdam.

Lazarus, C. M., and Küntzel, H. (1981). Anatomy of amplified mitochondrial DNA in "ragged" mutants of *Aspergillus amstelodami:* Excision points within protein genes and a common 215 bp segment containing a possible origin of replication. *Curr. Genet.* **4,** 99–107.

Lazarus, C. M., Earl, A. J., Turner, G., and Küntzel, H. (1980a). Amplification of a mitochondrial DNA sequence in the cytoplasmically inherited "ragged" mutant of *Aspergillus amstelodami. Eur. J. Biochem* **106,** 633–641.

Lazarus, C. M., Lünsdorf, H., Hahn, U., Stepien, P. P., and Küntzel, H. (1980b). Physical map of *Aspergillus nidulans* mitochondrial genes coding for ribosomal RNA: An intervening sequence in the large rRNA cistron. *Mol. Gen. Genet.* **177,** 389–397.

Lazowska, J., and Slonimski, P. P. (1977). Site-specific recombination in petite colony mutants of *Saccharomyces cerevisiae.* 1. Electron microscopic analysis of organization of recombinant DNA resulting from end to end joining of 2 mitochondrial segments. *Mol. Gen. Genet.* **156,** 163–175.

Lazowska, J., Jacq, C., and Slonimski, P. P. (1980). Sequence of introns and flanking exons in the wild type and *box3* mutants of the mitochondrial cytochrome *b* gene reveals an interlaced splicing protein coded by an intron. *Cell* **22,** 333–348.

Lazowska, J., Jacq, C., and Slonimski, P. P. (1981). Splice points of the third intron in the yeast mitochondrial cytochrome b gene. *Cell* **27,** 12–14.

Leff, J., and Eccleshall, T. R. (1978). Replication of bromodeoxyuridylate-substituted mitochondrial DNA in yeast. *J. Bacteriol.* **135,** 436–444.

Lemire, E. G., and Nargang, F. E. (1986). A missense mutation in the *oxi 3*-gene of the *mi-3* extranuclear mutant of *Neurospora crassa. J. Biol. Chem.* **261,** 5610–5615.

Levens, D., Edwards, J., Locker, J., Lustig, A., Merten, S., Morimoto, R., Synenki, R., and Rabinowitz, M. (1979). Transcription of yeast mitochondrial DNA. *In* "Extrachromosomal DNA" (D. J. Cummings, P. Borst, I. B. Dawid, S. M. Weissman, and C. F. Fox, eds.), pp. 287–304. Academic Press, New York.

Levens, D., Lustig, A., and Rabinowitz, M. (1981a). Purification of mitochondrial RNA polymerase from *Saccharomyces cerevisiae. J. Biol. Chem.* **256,** 1474–1481.

Levens, D. Morimoto, R., and Rabinowitz, M. (1981b). Mitochondrial transcription complex from *Saccharomyces cerevisiae. J. Biol. Chem.* **256,** 1466–1473.

Levens, D., Ticho, B., Ackermann, E., and Rabinowitz, M. (1981c). Transcriptional initiation and 5′ termini of yeast mitochondrial RNA. *J. Biol. Chem.* **256,** 5226–5232.

Lewin, A. S., Morimoto, R., Rabinowitz, M., and Fukuhara, H. (1978). Restriction enzyme analysis of mitochondrial DNAs of petite mutants of yeast: Classification of petites and deletion mapping of mitochondrial genes. *Mol. Gen. Genet.* **163,** 257–275.

Lewin, R. (1980). Alternatives for splicing: An intron-coded protein. *Cell* **22,** 645–646.

Lewin, R. (1982). RNA can be a catalyst. *Science* **218,** 872–874.

Li, M., and Tzagoloff, A. (1979). Assembly of the mitochondrial membrane system: Sequences of yeast mitochondrial valine and an unusual threonine tRNA gene. *Cell* **18,** 47–54.

Li, M., Tzagoloff, A., Underbrink-Lyon, K., and Martin, N. (1982). Identification of the paramomycin-resistance mutation in the 15S rRNA gene of yeast mitochondria. *J. Biol. Chem.* **257,** 5921–5928.

Linnane, A. W., and Nagley, P. (1978). Mitochondrial genetics in perspective. The derivation of a genetic and physical map of the yeast mitochondrial genome. *Plasmid* **1,** 324–345.

Linnane, A. W., Lamb, A. L., and Christodoulou, C. (1968a). The biogenesis of mitochondria. VI. Biochemical basis of resistance of *Saccharomyces cerevisiae* towards antibiotics. *Proc. Natl. Acad. Sci. U.S.A.* **59,** 1288–1293.

Linnane, A. W., Saunders, G. W., Gingold, E. B., and Lukins, H. B. (1968b). The biogenesis of mitochondria. V. Cytoplasmic inheritance of erythromycin resistance in *Saccharomyces cerevisiae. Proc. Natl. Acad. Sci. U.S.A.* **59,** 903–910.

Linnane, A. W., Lukins, H. B., Molloy, P. L., Nagley, P., Rytka, J., Sriprakash, K. S., and Trembath, M. K. (1976). Biogenesis of mitochondria—Molecular mapping of mitochondrial genome of yeast. *Proc. Natl. Acad. Sci. U.S.A.* **73,** 2082–2085.

Locker, J., and Rabinowitz, M. (1981). Transcription in yeast mitochondria: Analysis of the 21S rRNA region and its transcripts. *Plasmid* **6,** 302–314.

Lopez, I. C., Farelly, F., and Butow, R. A. (1981). *Trans* action of the *var1* determinant region on yeast mitochondrial DNA. Specific labeling of mitochondrial translation products in zygotes. *J. Biol. Chem.* **256,** 6496–6501.

Lopez-Perez, M., and Turner, G. (1975). Mitochondrial DNA from *Aspergillus nidulans. FEBS Lett.* **58,** 159–163.

Lückemann, G., Sipiczki, M., and Wolf, K. (1979). Transmission, segregation, and recombination of mitochondrial genomes in zygotic clones and protoplast fusion clones in yeast. *Mol. Gen. Genet.* **177,** 185–187.

Lückemann, G., Merlos-Lange, A. M., Del Giudice, L., and Wolf, K. (1988). Genetic and physical analysis of transmission, segregation and recombination of mitochondrial genomes in the fission yeast *Schizosaccharomyces pombe. Life Sci. Adv.,* in press.

Lukins, H. B., Devenish, R. J., and Linnane, A. W. (1984). Constants and variables of fungal mitochondrial genomes. *Microbiol. Sci.* **1,** 141–149.

Lustig, A., Levens, D., and Rabinowitz, M. (1982). The biogenesis and regulation of yeast mitochondrial RNA polymerase. *J. Biol. Chem.* **257,** 5800–5808.

Mabuchi, T., and Wakabayashi, K. (1984). Nucleotide sequence of an essential region for autonomous replication of cloned yeast mitochondrial DNA. *J. Biochem.* **95,** 589–592.

Mabuchi, T., Nishikawa, S., and Wakabayashi, K. (1984). Two autonomously replicating sequences near *oli-1* gene of yeast mitochondrial DNA. *J. Biochem.* **95,** 729–726.

McArthur, C. R., and Clark-Walker, G. D. (1983). Mitochondrial DNA size diversity in the *Dekkera/Brettanomyces* yeasts. *Curr. Genet.* **7,** 29–35.

McDougall, E. J., and Pittenger, T. H. (1966). A cytoplasmic variant of *Neurospora crassa. Genetics* **54,** 551–565.

Macino, G. (1980). Mapping of mitochondrial structural genes in *Neurospora crassa. J. Biol. Chem.* **255,** 10563–10565.

Macino, G., and Morelli, G. (1983). Cytochrome oxidase subunit 2 gene in *Neurospora crassa* mitochondria. *J. Biol. Chem.* **258,** 13230–13235.

Macino, G., and Tzagoloff, A. (1979a). Assembly of the mitochondrial membrane system. Two separate genes coding for threonyl-tRNA in the mitochondrial DNA of *Saccharomyces cerevisiae. Mol. Gen. Genet.* **169,** 183–188.

Macino, G., and Tzagoloff, A. (1979b). Assembly of the mitochondrial membrane system. The DNA sequence of a mitochondrial ATPase gene in *Saccharomyces cerevisiae. J. Biol. Chem.* **254,** 4617–4623.

Macino, G., and Tzagoloff, A. (1979c). Assembly of the mitochondrial membrane system—partial sequence of a mitochondrial ATPase gene in *Saccharomyces cerevisiae*. *Proc. Natl. Acad. Sci. U.S.A.* **76,** 131–135.

Macino, G., and Tzagoloff, A. (1980). Assembly of the mitochondrial membrane system: Sequence analysis of a yeast mitochondrial ATPase gene containing the *oli-2* and *oli-4* loci. *Cell* **20,** 507–517.

Macino, G., Scazzocchio, C., Waring, R. B., McPhail Berks, M., and Davies, R. W. (1980). Conservation and rearrangement of mitochondrial structural gene sequences. *Nature (London)* **288,** 404–406.

Macino, G., Coruzzi, G., Nobrega, F. G., Li, M., and Tzagoloff, A. (1979). Use of the UGA terminator as a tryptophan codon in mitochondria. *Proc. Natl. Acad. Sci. U.S.A.* **76,** 3784–3785.

Macreadie, I. G., Choo, W. M., Novitski, C. E., Marzuki, S., and Nagley, P. (1982). Novel mitochondrial mutations between the *oli2* and *oxi3* genes affect the yeast mitochondrial ATPase. *Biochem. Int.* **5,** 129–136.

Macreadie, I. G., Novitski, C. E., Maxwell, R. J., John, U., Ooi, B.-G., McMullen, G. L., Lukins, H. B., Linnane, A. W., and Nagley, P. (1983). Biogenesis of mitochondria: The mitochondrial gene (*aap1*) coding for mitochondrial ATPase subunit 8 in *Saccharomyces cerevisiae*. *Nucleic Acids Res.* **11,** 4435–4451.

Macreadie, I. G., Scott, R. M., Zinn, A. R., and Butow, R. A. (1985a). Transpositon of an intron in yeast mitochondria requires a protein encoded by that intron. *Cell* **41,** 395–402.

Macreadie, I. G., Zinn, A. R., and Butow, R. A. (1985b). The yeast mitochondrial *fit1* gene. *In* "Achievements and Perspectives of Mitochondrial Research" (E. Quagliariello, E. C. Slater, F. Palmieri, C. Saccone, and A. M. Kroon, eds.), Vol. 2, pp. 349–354. Elsevier, Amsterdam.

Maheshwari, K. K., Marzuki, S., and Linnane, A. W. (1982). The formation of defective small ribosomal subunit in yeast mitochondria in the absence of mitochondrial protein synthesis. *Biochem. Int.* **4,** 109–116.

Mahler, H. R., Hanson, D., Miller, D., Bilinski, T., Ellis, D. M., Alexander, N. J., and Perlman, P. S. (1977). Structural and regulatory mutations affecting mitochondrial gene products. *In* "Mitochondria 1977" (W. Bandlow, R. J. Schweyen, K. Wolf, and F. Kaudewitz, eds.), pp. 345–370. De Gruyter, Berlin.

Mahler, H. R., Hansen, D. K., Lamb, M. R., Perlman, P. S., Anziano, P. G., Glaus, K. R., and Haldi, M. L. (1982). Regulatory interactions between mitochondrial genes: Expressed introns—their function and regulation. *In* "Mitochondrial Genes" (P. Slonimski, P. Borst, and G. Attardi, eds.), pp. 185–200. Cold Spring Harbor Laboratory, Cold Spring Harbor, New York.

Maly, P., and Brimacombe, R. (1983). Refined secondary structure models for the 16S and 23S ribosomal RNA of *Escherichia coli*. *Nucleic Acids Res.* **11,** 7263–7286.

Manella, C. A., and Lambowitz, A. (1978). Interactions of wild-type and *poky* mitochondrial DNA in heterokaryons of *Neurospora*. *Biochem. Biophys. Res. Commun.* **80,** 673–679.

Manella, C. A., and Lambowitz, A. M. (1979). Unidirectional gene conversion associated with two insertions in *Neurospora crassa* mitochondrial DNA. *Genetics* **93,** 645–654.

Manella, C. A., Goewert, R. R., and Lambowitz, A. M. (1979). Characterization of variant *Neurospora crassa* mitochondrial DNAs which contain tandem reiterations. *Cell* **18,** 1197–1207.

Mangin, M., Faugeron-Fonty, G., and Bernardi, G. (1983). The *ori*^{r} to *ori*$^{+}$ mutation in spontaneous yeast petites is accompanied by a drastic change in mitochondrial genome replication. *Gene* **24,** 73–81.

Manna, F., Del Giudice, L., Schreil, W. H., and Wolf, K. (1981). Electron microscopy of native and cloned mitochondrial DNA. *Mol. Gen. Genet.* **184,** 469–470.

Marmiroli, N., and Puglisi, P. P. (1980). Antibiotic-resistant temperature-sensitive mutants in *Kluyveromyces lactis* as a tool for the analysis of nucleo–mitochondrial relationships in a petite negative yeast. Ethidium bromide mutagenesis and biochemical analysis. *Mutat. Res.* **72,** 405–422.

Marotta, R., Colin, Y., Goursot, R., and Bernardi, G. (1982). A region of extreme instability in the mitochondrial genome of yeast. *EMBO J.* **1,** 529–534.

Martin, N. C., and Rabinowitz, M. (1978). Mitochondrial transfer RNAs in yeast: Identification of isoaccepting transfer RNAs. *Biochemistry* **17,** 1628–1634.

Martin, N. C., and Underbrink-Lyon, K. (1981). A mitochondrial locus is necessary for the synthesis of mitochondrial tRNA in the yeast *Saccharomyces cerevisiae. Proc. Natl. Acad. Sci. U.S.A.* **78,** 4743–4747.

Martin, N. C., Rabinowitz, M., and Fukuhara, H. (1976). Isoaccepting mitochondrial glutamyl-tRNA species transcribed from different regions of the mitochondrial genome of *Saccharomyces cerevisiae. J. Mol. Biol.* **101,** 285–296.

Martin, N. C., Rabinowitz, M., and Fukuhara, H. (1977). Yeast mitochondrial DNA specifies tRNA for 19 amino acids. Deletion mapping of the tRNA genes. *Biochemistry* **16,** 4672–4677.

Martin, N. C., Miller, D. L., and Donelson, J. E. (1979a). Cloning of yeast mitochondrial DNA in the *Escherichia coli* plasmid *pBR322*. Identification of tRNA genes. *J. Biol. Chem.* **254,** 11729–11734.

Martin, N. C., Miller, D. L., Donelson, J. E., Sigurdson, C., Hartley, J. L., Moynihan, P. S., and Pham, H. D. (1979b). Identification and sequencing of yeast mitochondrial transfer RNA genes in mitochondrial DNA–*pBR322* recombinants. *In* "Extrachromosomal DNA" (D. J. Cummings, P. Borst, I. B. Dawid, S. M. Weissman, and C. F. Fox, eds.), pp. 357–376. Academic Press, New York.

Martin, N. C., Miller, D. L., Hartley, J., Moynihan, P., and Donelson, J. E. (1980a). The tRNA Ser <AGY> and tRNA Arg <CGY> genes form a gene cluster in yeast mitochondrial DNA. *Cell* **19,** 339–343.

Martin, N. C., Pham, H. D., Underbrink-Lyon, K. Miller, D. L., and Donelson, J. E. (1980b). Yeast mitochondrial tRNA Trp can recognize the nonsense codon UGA. *Nature (London)* **285,** 579–581.

Martin, N. C., Underbrink-Lyon, K., and Miller, D. L. (1982). Identification and characterization of a yeast mitochondrial locus necessary for tRNA biosynthesis. *In* "Mitochondrial Genes" (P. Slonimski, P. Borst, and G. Attardi, eds.), pp. 264–267. Cold Spring Harbor Laboratory, Cold Spring Harbor, New York.

Martin, N. C., Hollingsworth, M. J., Shu, H.-H., and Najarian, D. R. (1985a). Expression and function of the tRNA synthesis locus in *Saccharomyces cerevisiae. In* "Achievements and Perspectives of Mitochondrial Research" (E. Quagliariello, E. C. Slater, F. Palmieri, C. Saccone, and A. M. Kroon, eds.), Vol. 2, pp. 193–202. Elsevier, Amsterdam.

Martin, N. C., Miller, D. L., Underbrink, K., and Ming, X. (1985b). Structure of the primary transcript of the yeast mitochondrial tRNA fMet gene: Implications for the function of the tRNA synthesis locus. *J. Biol. Chem.* **260,** 1479–1483.

Martin, R. P., and Dirheimer, G. (1983). Two-dimensional polyacrylamide gel electrophoresis of tRNA: Its application to the study of yeast mitochondrial tRNAs. *Mol. Biol.* **17,** 1117–1125.

Martin, R. P., Schneller, J. M., Stahl, A. J. C., and Dirheimer, G. (1976a). Studies of odd bases in yeast mitochondrial tRNA: II. Characterization of rare nucleosides. *Biochem. Biophys. Res. Commun.* **70,** 997–1002.

Martin R. P., Schneller, J. M., Stahl, A. J. C., and Dirheimer, G. (1976b). Isoacceptor tRNA species in yeast mitochondria. Methionine and formylmethionine specific tRNAs coded by mitochondrial DNA. *In* Genetics and Biogenesis of Chloroplasts and Mitochondria" (T. Bücher, W. Neupert, W. Sebald, and S. Werner, eds.), pp. 755–758. Elsevier, Amsterdam.

Martin, R. P., Schneller, J. M., Stahl, A. J. C., and Dirheimer, G. (1977). Study of yeast mitochondrial tRNAs by two dimensional polyacrylamide electrophoresis: Characterization of isoaccepting species and search for imported cytoplasmic tRNAs. *Nucleic Acids. Res.* **4,** 3497–3510.

Martin, R. P., Sibler, A. P., Schneller, J. M., Keith, G., Stahl, A. J. C., and Dirheimer, G. (1978). Primary structure of yeast mitochondrial DNA-coded phenylalanine tRNA. *Nucleic Acids Res.* **5,** 4579–4592.

Martin, R. P., Schneller, J. M., Stahl, A. J. C., and Dirheimer, G. (1979). Import of nuclear deoxyribonucleic acid coded lysine-accepting transfer ribonucleic acid (anticodon CUU) into yeast mitochondria. *Biochemistry* **18,** 4600–4605.

Martin, R. P., Sibler, A.-P., Dirheimer, G., de Henau, S., and Grosjean, H. (1981). Yeast mitochondrial tRNA Trp injected with *E. coli* activating enzyme into *Xenopus* oocytes suppresses UGA termination. *Nature (London)* **293,** 235–237.

Martin, R. P. Sibler, A.-P., Bordonne, R., and Dirheimer, G. (1983). Yeast mitochondrial tRNAs: Structure, coding properties and gene organization. *Mol. Biol.* **17,** 1126–1146.

Mason, J. R., and Turner, G. (1975). Transmission and recombination of extranuclear genes during sexual crosses in *Aspergillus nidulans. Mol. Gen. Genet.* **143,** 93–100.

Mason, T., Boerner, P., and Biron, C. (1976). The use of double mutant strains containing both heat and cold sensitive mutations in studies of mitochondrial biogenesis. *In* "Genetics and Biogenesis of Chloroplasts and Mitochondria" (T. Bücher, W. Neupert, W. Sebald, and S. Werner, eds.), pp. 239–246. Elsevier, Amsterdam.

Mason, T., Breitbart, M., and Meyers, J. (1978). Temperature sensitive mutations of *Saccharomyces cerevisiae* with defects in mitochondrial functions. *In* "Methods in Enzymology" (S. Fleischer and L. Packer, eds.), Vol. 56, pp. 131–139, Academic Press, New York.

Massardo, D. R., Del Giudice, L., Manna, F., and Wolf, K. (1982). Extrachomosomal inheritance of nalidixic acid resistance in the petite negative yeast *Schizosaccharomyces pombe. Mol. Gen. Genet.* **187,** 96–100.

Mathews, S., Schweyen, R. J., and Kaudewitz, F. (1977). Preferential loss or retention of mitochondrial genes in *rho*$^-$ clones. *In* "Mitochondria 1977" (W. Bandlow, R. J. Schweyen, K. Wolf, and F. Kaudewitz, eds.), pp. 133–138. De Gruyter, Berlin.

Mattick, J. S., and Hall, R. M. (1977). Replicative deoxyribonucleic acid synthesis in isolated mitochondria from *Saccharomyces cerevisiae. J. Bacteriol.* **130,** 973–982.

Maxwell, R. J., Devenish, R. J., and Nagley, P. (1986). The nucleotide sequence of the mitochondrial DNA genome of an abundant petite mutant of *Saccharomyces cerevisiae* carrying the *ori 1* replication origin. *Biochem. Int.* **13,** 101–108.

Merlos-Lange, A. M., and Wolf, K. (1986). A mutant *tRNA-Met* gene in the mitochondrial genome of *Schizosaccharomyces pombe. Nucleic Acids Res.* **14,** 8687.

Merlos-Lange, A.-M., Kanbay, F., Zimmer, M., and Wolf, K. (1987). DNA-splicing of mitochondrial group I and II introns in *Schizosaccharomyces pombe. Mol. Gen. Genet.* **206,** 273.

Merten, S., Synenki, R. M., Locker, J., Christianson, T., Rabinowitz, M. (1980). Processing of precursors of 21S ribosomal RNA from yeast mitochondria. *Proc. Natl. Acad. Sci. U.S.A.* **77,** 1417–1421.

Meunier-Lemesle, D., Chevilotte-Brivet, P., and Pajot, P. (1980). Cytochrome *b-565* in *Saccharomyces cerevisiae:* Use of mutants in the *cob–box* region of the mitochondrial DNA to study the functional role of this spectral species of cytochrome *b*. 1. Measurements of cytochromes *b-562* and *b-565* and selection of revertants devoid of cytochrome *b-565*. *Eur. J. Biochem.* **111,** 151–159.

Michael, N. L., Rothbard, J. B., Shiurba, R. A., Linke, H. K., Schoolnik, G. K., and Clayton, D. A. (1984). All eight unassigned reading frames of mouse mitochondrial DNA are expressed. *EMBO J.* **3,** 3165–3175.

Michaelis, G. (1976). Cytoplasmic inheritance of antimycin A resistance in *Saccharomyces cerevisiae. Mol. Gen. Genet.* **146,** 133–137.

Michaelis, G., and Pratje, E. (1977). Mapping of the two mitochondrial antimycin A resistance loci in *Saccharomyces cerevisiae. Mol. Gen. Genet.* **156,** 79–85.

Michaelis, G., and Somlo, M. (1976). Genetic analysis of mitochondrial biogenesis and function in *Saccharomyces cerevisiae. J. Bioenerg.* **8,** 93–107.

Michaelis, G., Petrochilo, E., and Slonimski, P. P. (1973). Mitochondrial genetics. III. Recombined molecules of mitochondrial DNA obtained from crosses between cytoplasmic petite mutants of *Saccharomyces cerevisiae:* Physical and genetic characterization. *Mol. Gen. Genet.* **123,** 51–65.

Michaelis, G., Michel, F., Lazowska, J., and Slonimski, P. P. (1976). Recombined molecules of mitochondrial DNA obtained from crosses between cytoplasmic petite mutants of *Saccharomyces cerevisiae:* The stoichiometry of parental DNA repeats within the recombined molecule. *Mol. Gen. Genet.* **149,** 125–130.

Michel, F. (1984). A maturase-like sequence downstream of the *oxi2* gene of yeast mitochondrial DNA is interrupted by two GC clusters and a putative end-of-messenger signal. *Curr. Genet.* **8,** 307–317.

Michel, F., and Cummings, D. J. (1985). Analysis of class I introns in a mitochondrial plasmid associated with senescence of *Podospora anserina* reveals extraodrinary resemblance to the *Tetrahymena* ribosomal intron. *Curr. Genet.* **10,** 69–80.

Michel, F., and Dujon, B. (1983). Conservation of RNA secondary structure in two intron families including mitochondrial-, chloroplast- and nuclear-encoded members. *EMBO J.* **2,** 33–38.

Michel, F., and Lang, B. F. (1985). Mitochondrial class II introns encode proteins related to the reverse transcriptases of retroviruses. *Nature* (*London*) **316,** 641–643.

Michel, F., Grandchamp, C., and Dujon, B. (1979). Genetic and physical characterization of a segment of yeast mitochondrial DNA involved in the control of genetic recombination. *Biochimie* **61,** 985–991.

Michel, F., Jaquier, A., and Dujon, B. (1982). Comparison of fungal mitochondrial introns reveals extensive homologies in RNA secondary structure. *Biochimie* **64,** 867–881.

Mieszczak, M., and Zagorski, W. (1987). *Mim 3* and *nam 3* omnipotent suppressor genes similarly affect the polypeptide composition of yeast mitoribosomes. *Biochimie* **69,** 531–538.

Miller, D. L., and Martin, N. C. (1981). Organization and expression of a tRNA gene cluster in *Saccharomyces cerevisiae* mitochondrial DNA. *Curr. Genet.* **4,** 135–143.

Miller, D. L., and Martin, N. C. (1983). Characterization of the yeast mitochondrial locus necessary for tRNA biosynthesis: DNA sequence analysis and identification of a new transcript. *Cell* **34,** 911–917.

Miller, D. L., Martin, N. C., Pham, H. D., and Donelson, J. E. (1979). Sequence analysis of two yeast mitochondrial DNA fragments containing the genes for tRNA Ser <UCR> and tRNA Phe <UUY>. *J. Biol. Chem.* **254,** 11735–11740.

Miller, D. L., Sigurdson, C., Martin, N. C., and Donelson, J. F. (1980). Nucleotide sequence of the mitochondrial genes coding for tRNA Gly <GGR> and tRNA Val <GUR>. *Nucleic Acids Res.* **8,** 1435–1442.

Miller, D. L., Najarian, D. R., Folse, J. R., and Martin, N. C. (1981). A mutation in the tRNA Asp gene from yeast mitochondria. Effects on RNA and protein synthesis. *J. Biol. Chem.* **256,** 9774–9777.

Miller, D. L., Folse, J. R., Benson, P. J., Martin, N. C. (1983a). Identification and consequences of a guanosine-15 to adenosine-15 change in the yeast mitochondrial tRNA Ser <UCX> gene. *Biochemistry* **22,** 1709–1714.

Miller, D. L., Underbrink-Lyon, K., Najarian, D. R., Krupp, J., and Martin, N. C. (1983b). Transcription of yeast mitochondrial tRNA genes and processing of tRNA gene transcripts. *In* "Mitochondria 1983" (R. J. Schweyen, K. Wolf, and F. Kaudewitz, eds.), pp. 151–164. De Gruyter, Berlin.

Mitchell, M. B., and Mitchell, H. K. (1952). A case of maternal inheritance in *Neurospora crassa. Proc. Natl. Acad. Sci. U.S.A.* **38,** 442–449.

Mitchell, M. B., Mitchell, H. K., and Tissieres, A. (1953). Mendelian and nonmendelian factors affecting the cytochrome system in *Neurospora crassa. Proc. Natl. Acad. Sci. U.S.A.* **39,** 606–613.

Mol, J. N. M., Borst, P., Grosveld, F. G., and Spencer, J. H. (1974). The size of the repeating unit of the repetitive mitochondrial DNA from a "low density" petite mutant of yeast. *Biochem. Biophys. Acta* **374,** 115–128.

Molloy, P. L., Howell, N., Plummer, D. T., Linnane, A. W., and Lukins, H. B. (1973). Mitochondrial mutants of the yeast *Saccharomyces cerevisiae* showing resistance *in vitro* to chloramphenicol inhibition of mitochondrial protein synthesis. *Biochem. Biophys. Res. Commun.* **52,** 9–14.

Molloy, P. L., Linnane, A. W., and Lukins, H. B. (1975). Biogenesis of mitochondria: Analysis of deletion of mitochondrial antibiotic resistance markers in petite mutants of *Saccharomyces cerevisiae. J. Bacteriol.* **122,** 7–18.

Molloy, P. L., Linnane, A. W., and Lukins, H. B. (1976). Relative retention of mitochondrial markers in petite mutants: Mitochondrially determined differences between *rho*$^+$ strains. *Genet. Res.* **26,** 319–325.

Monnerot, M., Schweyen, R. J., and Fukuhara, F. (1977). Mapping of mutation *tsm-8* with respect to transfer RNA genes on the mitochondrial DNA of *Saccharomyces cerevisiae. Mol. Gen. Genet.* **152,** 307–309.

Montoya, J., Ojala, D., and Attardi, G. (1981). Distinct features of the 5′-terminal sequences of the human mitochondrial mRNAs. *Nature (London)* **290,** 465–470.

Montoya, J., Christiansen, T., Levens, D., Rabinowitz, M., and Attardi, G. (1982). Identification of initiation sites for heavy strand and light strand transcription in human mitochondrial DNA. *Proc. Natl. Acad. Sci. U.S.A.* **79,** 7195–7199.

Moorman, A. F. M., Van Ommen, G.-J. B., and Grivell, L. A. (1978). Transcription of yeast mitochondria: Isolation and physical mapping of messenger RNAs for subunits of cytochrome *c* oxidase and ATPase. *Mol. Gen. Genet.* **160,** 13–24.

Morelli, G., and Macino, G. (1984). Two intervening sequences in the ATPase subunit 6 gene of *Neurospora crassa.* A short intron (93 base-pairs) and a long intron that is stable after excision. *J. Mol. Biol.* **178,** 491–508.

Morgan, A. J., and Whittaker, P. A. (1978). Biosynthesis of yeast mitochondria. IV. Antibiotic effects of growth, cytochrome synthesis, and respiration in *Kluyveromyces lactis. Mol. Gen. Genet.* **164,** 185–193.

Morimoto, R., and Rabinowitz, M. (1979a). Physical mapping of the *Xba* I, *Hinc* II, *Bgl* II,

Xho I, *Sst* I, and *Pvu* II restriction endonuclease cleavage fragments of mitochondrial DNA of *S. cerevisiae. Mol. Gen. Genet.* **170,** 11–23.

Morimoto, R., and Rabinowitz, M. (1979b). Physical mapping of the yeast mitochondrial genome. *Mol. Gen. Genet.* **170,** 25–48.

Morimoto, R., Lewin, A., and Rabinowitz, M. (1977). Restriction cleavage map of mitochondrial DNA from the yeast *Saccharomyces cerevisiae. Nucleic Acids Res.* **4,** 2331–2351.

Morimoto, R., Merten, S., Lewin, A., Martin, N. C., and Rabinowitz, M. (1978). Physical mapping of genes on yeast mitochondrial DNA: Localization of antibiotic resistance loci, and rRNA and tRNA genes. *Mol. Gen. Genet.* **163,** 241–255.

Morimoto, R., Lewin, A., and Rabinowitz, M. (1979a). Physical mapping and characterization of the mitochondrial DNA and RNA sequences from *mit*$^-$ mutants defective in cytochrome oxidase peptide 1 (*OXI 3*). *Mol. Gen. Genet.* **170,** 1–9.

Morimoto, R., Locker, J., Synenki, R. M., and Rabinowitz, M. (1979b). Transcription, processing, and mapping of mitochondrial RNA from grande and petite yeast. *J. Biol. Chem.* **254,** 12461–12470.

Mueller, D. M., and Getz, G. S. (1986a). Transcriptional regulation of the mitochondrial genome of yeast *Saccharomyces cerevisiae. J. Biol. Chem.* **261,** 11756–11764.

Mueller, D. M., and Getz G. S. (1986b). Steady state analysis of mitochondrial RNA after growth of yeast *Saccharomyces cerevisiae* under catabolite repression and derepression. *J. Biol. Chem.* **261,** 11816–11822.

Munz, P., Kohli, J., Wolf, K., and Leupold, U. (1988). Genetics overview. *In* "The Molecular Biology of the Fission Yeast" (A. Nasim, P. Young, and B. F. Johnson, eds.). Academic Press, New York, in press.

Murphy, M., Gutowski, S. J., Marzuki, S., Lukins, H. B., and Linnane, A. W. (1978). Mitochondrial oligomycin-resistance mutations affecting the proteolipid subunit of the mitochondrial adenosine triphosphatase. *Biochem. Biophys. Res. Commun.* **85,** 1283–1290.

Murphy, M., Roberts, H., Choo, W. M., Macreadie, I., Marzuki, S., Lukins, H. B., and Linnane, A. W. (1980). Biogenesis of mitochondria. *Oli2* mutations affecting the coupling of oxidation to phosphorylation in *Saccharomyces cerevisiae. Biochim. Biophys. Acta* **592,** 431–444.

Myers, A. M., Pape, L. K., and Tzagoloff, A. (1985). Mitochondrial protein synthesis is required for maintenance of intact mitochondrial genomes in *Saccharomyces cerevisiae. EMBO J.* **4,** 2087–2092.

Nagley, P., and Linnane, A. W. (1970). Mitochondrial DNA deficient petite mutants of yeast. *Biochem. Biophys. Res. Commun.* **39,** 989–996.

Nagley, P., and Novitski, C. E. (1982). Coding capacity of yeast mitochondrial DNA. *TIBS* **7,** 281–285.

Nagley, P., Cobon, G. S., Linnane, A. W., and Beilharz, M. W. (1981). Transcription of the *oli2* region of yeast mitochondrial DNA shows strain-dependent variation. *Biochem. Int.* **3,** 473–481.

Nagley, P., Hall, R. M., and Ooi, B. G. (1986). Amino acid substitutions in mitochondrial ATPase subunit 9 of *Saccharomyces cerevisiae* leading to oligomycin or venturicidin resistance. *FEBS Lett.* **195,** 159–163.

Najarian, D., Shu, H. H., and Martin, N. C. (1986). Sequence and expression of four mutant aspartic acid tRNA genes from the mitochondria of *Saccharomyces cerevisiae. Nucleic Acids Res.* **14,** 9561–9594.

Nargang, F. E., Bell, J. B., Stohl, L. L. and Lambowitz, A. M. (1983). A family of repetitive palindromic sequences found in *Neurospora* mitochondrial DNA is also found in a mitochondrial plasmid DNA. *J. Biol. Chem.* **258,** 4257–4260.

Nargang, F. E., Bell, J. B., Stohl, L. L., and Lambowitz, A. M. (1984). The DNA sequence and genetic organization of a *Neurospora* mitochondrial plasmid suggest a relationship to introns and mobile elements. *Cell* **38,** 441–453.

Nelson, M. A., and Macino, G. (1985). Gene organization and expression in *Neurospora crassa* mitochondria. *In* "Achievements and Perspectives of Mitochondrial Research" (E. Quagliariello, E. C. Slater, F. Palmieri, C. Saccone, and A. M. Kroon, eds.), Vol. 2, pp. 293–304. Elsevier, Amsterdam.

Nelson, M. A., and Macino, G. (1987a). Structure and expression of the overlapping *ND4L* and *ND5* genes of *Neurospora crassa* mitochondria. *Mol. Gen. Genet.* **206,** 307–317.

Nelson, M. A., and Macino, G. (1987b). Three class I introns in the *ND4L*/*ND5* transcriptional unit of *Neurospora crassa* mitochondria. *Mol. Gen. Genet.* **206,** 318–325.

Netter, P., Petrochilo, E. Slonimski, P. P., Bolotin-Fukuhara, M., Coen, D., Deutsch, J., and Dujon, B. (1974). Mitochondrial genetics. VII. Allelism and mapping studies of ribosomal mutants resistant to chloramphenicol, erythromycin and spiramycin in *S. cerevisiae. Genetics* **78,** 1063–1100.

Netter, P., Carignani, G., Jacq, C., Groudinsky, O., Clavilier, L., and Slonimski, P. P. (1982). The cytochrome oxidase subunit I split gene in *Saccharomyces cerevisiae:* Genetic and physical studies of the mtDNA segments encompassing the "cytochrome *b*-homologous" intron. *Mol. Gen. Genet.* **188,** 51–59.

Netter, P., Jacq, C., Carignani, G., and Slonimski, P. P. (1982b). Critical sequences within mitochondrial introns: Cis-dominant mutations of the "cytochrome-*b*-like" intron of the oxidase gene. *Cell* **28,** 733–738.

Netzker, R., Köchel, H. G., Basak, N., and Küntzel, H. (1982). Nucleotide sequence of *Aspergillus nidulans* mitochondrial genes coding for ATPase subunit 6, cytochrome oxidase subunit 3, seven unidentified proteins, four tRNAs and L-rRNA. *Nucleic Acids Res.* **10,** 4783–4794.

Neupert, W., Massinger, P., and Pfaller, A. (1971). Amino acid incorporation into mitochondrial ribosomes of *Neurospora crassa* wild type and *MI-1*. *In* "Anatomy and Biogenesis of Mitochondria and Chloroplasts" (N. K. Boardman, A. W. Linnane, and R. M. Smillie, eds.), pp. 328–338. Elsevier, Amsterdam.

Newlon, C. S., and Fangman, W. L. (1975). Mitochondrial DNA synthesis in cell cycle mutants of *Saccharomyces cerevisiae. Cell* **5,** 423–428.

Newman, D., Pham, H. D., Underbrink-Lyon, K., and Martin, N. C. (1980). Characterization of tRNA genes in tRNA region II of yeast mitochondrial DNA. *Nucleic Acids Res.* **8,** 5007–5016.

Nielsen, H., and Engberg, J. (1985). Sequence comparison of the rDNA introns from six different species of *Tetrahymena. Nucleic Acids Res.* **13,** 7445–7456.

Nobrega, M. P., and Nobrega, F. G. (1986). Mapping and sequencing of the wild-type and mutant (*G 116-40*) alleles of the tyrosyl-tRNA mitochondrial gene in *Saccharomyces cerevisiae. J. Biol. Chem.* **261,** 3054–3059.

Nobrega, F. G., and Tzagoloff, A. (1980a). Assembly of the mitochondrial membrane system. Structure and location of the mitochondrial glutamic tRNA gene in *Saccharomyces cerevisiae. FEBS Lett.* **113,** 52–54.

Nobrega, F. G., and Tzagoloff, A. (1980b). Assembly of the mitochondrial membrane system. Complete restriction map of the cytochrome *b* region of mitochondrial DNA in *Saccharomyces cerevisiae* D273-10B. *J. Biol. Chem.* **255,** 9821–9827.

Nobrega, F. G., and Tzagoloff, A. (1980c). Assembly of the mitochondrial membrane system. DNA sequence and organization of the cytochrome *b* gene in *Saccharomyces cerevisiae* D273-10B. *J. Biol. Chem.* **255,** 9828–9837.

Noller, H. F. (1984). Structure of ribosomal DNA. *Annu. Rev. Biochem.* **53,** 119–162.

Novitski, C. E., Macreadie, I. G., Maxwell, R. J., Lukins, H. B., Linnane, A. W., and Nagley, P. (1983). Features of nucleotide sequences in the region of the *oli2*-genes and *aapi*-genes in the yeast mitochondrial genome. *In* "Manipulation and Expression of Genes in Eukaryotes" (P. Nagley *et al.,* eds.), pp. 257–268. Academic Press, Sidney.

Novitski, C. E., Macreadie, I. G., Maxwell, R. J., Lukins, H. B., Linnane, A. W., and Nagley, P. (1984). Biogenesis of mitochondria: Genetic and molecular analysis of the *oli2* region of mitochondrial DNA in *Saccharomyces cerevisiae. Curr. Genet.* **8,** 135–146.

O'Connor, R. M., McArthur, C. R., and Clark-Walker, G. D. (1975). Closed-cirular DNA from mitochondria-enriched fractions of four petite-negative yeasts. *Eur. J. Biochem.* **53,** 137–144.

O'Connor, R. M., McArthur, C. R., and Clark-Walker, G. D. (1976). Respiratory-deficient mutants of *Torulopsis glabrata,* a yeast with circular mitochondrial deoxyribonucleic acid of 6 μm. *J. Bacteriol.* **126,** 959–968.

Ojala, D., Merkel, C., Gelfand, R., and Attardi, G. (1980). The tRNA genes punctuate the reading of genetic information in human mitochondrial DNA. *Cell* **22,** 393–403.

Ojala, D., Montoya, J., and Attardi, G. (1981). tRNA punctuation model of RNA processing in human mitochondria. *Nature* (*London*) **290,** 470–474.

Ooi, B. G., and Nagley, P. (1986). The *oli 1* gene and flanking sequences in mitochondrial DNA of *Saccharomyces cerevisiae:* The complete nucleotide sequence of a 1.35 kilobase petite mitochondrial genome covering the *oli 1* gene. *Curr. Genet.* **10,** 713–724.

Ooi, B. G., McMullen, G. L., Linnane, A. W., Nagley, P., and Novitski, C. E. (1985). Biogenesis of mitochondria: DNA sequence analysis of *mit*$^-$ mutations in the mitochondrial *oli1* gene coding for mitochondrial ATPase subunit 9 in *Saccharomyces cerevisiae. Nucleic Acids Res.* **13,** 1327–1340.

Ooi, B. G., Lukins, H. B., Linnane, A. W., and Nagley, P. (1987). Biogenesis of mitochondria: A mutation in the 5′-untranslated region of yeast mitochondrial *oli1* mRNA leading to impairment in translation of subunit 9 of the mitochondrial ATPase complex. *Nucleic Acids Res.* **15,** 1965–1978.

Osiewacz, H. D., and Esser, K. (1983). DNA sequence analysis of the mitochondrial plasmid of *Podospora anserina. Curr. Genet.* **7,** 219–223.

Osiewacz, H. D., and Esser, K. (1984). The mitochondrial plasmid of *Podospora anserina:* A mobile intron of a mitochondrial gene. *Curr. Genet.* **8,** 299–305.

Osinga, K. A., and Tabak, H. F. (1982). Initiation of transcription of genes for mitochondrial ribosomal RNA in yeast: Comparison of the nucleotide sequence around the 5′ ends of both genes reveals a homologous stretch of 17 nucleotides. *Nucleic Acids Res.* **10,** 3617–3626.

Osinga, K. A., Evers, R. F., Van der Laan, J. C., and Tabak, H. F. (1981). A putative precursor for the small ribosomal RNA from mitochondria of *Saccharomyces cerevisiae. Nucleic Acids Res.* **9,** 1351–1364.

Osinga, K. A., De Haan, M., Christianson, T., and Tabak, H. F. (1982). A nonanucleotide sequence involved in promotion of ribosomal RNA synthesis and RNA priming of DNA replication in yeast mitochondria. *Nucleic Acids Res.* **10,** 7993–8006.

Osinga, K. A., De Vries, E., Van der Horst, G., and Tabak, H. F. (1984a). Processing of yeast mitochondrial messenger RNAs at a conserved dodecamer sequence. *EMBO J.* **3,** 829–834.

Osinga, K. A., De Vries, E., Van der Horst, G. T. J., and Tabak, H. F. (1984b). Initiation of transcription in yeast mitochondria: Analysis of origins of replication and of genes coding for a messenger RNA and a transfer RNA. *Nucleic Acids Res.* **12,** 1889–1900.

Pajot, P., Wambier-Kluppel, M. L., Kotylak, Z., and Slonimski, P. P. (1976). Regulation of cytochrome oxidase formation by mutations in a mitochondrial gene for cytochrome *b*. *In* "Genetics and Biogenesis of Chloroplasts and Mitochondria" (T. Bücher, W. Neupert, W. Sebald, and S. Werner, eds.), pp. 443–452. Elsevier, Amsterdam.

Pajot, P., Wambier-Kluppel, M. L., and Slonimski, P. P. (1977). Cytochrome *c* reductase and cytochrome *c* oxidase formation in mutants and revertants in the "*box*" region of the mitochondrial DNA. *In* "Mitochondria 1977" (W. Bandlow, R. J. Schweyen, K. Wolf, and F. Kaudewitz, eds.), pp. 173–184. De Gruyter, Berlin.

Palleschi, C., Francisci, C., Bianich, M. M., and Frontali, L. (1984a). Initiation of transcription of a mitochondrial tRNA gene cluster in *S. cerevisiae*. *Nucleic Acids Res.* **12,** 7317–7326.

Palleschi, C., Francisci, C., Zennaro, E., and Frontali, L. (1984b). Expression of the clustered mitochondrial tRNA gene in *Saccharomyces cerevisiae:* Transcription and processing of transcripts. *EMBO J.* **3,** 1389–1395.

Parker, J. H., Trimble, J. R., and Mattoon, J. R. (1968). Oligomycin resistance in normal and mutant yeast. *Biochem. Biophys. Res. Commun.* **33,** 590–595.

Partis, M. D., Bertoli, E., Zanders, E. D., and Griffiths, D. E. (1979). A modified subunit of mitochondrial ATPase in mutants of *Saccharomyces cerevisiae* with decreased sensitivity to dicyclohexylcarbodiimide. *FEBS Lett.* **105,** 167–170.

Peebles, C. L., Dietrich, R. C., Perlman, P. S., Petrillo, M. L., Mecklenburg, K. L., and Romiti, S. L. (1986). A self-splicing RNA excises an intron lariat. *Cell* **44,** 213–223.

Perea, J., and Jacq, C. (1985). Role of the 5′ hairpin structure in the splicing accuracy of the fourth intron of the yeast *cob-box* gene. *EMBO J.* **4,** 3281–3288.

Perlman, P. S., and Birky, C. W. (1974). Mitochondrial genetics in bakers' yeast: A molecular mechanism for recombinational polarity and suppressiveness. *Proc. Natl. Acad. Sci. U.S.A.* **71,** 4612–4616.

Perlman, P. S., Douglas, M. G., Strausberg, R. L., and Butow, R. A. (1977). Localization of genes for variant forms of mitochondrial proteins on mitochondrial DNA of *Saccharomyces cerevisiae. J. Biol. Chem.* **115,** 675–694.

Perlman, P. S., Birky, C. W., jr., and Strausberg, R. L. (1979). Segregation of mitochondrial markers in yeast *In* "Methods in Enzymology" (S. Fleischer and L. Packer, eds.), Vol. 56, pp. 139–154. Academic Press, New York.

Perlman, P. S., Alexander, N. J., Hanson, D: K., and Mahler, H. R. (1980). Mosaic genes in yeast mitochondria *In* "Gene Structure and Expression" (D. H. Dean *et al.,* eds.), pp. 221–253. Ohio State Univ. Press, Columbus, Ohio.

Petes, T. D., and Fangman, W. L. (1973). Preferential synthesis of yeast mitochondrial DNA in *alpha* factor arrested cells. *Biochem. Biophys Res. Commun.* **55,** 603–609.

Pinon, R., Salts, Y., and Simchen, G. (1974). Nuclear and mitochondrial DNA synthesis during yeast sporulation. *Exp. Cell Res.* **83,** 231–238.

Pittenger, T. H. (1956). Synergism of two cytoplasmically inherited mutants in *Neurospora crassa. Proc. Natl. Acad. Sci. U.S.A.* **42,** 747–752.

Pittenger, T. H., and West, D. J. (1979). Isolation and characterization of temperature-sensitive respiratory mutants of *Neurospora crassa. Genetics* **93,** 539–555.

Pratje, E., and Michaelis, G. (1976). Two mitochondrial antimycin A resistance loci in

Saccharomyces cerevisiae. In "Genetics and Biogenesis of Chloroplasts and Mitochondria" (T. Bücher, W. Neupert, W. Sebald, and S. Werner, eds.), pp. 467–471. Elsevier, Amsterdam.

Pratje, E., and Michaelis, G. (1977). Allelism studies of mitochondrial mutants resistant to antimycin A or funiculosin in *Saccharomyces cerevisiae. Mol. Gen. Genet.* **152,** 167–174.

Price, J. V. (1987). Origin of the phosphate at the ligation junction produced by self-splicing of *Tetrahymena thermophila* pre-ribosomal RNA. *J. Mol. Biol.* **196,** 217–221.

Price, J. V., Kieft, G. L., Kent, J. R., Sievers, E. L., and Cech, T. R. (1985). Sequence requirements for self-splicing of the *Tetrahymena thermophila* preribosomal RNA. *Nucleic Acids Res.* **13,** 1871–1890.

Price, J. V., Engberg, J., and Cech, T. R. (1987). 5′ exon requirement for self-splicing of the *Tetrahymena thermophila* pre-ribosomal RNA and identification of a cryptic 5′ splice site in the 3′ exon. *J. Mol. Biol.* **196,** 49–60.

Prunell, A., and Bernardi, G. (1974). The mitochondrial genome of wild-type yeast cells. IV. Genes and spacers. *J. Mol. Biol.* **86,** 825–841.

Prunell, A., and Bernardi, G. (1977). The mitochondrial genome of wild-type yeast cells. VI. Genome organization. *J. Mol. Biol.* **110,** 53–74.

Prunell, A., Kopecka, H., Strauss, F., and Bernardi, G. (1977). The mitochondrial genome of wild-type yeast cells. V. Genome evolution. *J. Mol. Biol.* **110,** 17–52.

Puhalla, J. E., and Srb, A. M. (1967). Heterokaryon studies of the cytoplasmic mutant *SG* in *Neurospora. Genet. Res.* **10,** 185–194.

Quagliariello, E., Slater, E. C., Palmieri, F., Saccone, C., and Kroon, A. M. eds. (1985). "Achievements and Perspectives of Mitochondrial Research," Vol. II. Elsevier, Amsterdam.

RajBhandary, U. L., Heckman, J. E., Yin, S., and Alzner-DeWeerd, B. (1979). Mitochondrial tRNAs and rRNAs of *Neurospora crassa:* Sequence studies, gene mapping and cloning. *In* "Extrachromosomal DNA" (D. J. Cummings, P. Borst, I. B. Dawid, S. M. Weissman, and C. F. Fox, eds.), pp. 377–394. Academic Press, New York.

Rank, G. H. (1970a). Genetic evidence for "Darwinian" selection at the molecular level. I. The effect of the suppressive factor on cytoplasmically-inherited erythromycin-resistance in *Saccharomyces cerevisiae. Can. J. Genet. Cytol.* **12,** 129–136.

Rank, G. H. (1970b). Genetic evidence for "Darwinian" selection at the molecular level. II. Genetic analysis of cytoplasmically-inherited high and low suppressivity in *Saccharomyces cerevisiae. Can. J. Genet. Cytol.* **12,** 340–346.

Rank, G. H., and Bech-Hansen, N. T. (1972a). Somatic segregation, recombination, asymmetrical distribution and complementation test of cytoplasmically-inherited antibiotic-resistance mitochondrial markers in *S. cerevisiae. Genetics* **72,** 1–15.

Rank, G. H., and Bech-Hansen, N. T. (1972b). Genetic evidence for "Darwinian" selection at the molecular level. III. The effect of the suppressive factor on nuclearly and cytoplasmically inherited chloramphenicol resistance in *S. cerevisiae. Can. J. Microbiol.* **18,** 1–7.

Rank, G. H., and Person, C. (1970). Reversion of cytoplasmically-inherited respiratory deficiency in *Saccharomyces cerevisiae. Can. J. Genet. Cytol.* **11,** 716–728.

Reijnders, L., and Borst, P. (1972). The number of 4S RNA genes on yeast mitochondrial DNA. *Biochim. Biophys. Acta* **47,** 126–133.

Reijnders, L., Sloof, P., and Borst, P. (1973). The molecular weights of the mitochondrial ribosomal RNAs of *Saccharomyces carlsbergensis. Eur. J. Biochem.* **35,** 266–269.

Rickwood, D., and Chambers, J. A. A. (1981). Evidence for protected regions of DNA in the mitochondrial nucleoid of *Saccharomyces cerevisiae*. *FEMS Microbiol. Lett.* **12,** 187–190.

Rickwood, D., Chambers, J. A. A., and Barat, M. (1981). Isolation and preliminary characterization of DNA–protein complexes from the mitochondria of *Saccharomyces cerevisiae*. *Exp. Cell Res.* **133,** 1–7.

Rieck, A., Griffiths, A. J. F., and Bertrand, H. (1982). Mitochondrial variants of *Neurospora intermedia* from nature. Can. J. Genet. Cytol. **24,** 741–748.

Rifkin, M. R., and Luck, D. J. L. (1971). Defective production of mitochondrial ribosomes in the *poky* mutant of *Neurospora crassa*. *Proc. Natl. Acad. Sci.* **68,** 287–290.

Rizet, G. (1952). Les phenomenes de barrage chez *Podospora anserina*. I. Analyse génétique des barrages entre souches *s* and *S*. *Rev. Cytol. Biol. Veg.* **13,** 51–92.

Rizet, G. (1953a). Sur la longevité des souches de *Podospora anserina*. *C. R. Acad. Sci. Paris* **237,** 838–840.

Rizet, G. (1953b). Sur la longevité des souches de *Podospora anserina*. *C. R. Acad. Sci. Paris* **237,** 1106–1109.

Roberts, H., Choo, W. M., Murphy, M., Marzuki, S., Lukins, H. B., and Linnane. A. W. (1979). *Mit*$^-$ mutations in the *oli2* region of mitochondrial DNA affecting the 20,000 dalton subunit of the mitochondrial ATPase in *Saccharomyces cerevisiae*. *FEBS Lett.* **108,** 501–504.

Roberts, H., Smith, S. C., Marzuki, S., and Linnane, A: W. (1980). Evidence that cytochrome *b* is the antimycin-binding component of yeast mitochondrial cytochrome bc_1 complex. *Arch. Biochem. Biophys.* **200,** 387–395.

Rödel, G., Holl, J., Schmelzer, C., Schmidt, C., Schweyen, R. J., Weiss-Brummer, G., and Kaudewitz, F. (1983). *Cob* intron 1 and 4; studies on mutants and revertants uncover functional domains and test the validity of predicted RNA-secondary structures. *In* "Mitochondria 1983" (R. J. Schweyen, K. Wolf, and F. Kaudewitz, eds.), pp. 191–201. De Gruyter, Berlin.

Rowlands, R. T., and Turner, G. (1973). Nuclear and extranuclear inheritance of oligomycin resistance in *Aspergillus nidulans*. *Mol. Gen. Genet.* **126,** 201–216.

Rowlands, R. T., and Turner, G. (1974a). Physical and biochemical studies of nuclear and extranuclear oligomycin-resistant mutants in *Aspergillus nidulans*. *Mol. Gen. Genet.* **132,** 73–88.

Rowlands, R. T., and Turner, G. (1974b). Recombination between the extranuclear genes conferring oligomycin resistance and cold-sensitivity in *Aspergillus nidulans*. *Mol. Gen. Genet.* **133,** 151–161.

Rowlands, R. T., and Turner, G. (1974c). Interaction and recombination between extranuclear genetic elements of *Aspergillus nidulans*. *Biochem. Soc. Trans.* **2,** 230–232.

Rowlands, R. T., and Turner, G. (1975). Three-marker extranuclear mitochondrial crosses in *Aspergillus nidulans*. *Mol. Gen. Genet.* **141,** 69–79.

Rowlands, R. T., and Turner, G. (1976). Maternal inheritance of extranuclear mitochondrial markers in *Aspergillus nidulans*. *Genet. Res.* **28,** 281–290.

Rowlands, R. T., and Turner, G. (1977). Nuclear–extranuclear interactions affecting oligomycin resistance in *Aspergillus nidulans*. *Mol. Gen. Genet.* **154,** 311–318.

Sainsard-Chanet, A., and Begel, O. (1986). Transformation of yeast and *Podospora:* Innocuity of senescence-specific DNAs. *Mol. Gen. Genet.* **204,** 443–451.

Sanders, J. P. M., Borst, P., and Weijers, P. J. (1975a). The organization of genes in yeast mitochondrial DNA. II. The physical maps of *EcoRI* and *Hind*II+III fragments. *Mol. Gen. Genet.* **143,** 53–64.

Sanders, J. P. M., Heyting, C., and Borst, P. (1975b). The organization of genes in yeast mitochondrial DNA. I. The genes for large and small ribosomal RNA are far apart. *Biochem. Biophys. Res. Commun.* **65,** 699–707.

Sanders, J. P. M., Heuyting, C., DiFranco, A., Borst, P., and Slonimski, P. P. (1976). The organization of genes in yeast mitochondrial DNA. *In* "The Genetic Function of Mitochondrial DNA" (C. Saccone and A. M. Kroon, eds.), pp. 259–272. Elsevier, Amsterdam.

Sanders, J. P. M., Verbeet, M. P., Meijlink, F. C. P. W., Heyting, C., and Borst, P. (1977). The construction of the physical maps of three different *Saccharomyces* mitochondrial DNAs. *Mol. Gen. Genet.* **157,** 271–280.

Scazzocchio, C., Brown, T. A., Waring, R. B., Ray, J. A., and Davies, R. W. (1983). Organization of the *Aspergillus nidulans* mitochondrial genome. *In* "Mitochondria 1983" (R. J. Schweyen, K. Wolf, and F. Kaudewitz, eds.), pp. 303–312. De Gruyter, Berlin.

Schinkel, A. H., Groot Koerkamp, M. J. A., Van der Horst, G. T. J., Touw, E. P. W., Osinga, K. A., Van der Bliek, A. M., Veeneman, G. H., Van Boom, J. H., and Tabak, H. F. (1986). Characterization of the promoter of the large ribosomal RNA gene in yeast mitochondria and separation of mitochondrial RNA polymerase into two different functional components. *EMBO J.* **5,** 1041–1047.

Schinkel, A. H., Groot Koerkamp, M. J. A., Stuiver, M. H., Van der Horst, G. T. J., and Tabak, H. F. (1987). Effect of point mutations on the *in vitro* transcription from the promoter for the large ribosomal RNA gene in yeast mitochondria. *Nucleic Acids Res.* **15,** 5597–5612.

Schmelzer, C., and Müller, M. W. (1987). Self-splicing of group II introns *in vitro:* Lariat formation and 3′ splice site selection in mutant RNAs. *Cell* **51,** 753–762.

Schmelzer, C., and Schweyen, R. J. (1982). Evidence for ribosomes involved in splicing of yeast mitochondrial transcripts. *Nucleic Acids Res.* **10,** 513–524.

Schmelzer, C., and Schweyen, R. J. (1986). Self-splicing of group II introns *in vitro:* Mapping of the branch point and mutational inhibition of lariat formation. *Cell* **46,** 557–565.

Schmelzer, C., Haid, A., Grosch, G., Schweyen, R. J., and Kaudewitz, F. (1981). Pathways of transcript splicing in yeast mitochondria. Mutations in intervening sequences of the split gene *cob* reveal a requirement for intervening sequence-encoded products. *J. Biol. Chem.* **2156,** 7610–7619.

Schmelzer, C., Schmidt, C., and Schweyen, R. J. (1982). Identification of splicing signals in introns of yeast mitochondrial split genes: Mutational alterations in intron bI1 and secondary structures in related introns. *Nucleic Acids Res.* **10,** 6797–6808.

Schmelzer, C., Schmidt, C., May, K., and Schweyen, R. J. (1983). Determination of functional domains in intron bI1 of yeast mitochondrial RNA by studies of mitochondrial mutations and a nuclear suppressor. *EMBO J.* **2,** 2047–2052.

Schmidt, U., Kosack, M., and Stahl, U. (1987). Lariat RNA of a group II intron in a filamentous fungus. *Curr. Genet.* **12,** 291–296.

Schneller, J. M., Accoceberry, B., and Stahl, A: J. C. (1975a). Fractionation of yeast mitochondrial tRNA tyr and tRNA leu. *FEBS Lett.* **53,** 44–48.

Schneller, J. M., Faye, G., Kujawa, C., and Stahl, A. J. C. (1975b). Number of genes and base composition of mitochondrial tRNA from *Saccharomyces cerevisiae. Nucleic Acids Res.* **2,** 831–838.

Schneller, J. M., Martin, R., Stahl, A., and Dirheimer, G. (1975c). Studies of odd bases in yeast mitochondrial tRNA: Absence of the fluorescent "Y" base in mitochondrial DNA coded tRNA phe, absence of 4-thiouridine. *Biochem. Biophys. Res. Commun.* **64,** 1046–1053.

Schneller, J. M., Stahl, A., and Fukuhara, H. (1975d). Coding origin of isoaccepting tRNA in yeast mitochondria. *Biochimie* **57,** 1051–1057.

Schroeder, R., Breitenbach, M., and Schweyen, R. J. (1983). Mitochondrial circular RNAs are absent in sporulating cells of *Saccharomyces cerevisiae. Nucleic Acids Res.* **11,** 1735–1746.

Schweyen, R. J., Steyrer, U., Kaudewitz, F., Dujon, B., and Slonimski, P. P. (1976a). Mapping of mitochondrial genes in *Saccharomyces cerevisiae.* Population and pedigree analysis of retention or loss of four genetic markers in *rho*$^-$ cells. *Mol. Gen. Genet.* **146,** 117–132.

Schweyen, R. J., Weiss-Brummer, B., Backhaus, B., and Kaudewitz, F. (1976b). Localization of seven gene loci on a circular map of the mitochondrial genome of *Saccharomyces cerevisiae. In* "Genetic Function of Mitochondrial DNA" (C. Saccone and A. M. Kroon, eds.), pp. 251–258. Elsevier, Amsterdam.

Schweyen, R. J., Weiss-Brummer, B., Backhaus, B., and Kaudewitz, F. (1977). The genetic map of the mitochondrial genome, including the fine structure of *cob* and *oxi* clusters. *In* "Mitochondria 1977" (W. Bandlow, R. J. Schweyen, K. Wolf, and F. Kaudewitz, eds.), pp. 139–148. De Gruyter, Berlin.

Schweyen, R. J., Weiss-Brummer, B., Backhaus, B., and Kaudewitz, F. (1978). The genetic map of the mitochondrial genome in yeast. Map positions of drug^r and *mit*$^-$ markers as revealed from population analyses of *rho*$^-$ clones in *Saccharomyces cerevisiae. Mol. Gen. Genet.* **159,** 151–160.

Schweyen, R. J., Francisci, S., Haid, A., Ostermayr, R., Rödel, G., Schmelzer, C., Schroeder, R., Weiss-Brummer, B., and Kaudewitz, F. (1982). Transcripts of yeast mitochondrial DNA: Processing of a split-gene transcript and expression of RNA species during adaptation and differentiation processes. *In* "Mitochondrial Genes" (P. Slonimski, P. Borst, and G. Attardi, eds.), pp. 201–212. Cold Spring Harbor Laboratory, Cold Spring Harbor, New York.

Schweyen, R. J., Wolf, K., and Kaudewitz, F., eds. (1983). Nucleo–mitochondrial interactions. "Mitochondria 1983." De Gruyter, Berlin.

Scragg, A. H. (1976). Origin of the mitochondrial RNA polymerase of yeast. *FEBS Lett.* **65,** 148–151.

Sebald, W., and Hoppe, J. (1981). On the structure and genetics of the proteolipid subunit of the ATP synthase complex. *Curr. Top. Bioenerg.* **12,** 1–64.

Sebald, W., and Wachter, E. (1979). Amino-acid sequence of the putative protonophore of the energy transducing ATPase complex. *29th Mosbach. Colloq. Energy Conserv. Biol. Membr.* pp. 228–236.

Sebald, W., Bücher, T., Olbrich, B., and Kaudewitz, F. (1968). Electrophoretic pattern of the amino acid incorporation *in vitro* into the insoluble mitochondrial protein of *Neurospora crassa* wild type and *mi-1* mutant. *FEBS Lett.* **1,** 235–240.

Sebald, W., Graff, T., and Lukins, H. B. (1979a). The dicyclohexylcarbodiimide-binding protein of the mitochondrial ATPase complex from *Neurospora crassa* and *Saccharomyces cerevisiae. Eur. J. Biochem.* **93,** 587–599.

Sebald, W., Wachter, E., and Tzagoloff, A. (1979b). Identification of amino acid substitutions in the dicyclohexylcarbodiimide binding subunit of the mitochondrial ATPase complex from oligomycin resistant mutants of *Saccharomyces cerevisiae. Eur. J. Biochem.* **100,** 599–607.

Sederoff, R. R. (1984). Structural variation in mitochondrial DNA. *Adv. Genet.* **22,** 1–108.

Seitz-Mayr, G., and Wolf, K. (1982). Extrachromosomal mutator inducing point mutations and deletions in mitochondrial genome of fission yeast. *Proc. Natl. Acad. Sci. U.S.A.* **79,** 2618–2622.

Seitz, G., Lückemann, G., Wolf, K., Kaudewitz, F., Boutry, M., and Goffeau, A. (1977a). Extrachromosomal inheritance in *Schizosaccharomyces pombe*. VI. Preliminary genetical and biochemical characterization of mitochondrially inherited respiratory-deficient mutants. *In* "Mitochondria 1977" (W. Bandlow, R. J. Schweyen, K. Wolf, and F. Kaudewitz, eds.), pp. 149–160. De Gruyter, Berlin.

Seitz, G., Wolf, K., and Kaudewitz, F. (1977b). Extrachromosomal inheritance in *Schizosaccharomyces pombe*. IV. Isolation and genetic characterization of mutants resistant to chloramphenicol and erythromycin using the mutator properties of mutant *ana*^{r}8. *Mol. Gen. Genet.* **155,** 339–346.

Seitz-Mayr, G., Wolf, K., and Kaudewitz, F. (1978). Extrachromosomal inheritance in *Schizosaccharomyces pombe*. VII. Studies by zygote clone analysis of transmission, segregation, recombination and uniparental inheritance of mitochondrial markers conferring resistance to antimycin, chloramphenicol, and erythromycin. *Mol. Gen. Genet.* **164,** 309–320.

Sena, E. P., Welch, J. W., Halvorson, H., and Fogel, S. (1975). Nuclear and mitochondrial DNA replication during mitosis in *Saccharomyces cerevisiae*. *J. Bacteriol.* **123,** 497–504.

Sena, E. P., Welch, J., and Fogel, S. (1978). Nuclear and mitochondrial DNA replication during zygote formation and maturation in yeast. *Science* **194,** 433–435.

Sena, E. P., Revet, B., and Moustacchi, E. (1986). *In vivo* homologous recombination intermediates of yeast mitochondrial DNA analyzed by electron microscopy. *Mol. Gen. Genet.* **202,** 421–428.

Seraphin, B., Simon, M., and Faye, G. (1985). A mitochondrial reading frame which may code for a maturase-like protein in *Saccharomyces cerevisiae*. *Nucleic Acids Res.* **13,** 3005–3014.

Seraphin, B., Boulet, A., Simon, M., and Faye, G. (1987a). Construction of a yeast strain devoid of mitochondrial introns and its use to screen nuclear genes involved in mitochondrial splicing. *Proc. Natl. Acad. Sci. U.S.A.* **84,** 6810–6814.

Seraphin, B., Simon, M., and Faye, G. (1987b). The mitochondrial reading frame *RF3* is a functional gene in *Saccharomyces uvarum*. *J. Biol. Chem.* **262,** 10146–10153.

Sevarino, K. A., and Poyton, R. O. (1980). Mitochondrial membrane biogenesis: Identification of a precursor yeast cytochrome *c* oxidase subunit II, an integral polypeptide. *Proc. Natl. Acad. Sci. U.S.A.* **77,** 142–146.

Shannon, C., Enns, R., Wheelis, L., Burchiel, K., and Criddle, R. S. (1973). Alterations in mitochondrial adenosine triphosphatase activity resulting from mutation of mitochondrial deoxyribonucleic acid. *J. Biol. Chem.* **248,** 3004–3001.

Sibler, A.-P., Martin, R. P., and Dirheimer, G. (1979). The nucleotide sequence of yeast mitochondrial histidine-tRNA. *FEBS Lett.* **107,** 182–186.

Sibler, A.-P., Dirheirmer, G., and Martin, R. P. (1983). The primary structure of yeast mitochondrial tyrosine tRNA. *FEBS Lett.* **152,** 153–156.

Sibler, A. P., Dirheimer, G., and Martin, R. P. (1985). Yeast mitochondrial tRNAIle and tRNAMetm: Nucleotide sequence and codon recognition patterns. *Nucleic Acids Res.* **13,** 1341–1346.

Sibler, A.-P., Dirheimer, G., and Martin, R. P. (1986). Codon reading patterns in *Saccharomyces cerevisiae* mitochondria based on sequences of mitochondrial tRNAs. *FEBS Lett.* **194,** 131–138.

Simon, M., and Faye, G. (1984). Organization and processing of the mitochondrial *oxi3/oli2* multigenic transcript in yeast. *Mol. Gen. Genet.* **196,** 266–274.

Singh, A., Mason, T. L., and Zimmermann, R. A. (1978). Cold-sensitive cytoplasmic mutation of *Saccharomyces cerevisiae* affecting assembly of mitochondrial 50S ribosomal subunit. *Mol. Gen. Genet.* **161,** 143–151.

Slonimski, P. P., and Lazowska, J. (1977). Transposable segments of mitochondrial DNA: A unitary hypothesis for the mechanism of mutation, recombination, sequence reiteration and suppressiveness of yeast "petite colony" mutants. *In* "Mitochondria 1977" (W. Bandlow, R. J. Schweyen, K. Wolf, and F. Kaudewitz, eds.), pp. 39–52. De Gruyter, Berlin.

Slonimski, P. P., and Tzagoloff, A. (1976). Localization in yeast mitochondrial DNA of mutations expressed in a deficiency of cytochrome oxidase and/or coenzyme QH_2–cytochrome c reductase. *Eur. J. Biochem.* **61,** 27–41.

Slonimski, P. P., Claisse, M. L., Foucher, M., Jacq, C., Kochko, A., Lamouroux, A., Pajot, P., Perrodin, G., Spyridakis, A., Wambier-Kluppel, M. L. (1978a). Mosaic organization and expression of the mitochondrial DNA region controlling cytochrome *c* reductase and oxidase. III. A model of structure and function. *In* "Biochemistry and Genetics of Yeast" (M. Bacila *et al.,* eds.), pp. 391–401. Academic Press, New York.

Slonimski, P. P., Pajot, P., Jacq, C., Foucher, M., Perroding, G., Kochko, A., and Lamouroux, A. (1978b). Mosaic organization and expression of the mitochondrial DNA region controlling cytochrome *c* reductase and oxidase. I. Genetic, physical and complementation maps of the *box* region. *In* "Biochemistry and Genetics of Yeast" (M. Bacila *et al.,* eds.), pp. 339–368. Academic Press, New York.

Slonimski, P., Borst, P., and Attardi, G., eds. (1982). "Mitochondrial Genes." Cold Spring Harbor Laboratory, Cold Spring Harbor, New York.

Smith, A. E., and Marcker, K. A. (1968). *N*-Formylmethionyl transfer RNA in mitochondria from yeast and rat liver. *J. Mol. Biol.* **38,** 241–243.

Smith, D., Tauro, P., Schweizer, E., and Halvorson, H. O. (1968). The replication of mitochondrial DNA during the cell cycle in *Saccharomyces lactis. Proc. Natl. Acad. Sci. U.S.A.* **60,** 936–942.

Smith, J. R., and Rubenstein, I. (1973). Cytoplasmic inheritance of the timing of "senescence" in *Podospora anserina. J. Gen. Microbiol.* **76,** 297–304.

Solioz, M., and Schatz, G. (1979). Mutations in putative intervening sequences of the mitochondrial cytochrome *b* gene of yeast produce abnormal cytochrome *b* polypeptides. *J. Biol. Chem.* **254,** 9331–9334.

Somlo, M., and Cosson, J. (1976). Mitochondrially encoded oligomycin-resistant mutants in *S. cerevisiae:* Structural integration of ATPase and phenotype. *In* "Genetics and Biogenesis of Chloroplasts and Mitochondria" (T. Bücher, W. Neupert, W. Sebald, and S. Werner, eds.), pp. 143–150. Elsevier, Amsterdam.

Somlo, M., Avner, P. R., Cosson, J., Dujon, B., and Krupa, M. (1974). Oligomycin sensitivity of ATPase studied as a function of mitochondrial biogenesis, using mitochondrially determined oligomycin-resistant mutants of *Saccharomyces cerevisiae. Eur. J. Biochem.* **42,** 439–445.

Somlo, M., Clavilier, L., Dujon, B., and Kermorgant, M. (1985). The *pho1* mutation. A frameshift, and its compensation, producing altered forms of physiologically efficient ATPase in yeast mitochondria. *Eur. J. Biochem.* **150,** 89–94.

Sor, F., and Faye, G. (1979). Mitochondrial and nuclear mutations that affect the biogenesis of the mitochondrial ribosomes of yeast. 2. Biochemistry. *Mol. Gen. Genet.* **177,** 47–56.

Sor, F., and Fukuhara, H. (1980). Séquence nucléotidique du gène de l'ARN ribosomique 15S mitochondrial de la levure. *C. R. Acad. Sci. Paris Ser. D* **291,** 933–936.

Sor, F., and Fukuhara, H. (1982a). Nature of an inserted sequence in the mitochondrial gene coding for the 15S ribosomal RNA of yeast. *Nucleic Acids Res.* **10,** 1625–1633.

Sor, F., and Fukuhara, H. (1982b). Identification of two erythromycin resistance mutations in the mitochondrial gene coding for the large ribosomal RNA in yeast. *Nucleic Acids Res.* **10,** 6571–6577.

Sor, F., and Fukuhara, H. (1982c). Nucleotide sequence of the small ribosomal RNA gene from the mitochondria of *Saccharomyces cerevisiae*. *In* "Mitochondrial Genes" (P. Slonimski, P. Borst, and G. Attardi, eds.), pp. 255–262. Cold Spring Harbor Laboratory, Cold Spring Harbor, New York.

Sor, F., and Fukuhara, H. (1983a). Complete DNA sequence coding for the large ribosomal RNA of yeast mitochondria. *Nucleic Acids Res.* **11,** 339–348.

Sor, F., and Fukuhara, H. (1983b). Unequal excision of complementary strands is involved in the generation of palindromic repetitions of *rho*⁻ mitochondrial DNA in yeast. *Cell* **32,** 391–396.

Spithill, T. W., Trembath, M. K., Lukins, H. B., and Linnane, A. W. (1978). Mutations of the mitochondrial DNA of *Saccharomyces cerevisiae* which affect the interaction between mitochondrial ribosomes and the inner mitochondrial membrane. *Mol. Gen. Genet.* **164,** 155–162.

Spithill, T. W., Nagley, P., and Linnane, A. W. (1979). Biogenesis of mitochondria. 51. Biochemical characterization of a mitochondrial mutation in *Saccharomyces cerevisiae* affecting the mitochondrial ribosome by conferring resistance to aminoglycoside antibiotics. *Mol. Gen. Genet.* **173,** 159–170.

Spyridakis, A., and Claisse, M. (1978). Yeast cytochrome *b:* structural modifications in yeast mitochondrial mutants. *In* "Plant Mitochondria" (G. Doucet and C. Lance, eds.), pp. 11–18. Elsevier, Amsterdam.

Sriprakash, K. S., and Batum, C. (1981). Segregation and transmission of mitochondrial markers in fusion products of the asporogenous yeast *Torulopsis glabrata. Curr. Genet.* **4,** 73–80.

Sriprakash, K. S., and Clark-Walker, G. D. (1980). The size of yeast mitochondrial ribosomal RNAs. *Biochem. Biophys. Res. Commun.* **93,** 186–193.

Sriprakash, K. S., and Clark-Walker, G. D. (1983). Multiple start-transcription sites in *Torulopsis glabrata* mitochondrial DNA. *In* "Manipulation and Expression of Genes in Eukaryotes" (P. Nagley *et al.,* eds.), pp. 303–310. Academic Press, Sidney.

Sriprakash, K. S., Choo, K. B., Nagley, P., and Linnane, A. W. (1976a). Physical mapping of mitochondrial rRNA genes in *Saccharomyces cerevisiae. Biochem. Biophys. Res. Commun.* **69,** 85–91.

Sriprakash, K. S., Molloy, P. L., Nagley, P., Lukins, H. B., and Linnane, A. W. (1976b). Biogenesis of mitochondria. XLI. Physical mapping of mitochondrial genetic markers in yeast. *J. Mol. Biol.* **4,** 485–503.

Stahl, U., Lemke, P. A., Tudzynski, P., Kück, U., and Esser, K. (1978). Evidence for plasmid-like DNA in a filamentous fungus, ascomycete *Podospora anserina. Mol. Gen. Genet.* **162,** 341–343.

Stahl, U., Tudzynski, P., Kück, U., and Esser, K. (1979). Plasmid-like DNA in senescent cultures of the ascomycetous fungus *Podospora anserina. Hoppe-Seylers Z. Physiol. Chem.* **360,** 1045–1050.

Stahl, U., Kück, U., Tudzynski, P., and Esser, K. (1980). Characterization and cloning of plasmid like DNA of the ascomycete *Podospora anserina. Mol. Gen. Genet.* **178,** 639–646.

Stahl, U., Tudzynski, P., Kück, U., and Esser, K. (1982). Replication and expression of a bacterial–mitochondrial hybrid plasmid in the fungus *Podospora anserina. Proc. Natl. Acad. Sci. U.S.A.* **79,** 36141–3645.

Steinhilber, W., and Cummings, D. J. (1986). A DNA polymerase activity with characterisitics of a reverse transcriptase in *Podospora anserina. Curr. Genet.* **10,** 389–392.

Stephenson, G., Marzuki, S., and Linnane, A. W. (1981a). Biogenesis of mitochondria. Defective assembly of the proteolipid into the mitochondrial adenosine triphosphatase complex in an *oli2mit*⁻ mutant of *Saccharomyces cerevisiae*. *Biochim. Biophys. Acta* **636,** 104–112.

Stephenson, G., Marzuki, S., and Linnane, A. W. (1981b). *Mit*⁻ mutants in the structural gene of subunit III of cytochrome oxidase. *Biochim. Biophys. Acta* **653,** 416–422.

Stepien, P. P. (1982). DNA sequence of the region adjacent to the *oxi1* and *oxi2* genes from *Aspergillus nidulans* mitochondrial DNA. *Acta Biochem. Pol.* **29,** 143–149.

Stepien, P. P., Bernard, U., Cooke, H. J., and Küntzel, H. (1978). Restriction endonuclease cleavage map of mitochondrial DNA from *Aspergillus nidulans*. *Nucleic Acids Res.* **5,** 317–330.

Stiegler, P., Carbon, P., Zuker, M., Ebel, J. B., and Ehresmann, C. (1980). Structure secondaire et topographie du RNA ribosomique 16S d'*Escherichia coli*. *C. R. Acad. Sci Paris Ser. D* **291,** 937–940.

Stiegler, P., Carbon, P., Ebel, P., and Ehresmann, C. (1981). A general secondary-structure model for procaryotic and eucaryotic RNAs of the small ribosomal subunits. *Eur. J. Biochem.* **120,** 487–495.

Stohl, L. L., Akins, R. A., and Lambowitz, A. M. (1984). Characterization of deletion derivatives of an autonomously replicating *Neurospora* plasmid. *Nucleic Acids Res.* **12,** 6169–6178.

Storm, E. M., and Marmur, J. (1975). A temperature sensitive mitochondrial mutation of *Saccharomyces cerevisiae*. *Biochem. Biophys. Res. Commun.* **64,** 752–759.

Strausberg, S. L., and Birky, C. W., Jr. (1979). Recombination of yeast mitochondrial DNA does not require mitochondrial protein synthesis. *Curr. Genet.* **1,** 21–31.

Strausberg, R. L., and Butow, R. A. (1977). Expression of petite mitochondrial DNA *in vivo*—zygotic gene rescue. *Proc. Natl. Acad. Sci. U.S.A.* **74,** 2715–2719.

Strausberg, R. L., and Butow, R. A. (1981). Gene conversion at the *var1* locus on yeast mitochondrial DNA. *Proc. Natl. Acad. Sci. U.S.A.* **78,** 494–498.

Strausberg, R. L., Vincent, R. D., Perlman, P. S., and Butow, R. A. (1978). Asymmetric gene conversion at inserted segments on yeast mitochondrial DNA. *Nature (London)* **276,** 577–583.

Stuart, K. D. (1970). Cytoplasmic inheritance of oligomycin and rutamycin resistance in yeast. *Biochem. Biophys. Res. Commun.* **39,** 1045–1051.

Subik, J. (1975). Mucidin-resistant antimycin A-sensitive mitochondrial mutant of *Saccharomyces cerevisiae*. *FEBS Lett.* **59,** 273–276.

Subik, J. (1976). Mitochondrial inheritance of mucidin resistance in yeast. *In* "Genetics and Biogenesis of Chloroplasts and Mitochondria" (T. Bücher W. Neupert, W. Sebald, and S. Werner, eds.), pp. 473–478. Elsevier, Amsterdam.

Subik, J., and Goffeau, A. (1980). Cytochrome *b*-deficiency in a mitochondrial *muc1muc2* recombinant of *Saccharomyces cerevisiae*. *Mol. Gen. Genet.* **178,** 603–610.

Subik, J., and Goffeau, A. (1981). Mitochondrial translation products of yeast mutants resistant to mucidin. *Folia Microbiol.* **26,** 300–306.

Subik, J., and Takacsova, G. (1978). Genetic determination of ubiquinol–cytochrome *c* reductase—mitochondrial locus *muc3* specifying resistance of *Saccharomyces cerevisiae* to mucidin. *Mol. Gen. Genet.* **161,** 99–108.

Subik, J., Behun, M., and Musilek, V. (1974). Antibiotic mucidin, a new antimycin A-like inhibitor of electron transport in rat liver mitochondria. *Biochem. Biophys. Res. Commun.* **57,** 17–22.

Subik, J., Kovacova, V., and Takacsova, G. (1977). Mucidin resistance in yeast. Isolation, characterization and genetic analysis of nuclear and mitochondrial mucidin-resistant mutants of *Saccharomyces cerevisiae*. *Eur. J. Biochem.* **73,** 275–286.

Subik, J., Briquet, M., and Goffeau, A. (1981). Spectral properties of cytochrome *b-561* and cytochrome *b-565* in mucidin-resistant mutants of *Saccharomyces cerevisiae*. *Eur. J. Biochem.* **119,** 613–618.

Szostak, J., Orr-Weaver, T. L., and Rothstein, R. J. (1983). The double strand break repair model for recombination. *Cell* **33,** 25–35.

Tabak, H. F., and Arnberg, A. C. (1986). Splicing of yeast mitochondrial precursor RNAs. *In* "Oxford Surveys on Eukaryotic Genes" (N. Maclean, ed.), Vol. 3, pp. 161–182. Oxford Univ. Press, London and New York.

Tabak, H. F., and Grivell, L. A. (1986). RNA catalysis in the excision of yeast mitochondrial introns. *Trends Genet.* **2,** 51–54.

Tabak, H. F., Hecht, N. B., Menke, H. H., and Hollenberg, C. P. (1979). The gene for the small ribosomal RNA on yeast mitochondrial DNA: Physical map, direction of transcription and absence of an intervening sequence. *Curr. Genet.* **1,** 33–43.

Tabak, H. F., Van der Laan, J. C., Osinga, K. A., Schooten, J. P., Van Boom, J. H., and Veeneman, G. H. (1981). Use of a synthetic DNA oligonucleotide to probe the precision of RNA splicing in a yeast mitochondrial petite mutant. *Nucleic Acids Res.* **9,** 4475–4483.

Tabak, H. F., Van der Laan, J. C., Landegent, J. E., Evers, R. F., and Wassenaar, G. M. (1982). Mitochondrially encoded resistance to paromomycin in *Saccharomyces cerevisiae:* Reinvestigation of a controversy. *Plasmid* **8,** 261–275.

Tabak, H. F., Grivell, L. A., and Borst, P. (1983a). Transcription of mitochondrial DNA. *CRC Crit. Rev. Biochem.* **14,** 297–302.

Tabak, H. F., Osinga, K. A., De Vries, E., Van der Bliek, A. M., Van der Horst, G. T. J., Groot Koerkamp, M. J. A., Van der Horst, E. C., Zwarthoff, E. V., and MacDonald, M. E. (1983b). Initiation of transcription of yeast mitochondrial DNA. *In* "Mitochondria 1983" (R. J. Schweyen, K. Wolf, and F. Kaudewitz, eds.), pp. 79–94. De Gruyter, Berlin.

Tabak, H. F., Van der Horst, G., Osinga, K., and Arberg, A. C. (1984). Splicing of large ribosomal precursor RNA and processing of intron RNA in yeast mitochondria. *Cell* **39,** 623–629.

Tabak, H. F., Schinkel, A. H., Groot Koerkamp, M. J. A., Van der Horst, G. T. J., Van der Horst, G., and Arnberg, A. C. (1985). The large ribosomal RNA gene of mitochondrial DNA of *Saccharomyces cerevisiae: in vitro* initiation of transcription and self-splicing of precursor RNA. *In* "Achievements and Perspectives of Mitochondrial Research" (E. Quagliariello, E. C. Slater, F. Palmieri, C. Saccone, and A. M. Kroon, eds.), Vol. 2, pp. 183–192. Elsevier, Amsterdam.

Tabak, H. F., Van der Horst, G., Kamps, A. M. J. E., and Arnberg, A. C. (1987). Interlocked RNA circle formation by a self-splicing yeast mitochondrial group I intron. *Cell* **48,** 101–110.

Takacsova, G., Subik J., and Kotylak, Z. (1980). Localization of mucidin-resistant locus *muc3* on mitochondrial DNA with respect of ubiquinol–cytochrome *c* reductase deficient *box* loci. Locus *muc3* is allelic to *box2*. *Mol. Gen. Genet.* **179,** 141–146.

Tanner, N. K., and Cech, T. R. (1985a). Self-catalyzed cyclization of the intervening sequence RNA of *Tetrahymena:* Inhibition by intercalating dyes. *Nucleic Acids Res.* **13,** 7741–7758.

Tanner, N. K., and Cech, T. R. (1985b). Self-catalyzed cyclization of the intervening sequence RNA of *Tetrahymena:* Inhibition by methidiumpropyl·EDTA and location of the major dye binding sites. *Nucleic Acids Res.* **13,** 7759–7779.

Taylor, J. W., and Smolich, B. D. (1985). Molecular cloning and physical mapping of the *Neurospora crassa* 74-OR23-1A mitochondrial genome. *Curr. Genet.* **9,** 597–604.

Terpstra, P., and Butow, R. A. (1979). The role of *var1* in the assembly of yeast mitochondrial ribosomes. *J. Biol. Chem.* **254,** 12662–12669.

Terpstra, P., Holtrop, M., and Kroon, A. M. (1976). Restriction fragment map of *Neurospora crassa* mitochondrial DNA. *In* "The Genetic Functions of Mitochondrial DNA" (C. Saccone and A. M. Kroon, eds.), pp. 111–118. Elsevier, Amsterdam.

Terpstra, P., De Vries, H., and Kroon, A. M. (1977a). Properties and genetic localization of mitochondrial transfer RNAs of *Neurospora crassa. In* "Mitochondria 1977" (W. Bandlow, R. J. Schweyen, K. Wolf, and F. Kaudewitz, eds.), pp. 291–302. De Gruyter, Berlin.

Terpstra, P., Holtrop, M., and Kroon, A. M. (1977b). The ribosomal RNA genes on *Neurospora crassa* mitochondrial DNA are adjacent. *Biochim. Biophys. Acta* **478,** 146–155.

Terpstra, P., Holtrop, M., and Kroon, A. M. (1977c). A complete cleavage map of *Neurospora crassa* mtDNA obtained with endonucleases *Eco*RI and *Bam*HI. *Biochim. Biophys. Acta* **475,** 571–588.

Terpstra, P., Zanders, E., and Butow, R. A. (1979). The association of *var1* with the 38S mitochondrial ribosomal subunit of yeast. *J. Biol. Chem.* **254,** 12653–12661.

Thalenfeld, B. E., and Tzagoloff, A. (1980). Assembly of the mitochondrial membrane system. Sequence of the *oxi2* gene of yeast mitochondrial DNA. *J. Biol. Chem.* **255,** 6173–6180.

Thalenfeld, B., Hille, E., and Tzagoloff, A. (1983). Assembly of the mitochondrial membrane system. Characterization of the *oxi2* transcript and localization of its promoter in *Saccharomyces cerevisiae. J. Biol. Chem.* **258,** 610–615.

Thierbach, G., and Michaelis, G. (1982). Mitochondrial and nuclear myxothiazol resistance in *Saccharomyces cerevisiae. Mol. Gen. Genet.* **186,** 501–506.

Thierbach, G., and Reichenbach, H. (1981). Myxothiazol, a new inhibitor of the cytochrome b-c_1 segment of the respiratory chain. *Biochim. Biophys. Acta* **638,** 282–289.

Thomas, D. Y., and Wilkie, D. (1968). Inhibition of mitochondrial synthesis in yeast by erythromycin: Cytoplasmic and nuclear factors controlling resistance. *Genet. Res.* **11,** 33–41.

Thrailkill, K., Birky, C. W., Jr., Lückemann, G., and Wolf, K. (1980). Intracellular population genetics: Evidence for random drift of mitochondrial allele frequencies in *Saccharomyces cerevisiae* and *Schizosaccharomyces pombe. Genetics* **96,** 237–262.

Tikomirova, L. P., Krynkoy, V. M., Stizkov, N. I., and Bayev, A. A. (1983). Mt DNA sequences of *Candida utilis* capable of supporting autonomous replication of plasmids in *Saccharomyces cerevisiae. Mol. Gen. Genet.* **189,** 479–484.

Tissieres, H., and Mitchell, H. K. (1954). Cytochromes and respiratory activities in some slow growing strains of *Neurospora. J. Biol. Chem.* **208,** 241–249.

Tissieres, H., Mitchell, H. K., and Haskins, F. A. (1953). Studies on the respiratory system of the *poky* strain of *Neurospora. J. Biol. Chem.* **205,** 423–433.

Trembath, M. K., Bunn, C. L., Lukins, H. B., and Linnane, A. W. (1973). Biogenesis of mitochondria. 27. Genetic and biochemical characterization of cytoplasmic and nuclear mutations to spiramycin resistance in *Saccharomyces cerevisiae. Mol. Gen. Genet.* **121,** 35–48.

Trembath, M. K., Molloy, P. L., Sriprakash, K. S., Cuttin, G. J., Linnane, A. W., and Lukins, H. B. (1976). Biogenesis of mitochondria. 44. Comparative studies and mapping of mitochondrial oligomycin resistance mutations in yeast based on gene recombination and petite deletion analysis. *Mol. Gen. Genet.* **145,** 43–52.

Trembath, M. K., Monk, B. C., Kellerman, G. M., and Linnane, A. W. (1975). Biogenesis of mitochondria. 36. Genetic and biochemical analysis of a mitochondrially determined cold sensitive oligomycin resistant mutant of *Saccharomyces cerevisiae* with affected mitochondrial ATPase assembly. *Mol. Gen. Genet.* **141,** 9–22.

Trembath, M. K., Macino, G., and Tzagoloff, A. (1977). The mapping of mutations in tRNA and cytochrome oxidase genes located in the *cap–par* segment of the mitochondrial genome of *S. cerevisiae. Mol. Gen. Genet.* **158,** 35–45.

Trinkl, H., and Wolf, K. (1986). The mosaic *cox1* gene in the mitochondrial genome of *Schizosaccharomyces pombe:* Minimal structural requirements and evolution of group I introns. *Gene* **45,** 289–297.

Trinkl, H., and Wolf, K. (1987). Submitted.

Trinkl, H., Lang, B. F., and Wolf, K. (1985). The mitochondrial genome of the fission yeast *Schizosaccharomyces pombe.* 7. Continuous gene for apocytochrome *b* in strain EF1 (CBS 356) and sequence variation in the region of intron insertion in strain ade7-50h−. *Mol. Gen. Genet.* **198,** 360–363.

Tsai, M.-J., Michaelis, G., and Criddle, R. S. (1971). DNA-dependent RNA polymerase from yeast mitochondria. *Proc. Natl. Acad. Sci. U.S.A.* **68,** 473–477.

Tudzynski, P., and Esser, K. (1977). Inhibitors of mitochondrial function prevent senescence in the ascomycete *Podspora anserina. Mol. Gen. Genet.* **153,** 111–113.

Tudzynski, P., and Esser, K. (1979). Chromosomal and extrachromosomal control of senescence in the ascomycete *Podospora anserina. Mol. Gen. Genet.* **173,** 71–84.

Tudzynski, P., Stahl, U., and Esser, K. (1980). Transformation to senescence with plasmid like DNA in the ascomycete *Podospora anserina. Curr. Genet.* **2,** 181–184.

Tudzynski, P., Stahl, U., and Esser, K. (1982). Development of a eukaryotic cloning system in *Podospora anserina. Curr. Genet.* **6,** 219–222.

Tuite, M. F., and McLaughlin, C. S. (1982). Endogenous read-through of a UGA termination codon in a *Saccharomyces cerevisiae* cell-free system: Evidence for involvement of both mitochondrial and nuclear tRNA. *Mol. Cell. Biol.* **2,** 490–497.

Turker, M. A., and Cummings, D. J. (1987). *Podospora anserina* does not senesce when serially passaged in liquid culture. *J. Bacteriol.* **169,** 454–460.

Turker, M. S., Nelson, J. G., and Cummings, D. J. (1987). A *Podospora anserina* longevity mutant with a temperature-sensitive phenotype for senescence. *Mol. Cell Biol.* **7,** 3199–3204.

Turner, G., and Rowlands, R. T. (1977). Mitochondrial genetics of *Aspergillus nidulans. In* "Genetics and Physiology of *Aspergillus* (J. E. Smith and J. A. Pateman, eds.), pp. 319–337. Academic Press, New York.

Turner, G., and Watson, C. M. J. (1976). A new extranuclear oligomycin-resistance locus in *Aspergillus nidulans. Heredity* **37,** 151–158.

Turner, G., Imam, G., and Küntzel, H. (1979). Mitochondrial ATPase complex of *Aspergillus nidulans* and the dicyclohexylcarbodiimide-binding protein. *Eur. J. Biochem.* **97,** 565–571.

Turner, G., Earl, A. J., and Greaves, D. R. (1982). Interspecies variation and recombination of mitochondrial DNA in the *Aspergillus nidulans* species group and the selection of species-specific sequences by nuclear background. *In* "Mitochondrial Genes" (P. Slonimski, P. Borst, and G. Attardi, eds.), pp. 411–414. Cold Spring Harbor Laboratory, Cold Spring Harbor, New York.

Tzagoloff, A., and Myers, A. M. (1986). Genetics of mitochondrial biosynthesis. *Annu. Rev. Biochem.* **55,** 249–286.

Tzagoloff, A., and Nobrega, F. G. (1980). Structure and nucleotide sequence of the cytochrome *b* gene in yeast mitochondrial DNA. *In* "Biological Chemistry of Organelle Formation" (T. Bücher *et al.,* eds.), pp. 2–9. Springer-Verlag, Berlin.

Tzagoloff, A., Akai, A., and Needleman, R. B. (1975a). Properties of cytoplasmic mutants of *Saccharomyces cerevisiae* with specific lesions in cytochrome oxidase. *Proc. Natl. Acad. Sci. U.S.A.* **72,** 2054–2057.

Tzagoloff, A., Akai, A., and Needleman, R. B. (1975b). Assembly of the mitochondrial membrane system: Isolation of nuclear and cytoplasmic mutants of *Saccharomyces cerevisiae* with specific defects in mitochondrial functions. *J. Bacteriol.* **122,** 826–831.

Tzagoloff, A., Akai, A., and Foury, F. (1976a). Assembly of the mitochondrial membrane system. XVI. Modified form of the ATPase proteolipid in oligomycin-resistant mutants of *Saccharomyces cerevisiae. FEBS Lett.* **65,** 391–395.

Tzagoloff, A., Foury, F., and Akai, A. (1976b). Assembly of the mitochondrial membrane system. XVIII. Genetic loci on mitochondrial DNA involved in cytochrome *b* biosynthesis. *Mol. Gen. Genet.* **149,** 33–42.

Tzagoloff, A., Macino, G., and Sebald, W. (1979). Mitochondrial genes and translation products. *Annu. Rev. Biochem.* **48,** 419–441.

Tzagoloff, A., Bonitz, S., Coruzzi, G., Thalenfeld, B., and Macino, G. (1980a). Yeast mitochondrial cytochrome oxidase gene. *In* "The Organization and Expression of the Mitochondrial Genome" (A. M. Kroon and C. Saccone, eds.), pp. 181–190. Elsevier, Amsterdam.

Tzagoloff, A., Nobrega, M., Akai, A., and Macino, G. (1980b). Assembly of the mitochondrial membrane system. Organization of yeast mitochondrial DNA in the *oli1* region. *Curr. Genet.* **2,** 149–157.

Underbrink-Lyon, K., Miller, D. L., Ross, N. A., Fukuhara, H., and Martin, N. C. (1983). Characterization of a yeast mitochondrial locus necessary for tRNA biosynthesis. Deletion mapping and restriction mapping studies. *Mol. Gen. Genet.* **191,** 512–518.

Van den Boogaart, P., Van Dijk, S., and Agsteribbe, E. (1982a). The mitochondrially made subunit 2 of *Neurospora crassa* cytochrome aa_3 is synthesized as a precursor protein. *FEBS Lett.* **147,** 97–100.

Van den Boogaart, P., Samallo, J., and Agsteribbe, E. (1982b). Similar genes for a mitochondrial ATPase subunit in the nuclear and mitochondrial genome of *Neurospora crassa. Nature* (*London*) **298,** 187–189.

Van den Boogaart, P., Samallo, J., van Dijk, S., and Agsteribbe, E. (1982c). Structural and functional analysis of the genes for subunit II of cytochrome aa_3 and for a dicyclohexylcarbodiimide-binding protein in *Neurospora crassa* mitochondrial DNA. *In* "Mitochondria Genes" (P. Slonimski, P. Borst, and G. Attardi, eds.), pp. 375–380. Cold Spring Harbor Laboratory, Cold Spring Harbor, New York.

Van der Horst, G., and Tabak, H. F. (1985). Self-splicing of yeast mitochondrial ribosomal and messenger RNA precursors. *Cell* **40,** 759–766.

Van der Horst, G., and Tabak, H. F. (1987). New RNA-mediated reactions by yeast mitochondrial group I introns. *EMBO J.* **6,** 2139–2144.

Van der Veen, R., Arnberg, A. C., Van der Horst, G., Bonen, L., Tabak, H. F., and Grivell, L. A. (1986). Excised group II introns in yeast mitochondria are lariats and can be formed by self-splicing *in vitro. Cell* **44,** 225–234.

Van der Veen, R., Arnberg, A. C., and Grivell, L. A. (1987). Self-splicing of a group II intron in yeast mitochondria: Dependence on 5′ exon sequences. *EMBO J.* **6,** 1079–1084.

Van Kreijl, C. F., and Bos, J. L. (1977). Repeating nucleotide sequence in repetitive mitochondrial DNA from a low-density petite mutant of yeast. *Nucleic Acids Res.* **4,** 2369–2376.

Van Ommen, G.-J. B., Groot, G. S. P., and Borst, P. (1977). Fine structure physical mapping of 4S RNA genes on mitochondrial DNA of *Saccharomyces cerevisiae. Mol. Gen. Genet.* **154,** 255–262.

Van Ommen, G.-J. B., Groot, G. S. P., and Grivell, L. A. (1979). Transcription maps of mtDNAs of two strains of *Saccharomyces:* Transcription of strain-specific insertions: Complex RNA maturation and splicing. *Cell* **18,** 511–523.

Van Ommen, G.-J. B., De Boer, P. H., Groot, G. S. P., de Haan, M., Roosendaal, E., Grivell, L. A., Haid, A., and Schweyen, R. J. (1980). Mutations affecting RNA splicing and the interaction of gene expression of the yeast mitochondrial loci *cob* and *oxi-3*. *Cell* **20,** 173–183.

Velours, J., Esparza, M., Hoppe, J., Sebald, W., and Guerin, B. (1984). Amino acid sequence of a new mitochondrially synthesized proteolipid of the ATP synthase of *Saccharomyces cerevisiae. EMBO J.* **3,** 207–212.

Verma, I. M., Edelman, M., Herzog, M., and Littauer, U. (1970). Size determination of mitochondrial ribosomal RNA from *Aspergillus nidulans* by electron microscopy. *J. Mol. Biol.* **52,** 137–140.

Verma, I. M., Edelman, M., and Littauer, U. Z. (1971). A comparison of nucleotide sequences from mitochondrial and cytoplasmic RNA of *Aspergillus nidulans. Eur. J. Biochem.* **19,** 124–129.

Vierny, C., Begel, O., Keller, A.-M., Raynal, A., and Belcour, L. (1980). Senescence specific DNA of *Podospora anserina:* Its variability and its relation with mitochondria. *In* "The Organization and Expression of the Mitochondrial Genome" (A. M. Kroon and C. Saccone, eds.), pp. 91–95. Elsevier, Amsterdam.

Vierny, C., Keller, A.-M., Begel, O., and Belcour, L. (1982). A sequence of mitochondrial DNA is associated with the onset of senescence in a fungus. *Nature (London)* **297,** 157–159.

Vincent, R. D., Perlman, P. S., Strausberg, R. L., and Butow, R. A. (1980). Physical mapping of genetic determinant on yeast mitochondrial DNA affecting the apparent size of the *var1* polypeptide. *Curr. Genet.* **2,** 27–38.

Visconti, N., and Delbrück, M. (1953). The mechanism of genetic recombination in phage. *Genetics* **38,** 5–33.

Wachter, E., Sebald, W., and Tzagoloff, A. (1977). Altered amino acid sequence of the DCCD-binding protein in the *oli-1* resistant mutant D273-10b/A21 of *Saccharomyces cerevisiae. In* "Mitochondria 1977" (W. Bandlow, R. J. Schweyen, K. Wolf, and F. Kaudewitz, eds.), pp. 441–450. De Gruyter, Berlin.

Wakabayashi, K. (1972). Oligomycin resistance in yeast. II. Change in mitochondrial ATPase of a mutant and its genetic character. *J. Antibiot.* **8,** 475–476.

Wakabayashi, K. (1976). A segment of mitochondrial DNA carrying oligomycin resistance. *In* "Genetics and Biogenesis of Chloroplasts and Mitochondria" (T. Bücher, W. Neupert, W. Sebald, and S. Werner, eds.), pp. 535–538. Elsevier, Amsterdam.

Wakabayashi, K. (1978). Segments of mitochondrial DNA of yeast carrying antibiotic resistances. *Mol. Gen. Genet.* **159,** 229–238.

Wakabayashi, K., and Furutani, Y. (1984). Inversions in mitochondrial DNA of a petite mutant of yeast. *J. Biochem.* **96,** 1559–1564.

Wakabayashi, K., and Kamei, S. (1973). Oligomycin resistance in yeast. Linkage of the mitochondrial drug resistance. *FEBS Lett.* **33,** 263–265.

Wakabayashi, K., and Kamei, S., (1974). Oligomycin resistance in yeast. IV. Loss of mitochondrial genome in a nuclear petite. *J. Antibiot.* **27,** 729–730.

Wakabayashi, K., and Mabuchi, T. (1984). Nucleotide sequence involved in the replication of cloned yeast mitochondrial DNA. *J. Biochem* **96,** 171–177.

Waldron, C., and Roberts, C. F. (1973). Cytoplasmic inheritance of cold-sensitive mutants in *Aspergillus nidulans. J. Gen. Microbiol.* **78,** 379–381.

Waldron, C., and Roberts, C. F. (1974a). Cold-sensitive mutants in *Aspergillus nidulans*. I. Isolation and general characterization. *Mol. Gen. Genet.* **134,** 99–113.

Waldron, C., and Robert, C. F. (1974b). Cold-sensitive mutants in *Aspergillus nidulans*. II. Mutations affecting ribosome production. *Mol. Gen. Genet.* **134,** 115–132.

Wallace, D. C. (1982). Structure and evolution of organelle genomes. *Microbiol. Rev.* **46,** 208–240.

Waring, R. B., and Davies, R. W. (1984). Assessment of a model for intron RNA secondary structure relevant to RNA self-splicing—a review. *Gene* **28,** 277–291.

Waring, R. B., and Scazzocchio, C. (1980). Nuclear and mitochondrial suppression of a mitochondrially inherited cold-sensitive mutation in *Aspergillus nidulans. J. Gen. Microbiol.* **119,** 297–311.

Waring, R. B., and Scazzocchio, C. (1983). Mitochondrial four-point crosses in *Aspergillus nidulans:* Mapping of a suppressor of a mitochondrially inherited cold-sensitive mutation. *Genetics 103,* 409–428.

Waring, R. B., Davies, R. W., Lee, S., Grisi, E., Berks, M. M., and Scazzocchio, C. (1981). The mosaic organization of the apocytochrome *b* gene of *Aspergillus nidulans* revealed by DNA sequencing. *Cell* **27,** 4–11.

Waring, R. B., Davies, R. W., Scazzocchio, C., and Brown, T. A. (1982). Internal structure of a mitochondrial intron of *Aspergillus nidulans. Proc. Natl. Acad. Sci. U.S.A.* **79,** 6332–6336.

Waring, R. B., Scazzocchio, C., Brown, T. A., and Davies, R. W. (1983). Close relationship between certain nuclear and mitochondrial introns. Implications for the mechanism of RNA splicing. *J. Mol. Biol.* **167,** 595–605.

Waring, R. B., Brown, T. A., Ray, J. A., Scazzocchio, C., and Davies, R. W. (1984). Three variant introns of the same general class in the mitochondrial gene for cytochrome oxidase subunit 1 in *Aspergillus nidulans. EMBO J.* **3,** 2121–2128.

Waring, R. B., Ray, J. A., Edwards, S. W., Scazzocchio, C., and Davies, R. W. (1985). The *Tetrahymena* rRNA intron self-splices in *E. coli: In vivo* evidence for the importance of key base-paired regions of RNA for RNA enzyme function. *Cell* **40,** 371–380.

Waring, R. B., Towner, P., Minter, S. J., and Davies, R. W. (1986). Splice-site selection by a self-splicing RNA of *Tetrahymena. Nature (London)* **321,** 133–138.

Weiss-Brummer, B., Guba, R., Haid, A., and Schweyen, R. J. (1979). Fine structure of *OXI1,* the mitochondrial gene coding for subunit II of yeast cytochrome *c* oxidase. *Curr. Genet.* **1,** 75–83.

Weiss-Brummer, B., Rödel, G., Schweyen, R. J., and Kaudewitz, F. (1982). Expression of the split gene *cob* in yeast: Evidence for a precursor of a "maturase" protein translated from intron 4 and preceding exons. *Cell* **29,** 527–536.

Weiss-Brummer, B., Holl, J., Schweyen, R. J., Rödel, G., and Kaudewitz, F. (1983). Processing of yeast mitochondrial RNA: Involvement of intramolecular hybrids in splicing of *cob* intron 4 RNA by mutation and reversion. *Cell* **33,** 195–202.

Weiss-Brummer, B., Sakai, H., and Kaudewitz, F. (1987). A mitochondrial frameshift-suppressor (+) of the yeast *S. cerevisiae* maps in the mitochondrial 15S rRNA locus. *Curr. Genet.* **11,** 295–302.

Wells, J. R. (1974). Mitochondrial DNA synthesis during the cell cycle of *Saccharomyces cerevisiae. Exp. Cell Res.* **85,** 278–286.

Wesolowski, M., and Fukuhara, H. (1979). The genetic map of transfer RNA genes of yeast mitochondria: Correction and extension. *Mol. Gen. Genet.* **170,** 261–275.

Wesolowski, M., and Fukuhara, H. (1981). Linear mitochondrial deoxyribonucleic acid from the yeast *Hansenula mrakii. Mol. Cell. Biol.* **1,** 387–393.

Wesolowski, M., Monnerot, M., and Fukuhara, H. (1980). *Mbo*I, *Th*I and *Hin*fI endonuclease cleavage maps of the yeast mitochondrial DNA. Localization of transfer RNA genes. *Curr. Genet.* **2,** 121–129.

Wesolowski, M., Algeri, A., and Fukuhara, H. (1981a). Gene organization of the mitochondrial DNA of yeasts: *Kluyveromyces lactis* and *Saccharomycopsis lipolytica. Curr. Genet.* **3,** 157–162.

Wesolowski, M. Palleschi, C., Agostinelli, M., Frontali, L., and Fukuhara, H. (1981b). Two genes for mitochondrial tyrosine transfer RNA in yeast: Localization and expression. *FEBS Lett.* **125,** 180–182.

Wettstein-Edwards, J., Ticho, B. S., Martin, N. C., Najarian, D., and Getz, G. S. (1986). *In vitro* transcription and promoter strength analysis of five mitochondrial tRNA promoters in yeast. *J. Biol. Chem.* **261,** 2905–2911.

Williamson, D. H. (1970). The effect of environmental and genetic factors on the replication of mitochondrial DNA in yeast. *In* "Control of Organelle Development," pp. 247–276. Cambridge Univ. Press, London.

Williamson, D. H. (1976). Packaging and recombination of mitochondrial DNA in vegetatively growing yeast cells. *In* "Genetic, Biogenesis and Bioenergetics of Mitochondria" (W. Bandlow, R. J. Schweyen, D. Y. Thomas, K. Wolf, and F. Kaudewitz, eds.), pp. 117–124. De Gruyter, Berlin.

Williamson, D. H. (1978). Gene conversion in mitochondria. *Nature (London)* **276,** 562.

Williamson, D. H., and Fennell, D. J. (1974). Apparent dispersive replication of yeast mitochondrial DNA as revealed by density labelling experiments. *Mol. Gen. Genet.* **131,** 193–207.

Williamson, D. H., and Fennell, D. J. (1975). The use of fluorescent DNA binding agent for detecting and separating yeast mitochondrial DNA. *In* "Methods in Cell Biology" (D. M. Prescott, ed.), pp. 335–351. Academic Press, New York.

Williamson, D. H., and Moustacchi, E. (1971). The synthesis of mitochondrial DNA during the cell cycle in the yeast *Saccharomyces cerevisiae. Biochem. Biophys. Res. Commun.* **142,** 195–201.

Williamson, D. H., Johnston, L. H., Richmond, K. M. V., and Game, J. C. (1977). Mitochondrial DNA and the heritable unit of the yeast mitochondrial genome: A review. *In* "Mitochondria 1977" (W. Bandlow, R. J. Schweyen, K. Wolf, and F. Kaudewitz, eds.), pp. 1–24. De Gruyter, Berlin.

Wills, J. W., Troutman, W. B., and Riggsby, W. S. (1985). Circular mitochondrial genome of *Candida albicans* contains a large inverted duplication. *J. Bacteriol.* **164,** 7–13.

Willson, T. A., and Nagley, P. (1987). Amino acid substitutions in subunit 9 of the mitochondrial ATPase complex of *Saccharomyces cerevisiae*—sequence analysis of a series of revertants of an *oli 1* mit$^-$ mutant carrying an amino acid substitution in the hydrophilic loop of subunit 9. *Eur. J. Biochem.* **167,** 291–297.

Willson, T. A., Ooi, B. G., Lukins, H. B., Linnane, A. W., and Nagley, P. (1986). Mutations in the mitochondrial *oli 1* gene of *Saccharomyces cerevisiae* affecting subunit 9 of the mitochondrial ATPase complex. *Nucleic Acids Res.* **14,** 8228.

Winkley, C. S., Keller, M. J., and Jaehning, J. A. (1985). A multicomponent mitochondrial RNA polymerase from *Saccharomyces cerevisiae. J. Biol. Chem.* **260,** 14214–14223.

Wintersberger, E. (1970). DNA-dependent RNA polymerase from mitochondria of a cytroplasmic "petite" mutant of yeast. *Biochem. Biophys. Res. Commun.* **40,** 1179–1184.

Wintersberger, E. (1972). Isolation of a distinct rifampicin-resistant RNA polymerase from mitochondria of yeast, *Neurospora* and liver. *Biochem. Biophys. Res. Commun.* **48,** 1287–1294.

Wintersberger, U., and Blutsch, H. (1976). DNA-dependent DNA polymerase from yeast mitochondria. Dependence of enzyme activity on conditions of cell growth, and properties of the highly purified polymerase. *Eur. J. Biochem.* **68,** 199–207.

Wintersberger, U., and Wintersberger, W. (1970). Studies on DNA polymerase from yeast. 2. Partial purification and characterization of mitochondrial DNA polymerase from wild type and respiratory deficient yeast cells. *Eur. J. Biochem.* **13,** 20–27.

Woese, C. L., Magrum, L. J., Gupta, R., Siegel, B., Stahl, A., Kop, J., Crawford, N., Brosius, J., Gutell, R., Hogan, J. J., and Noller, H. F. (1980). Secondary structure model for bacterial 16S ribosomal RNA: Phylogenetic, enzymatic and chemical evidence. *Nucleic Acids Res.* **8,** 2275–2293.

Wolf, K. (1983). Mitochondrial genome divergence in the petite negative yeast *Schizosaccharomyces pombe. Endocytobiology* II, 229–240.

Wolf, K. (1987a). Mitochondrial genes of the budding yeast *Saccharomyces cerevisiae. In* "Gene Structure in Eukaryotic Microbes. SGM Special Publication 22" (J. R. Kinghorn, ed.), pp. 41–62. IRL Press, Oxford.

Wolf, K. (1987b). Mitochondrial genes of the fission yeast *Schizosaccharomyces pombe. In* "Gene Structure in Eukaryotic Microbes. SGM Special Publication 22" (J. R. Kinghorn, ed.), pp. 69–91. IRL Press, Oxford.

Wolf, K., and Kaudewitz, F. (1977). Extrachromosomal inheritance in *Schizosaccharomyces pombe.* V. On a possible correlation between bias in transmission, spontaneous *mit*$^-$ production, and formation of uniparental zygotic clones. *In* "Mitochondria 1977" (W. Bandlow, R. J. Schweyen, K. Wolf, and F. Kaudewitz, eds.), pp. 65–81. De Gruyter, Berlin.

Wolf, K., Dujon, B., and Slonimski, P. P. (1973). Mitochondrial genetics. V. Multifactorial mitochondrial crosses involving a mutation conferring paromomycin resistance in *Saccharomyces cerevisiae. Mol. Gen. Genet.* **125,** 53–90.

Wolf, K., and Del Giudice, L. (1987). Horizontal gene transfer between mitochondrial genomes. *Endocyt. C. Res.* **4,** 103–120.

Wolf, K., Burger, G., Lang, B., and Kaudwitz, F. (1976a). Extrachromosomal inheritance in *Schizosaccharomyces pombe* I. Evidence for extrakaryotically inherited mutation conferring resistance to antimycin. *Mol. Gen. Genet.* **144,** 67–73.

Wolf, K., Lang, B., Burger, G., and Kaudewitz, F. (1976b). Extrachromosomal inheritance in *Schizosaccharomyces pombe.* II. Evidence for extrakaryotically inherited respiration deficient mutants. *Mol. Gen. Genet.* **144,** 75–81.

Wolf, K., Seitz, G., Lang, B., Burger, G., and Kaudewitz, F. (1976c). Extrachromosomal inheritance of drug resistance and respiratory deficiency in the fission yeast *Schizosaccharomyces pombe. In* "Genetics, Biogenesis and Bioenergetics of Mitochondria" (W. Bandlow, R. J. Schweyen, D. Y. Thomas, K. Wolf, and F. Kaudewitz, eds.), pp. 23–37. De Gruyter, Berlin.

Wolf, K., Seitz, G., Lückemann, G., Lang, B., Burger, G., Bandlow, W., and Kaudewitz, F. (1976d). Extrachromosomal inheritance in a petite-negative yeast, *Schizosaccharomyces pombe. In* "Genetics and Biogenesis of Chloroplasts and Mitochondria" (T. Bücher, W. Neupert, W. Sebald, and J. Werner, eds.), pp. 497–502. Elsevier, Amsterdam.

Wolf, K., Ahne, A., Del Giudice, L., Oraler, G., Kanbay, F., Merlos-Lange, A. M., Welser, F., and Zimmer, M. (1987). Introns as key elements in the evolution of mitochondrial genomes in lower eukaryotes. *In "Cytochrome Systems: Molecular Biology and Bioenergetics"* (Papa *et al.*, eds.). Plenum, New York, in press.

Wolf, K., Del Giudice, L., and Kaudewitz, F. (1979). Promotion of uniparental inheritance of mitochondrial drug resistance by delayed division of yeast zygotes. *Mol. Gen. Genet.* **176,** 301–302.

Wolf, K., Lang, B., Del Giudice, L., Anziano, P. Q., and Perlman, P. S. (1982). *Schizosaccharomyces pombe*—a short review of a short mitochondrial genome. *In* "Mitochondrial Genes" (P. Slonimski, P. Borst, and G. Attardi, eds.), pp. 355–360. Cold Spring Harbor Laboratory, Cold Spring Harbor, New York.

Wollenzien, P. L., Cantor, C. R., Grant, D. M., and Lambowitz, A. M. (1983). RNA splicing in *Neurospora* mitochondria: Structure of the unspliced 35 S precursor ribosomal RNA detected by psoralen cross-linking. *Cell* **32,** 397–407.

Woodward, D. O., and Munkres, K. D. (1966). Alterations of a maternally inherited mitochondrial structural protein in respiratory-deficient strains of *Neurospora. Proc. Natl. Acad. Sci. U.S.A.* **55,** 872–880.

Wright, R. M., and Cummings, D. J. (1983a). Transcription of a mitochondrial plasmid during senescence in *Podospora anserina. Curr. Genet.* **7,** 457–464.

Wright, R. M., and Cummings, D. J. (1983b). Integration of mitochondrial gene sequences within the nuclear genome during senescence in a fungus. *Nature (London)* **302,** 86–88.

Wright, R. M., and Cummings, D. J. (1983c). Mitochondrial DNA from *Podospora anserina.* IV. The large ribosomal RNA gene contains two long intervening sequences. *Curr. Genet.* **7,** 151–157.

Wright, R. M., Horrum M. A., and Cummings, D. J. (1982a). Are mitochondrial structural genes selectively amplified during senescence in *Podospora anserina? Cell* **29,** 505–515.

Wright, R. M., Laping, J. L., Horrum, M. A., and Cummings, D. J. (1982b). Mitochondrial DNA from *Podospora anserina.* III. Cloning, physical mapping, and localization of the ribosomal RNA genes. *Mol. Gen. Genet.* **185,** 56–64.

Yin, S., Heckman, J., Sarnoff, J., and RajBhandary, U. L. (1980). *Neurospora crassa* mitochondrial tRNAs: Structure, codon reading patterns, gene organization and unusual flanking of the tRNAs. *In* "The Organization and Expression of the Mitochondrial Genome" (A. M. Kroon and C. Saccone, eds.), pp. 307–310. Elsevier, Amsterdam.

Yin, S., Heckman, J., and RajBhandary, U. L. (1981). Highly conserved GC rich palindromic DNA sequences which flank tRNA genes in *Neurospora crassa* mitochondria. *Cell* **26,** 326–332.

Yin, S., Burke, J., Chang, D. D., Browning, K. S., Heckman, J. E., Alzner-De Weerd, B., Potter, M. J., and RajBhandary, U. L. (1982). *Neurospora crassa* mitochondrial tRNAs and rRNAs: Structure, gene organization, and DNA sequences. *In* "Mitochondrial Genes" (P. Slonimski, P. Borst, and G. Attardi, eds.), pp. 361–373. Cold Spring Harbor Laboratory, Cold Spring Harbor, New York.

Young, I. G., and Anderson, S. (1980). The genetic code in bovine mitochondria: Sequence of genes for the cytochrome oxidase subunit II and two tRNAs. *Gene* **12,** 257–265.

Zagorski, W., Kozlowski, M., Mieszczak, M., Spyridakis, A., Claisse, M., and Slonimski, P. P. (1987). Protein synthesis in mitochondria from yeast strains carrying *nam* and *mim* suppressor genes. *Biochimie* **69,** 517–530.

Zassenhaus, H. P., and Butow, R. A. (1984). Expression of GC clusters in the yeast mitochondrial *var1* gene. Transcription into stable RNAs. *J. Biol. Gen.* **259,** 8417–8421.

Zassenhaus, H. P., and Perlman, P. S. (1982). Respiration deficient mutants in the A+T−rich region of yeast mitochondrial DNA containing the *var1* gene. *Curr. Genet.* **6,** 179–188.

Zassenhaus, H. P., Farrelly, F., Hudspeth, M. E. S., Grossman, L. I., and Butow, R. A. (1983). Transcriptional analysis of the *Saccharomyces cerevisiae* mitochondrial *var1* gene: Anomalous hybridization of RNA from AT-rich regions. *Mol. Cell. Biol.* **3,** 1615–1624.

Zassenhaus, H. P., Martin, N. C., and Butow, R. A. (1984). Origins of transcripts of the yeast mitochondrial *var1* gene. *J. Biol. Chem.* **259,** 6019–6027.

Zaug, A. J., Cech, T. R. (1982). The intervening sequence excised from the ribosomal RNA precursor of *Tetrahymena* contains a 5′-terminal guanosine residue not encoded by the DNA. *Nucleic Acids Res.* **10,** 2823–2838.

Zaug, A. J., and Cech, T. R. (1986). The intervening sequence RNA of *Tetrahymena* is an enzyme. *Science* **231,** 470–475.

Zaug, A. J., Grabowski, P. J., and Cech, T. R. (1983). Autocatalytic cyclization of an excised intervening sequence RNA is a cleavage–ligation reaction. *Nature (London)* **301,** 578–583.

Zaug, A. J., Been, M. D., and Cech, T. R. (1986). The *Tetrahymena* ribozyme acts like an RNA restriction endonuclease. *Nature (London)* **324,** 429–432.

Zauner, R., Christner, J., Jung, G., Borchart, U., Machleidt, Videira, A., and Werner, S. (1985). Identification of the polypeptide encoded by the *URF-1* gene of *Neurospora crassa* mtDNA. *Eur. J. Biochem.* **150,** 447–454.

Zennaro, E., Grimaldi, L., Baldacci, G., and Frontali, L. (1985). Mitochondrial transcription and processing of transcripts during release from glucose repression in "resting cells" of *Saccharomyces cerevisiae. Eur. J. Biochem.* **146,** 191–196.

Zhu, H., Macreadie, I. G., and Butow, R. A. (1987). RNA processing and expression of an intron-encoded protein in yeast mitochondria: Role of a conserved dodecamer sequence. *Mol. Cell Biol.* **7,** 2530–2537.

Zimmer, M., Lückemann, G., Lang, B. F., and Wolf, K. (1984). The mitochondrial genome of the fission yeast *Schizosaccharomyces pombe*. 3. Gene mapping in strain EF1 (CBS 356) and analysis of hybrids between the strains EF1 and ade 7-50h$^-$. *Mol. Gen. Genet.* **196,** 473–481.

Zimmer, M., Welser, F., Oraler, G., and Wolf, K. (1987). Distribution of mitochondrial introns in the species *Schizosaccharomyces pombe* and the origin of the group II intron in the gene encoding apocytochrome b. *Curr. Genet.* **12,** 329–336.

Zimmermann, W., Chen, S. M., Bolden, A., and Weissbach, A. (1980). Mitochondrial DNA replication does not involve DNA polymerase alpha. *J. Biol. Chem.* **255,** 11847–11852.

Zimmern, D. (1983). Homologous proteins encoded by yeast mitochondrial introns and by a group of RNA viruses from plants. *J. Mol. Biol.* **171,** 345–352.

Zinn, A. R., and Butow, R. A. (1984). Kinetics and intermediates of yeast mitochondrial DNA recombination. *Cold Spring Harbor Symp.* **49,** 115–122.

Zinn, A. R., and Butow, R. A. (1985). Nonreciprocal exchange between alleles of the yeast mitochondrial 21S rRNA gene: Kinetics and the involvement of a double-strand break. *Cell* **40,** 887–895.

Zinn, A. R., Pohlmann, J. K., Perlman, P. S., and Butow, R. A. (1987). Kinetic and segregational analysis of mitochondrial recombination in yeast. *Plasmid* **17,** 248–256.

Zollinger, W. D., and Woodward, D. O. (1972). Comparison of cysteine and tryptophane content of insoluble proteins derived from wild-type and *mi-1* strains of *Neurospora crassa. J. Bacteriol.* **109,** 1001–1013.

INDEX

A

B

C

D

E

F

G

H

I

J

K

L

M

P

R

S

T

U

W

X

Y

Z